Deep Learning for Multimedia Processing Applications

Deep Learning for Multimedia Processing Applications is a comprehensive guide that explores the revolutionary impact of deep learning techniques in the field of multimedia processing. Written for a wide range of readers, from students to professionals, this book offers a concise and accessible overview of the application of deep learning in various multimedia domains, including image processing, video analysis, audio recognition, and natural language processing.

Divided into two volumes, Volume One begins by introducing the fundamental concepts of deep learning, providing readers with a solid foundation to understand its relevance in multimedia processing. Readers will discover how deep learning techniques enable accurate and efficient image recognition, object detection, semantic segmentation, and image synthesis. The book also covers video analysis techniques, including action recognition, video captioning, and video generation, highlighting the role of deep learning in extracting meaningful information from videos.

Furthermore, the book explores audio processing tasks such as speech recognition, music classification, and sound event detection using deep learning models. It demonstrates how deep learning algorithms can effectively process audio data, opening up new possibilities in multimedia applications. Lastly, the book explores the integration of deep learning with natural language processing techniques, enabling systems to understand, generate, and interpret textual information in multimedia contexts.

Throughout the book, practical examples, code snippets, and real-world case studies are provided to help readers gain hands-on experience in implementing deep learning solutions for multimedia processing. *Deep Learning for Multimedia Processing Applications* is an essential resource for anyone interested in harnessing the power of deep learning to unlock the vast potential of multimedia data.

Uzair Aslam Bhatti was born in 1986. He received his PhD degree in information and communication engineering, Hainan University, Haikou, Hainan, in 2019. He completed his postdoctoral from Nanjing Normal University, Nanjing, China, in implementing Clifford algebra algorithms in analysing the geospatial data using artificial intelligence (AI). He is currently working as an associate professor in the School of Information and Communication Engineering in Hainan University. His areas of specialty include AI, machine learning, and image processing. He is serving as guest editor of various journals including *Frontier in Plant Science, Frontier in Environmental Science, Computer Materials and Continua, PLoS One, IEEE Access,* etc., and has reviewed many IEEE Transactions and Elsevier journals.

Jingbing Li is a doctor, professor, doctoral supervisor, and the vice president of Hainan Provincial Invention Association. He has been awarded honorary titles of Leading Talents in Hainan Province, Famous Teaching Teachers in Hainan Province, Outstanding Young and Middle-aged Backbone Teachers in Hainan Province, and Excellent Teachers in Baosteel. He has also won the second prize of the Hainan Provincial Science and Technology Progress Award three times (the first completer twice, the second completer once). He has obtained 13 authorized national invention patents, published 5 monographs such as medical image digital watermarking, and published more than 80 SCI/EI retrieved academic papers (including 22 SCI retrieved papers) as the first author or corresponding author. He has presided over two projects of the National Natural Science Foundation of China, and five projects of Hainan Province's key research and development projects and Hainan Province's international scientific and technological cooperation projects.

Dr. Mengxing Huang is dean of the School of Information, at Hainan University. He has occupied many roles, such as the leader of the talent team of "Smart Service", the chief scientist of the National Key R&D Program, a member of the Expert Committee of Artificial Intelligence and Blockchain of the Science and Technology Committee of the Ministry of Education, the executive director of the Postgraduate Education Branch of the China Electronics Education Society, and the Computer Professional Teaching Committee of the Ministry of Education, among others. His main research areas include big data and intelligent information processing, multi-source information perception and fusion, artificial intelligence and intelligent services, etc. In recent years, he has published more than 230 academic papers as the first author and corresponding author, has obtained 36 invention patents authorized by the state, 96 software copyrights, published 4 monographs, and translated 2 books. He has won first prize and second prize in the Hainan Provincial Science and Technology Progress Award as the first person who completed it; and won two Hainan Provincial Excellent Teaching Achievement Awards and Excellent Teacher Awards. He has presided over and undertaken more than 30 national, provincial, and ministerial-level projects, such as national key research and development plan projects, national science and technology support plans, and National Natural Science Foundation projects.

Sibghat Ullah Bazai completed his undergraduate and graduate studies in computer engineering at the Balochistan University of Information Technology, Engineering, and Management Sciences (BUITEMS) in Quetta, Pakistan. He received his PhD (IT) in cybersecurity from Massey University in Auckland, New Zealand, in 2020. As part of his research, he is interested in applying cybersecurity, identifying diseases with deep learning, automating exams with natural language processing, developing local language sentiment data sets, and planning smart cities. Sibghat is a guest editor and reviewer for several journals' special issues in *MDPI, Hindawi, CMC, PLoS One, Frontier*, and others.

Muhammad Aamir received the bachelor of engineering degree in computer systems engineering from Mehran University of Engineering & Technology Jamshoro, Sindh, Pakistan, in 2008, the master of engineering degree in software engineering from Chongqing University, China, in 2014, and the PhD degree in computer science and technology from Sichuan University, Chengdu, China, in 2019. He is currently an associate professor at the Department of Computer, Huanggang Normal University, China. His main research interests include pattern recognition, computer vision, image processing, deep learning, and fractional calculus.

Deep Learning for Multimedia Processing Applications

Volume One:
Image Security and Intelligent Systems for Multimedia Processing

Edited by
Uzair Aslam Bhatti, Jingbing Li, Mengxing Huang,
Sibghat Ullah Bazai, and Muhammad Aamir

CRC Press
Taylor & Francis Group
Boca Raton London New York

CRC Press is an imprint of the
Taylor & Francis Group, an **informa** business

Designed cover image: Uzair Aslam Bhatti

First edition published [2024]
by CRC Press
2385 Executive Center Drive, Suite 320, Boca Raton, FL 33431

and by CRC Press
4 Park Square, Milton Park, Abingdon, Oxon, OX14 4RN

CRC Press is an imprint of Taylor & Francis Group, LLC

ISBN: 978-1-032-54824-1 (hbk)
ISBN: 978-1-032-54826-5 (pbk)
ISBN: 978-1-003-42767-4 (ebk)

DOI: 10.1201/9781003427674

Typeset in Minion
by MPS Limited, Dehradun

Contents

LEI CAO, JINGBING LI, AND UZAIR ASLAM BHATTI

FANGCHUN DONG, JINGBING LI, AND UZAIR ASLAM BHATTI

IQRA TABASSUM AND SIBGHAT ULLAH BAZAI

Contributors

Raza Muhammad Ahmad
School of Cyberspace Security
Hainan University
Haikou, China

Saira Akram
Balochistan University of Information
    Technology, Engineering, and
    Management Sciences (BUITEMS)
Quetta, Pakistan

Hooman Alavizadeh
Department of Computer Science and
    Information Technology, School of
    Computing, Engineering and
    Mathematical Sciences
La Trobe University
Melbourne, Australia

Hootan Alavizadeh
Department of Computer Science and
    Engineering
Wright State University
Dayton, OH, USA

Emadalden Alhatami
School of Information and
    Communication Engineering
Hainan University
Haikou, China

Sibghat Ullah Bazai
Department of Computer Science
Balochistan University of Information
    Technology, Engineering, and
    Management Sciences (BUITEMS)
Quetta, Pakistan

Uzair Aslam Bhatti
School of Information and
    Communication Engineering
and
State Key Laboratory of Marine Resource
    Utilization in the South China Sea
Hainan University
Haikou, China

Lei Cao
School of Information and
    Communication Engineering
Hainan University
Haikou, China

Muhammad Umar Chaudhry
Department of Computer Engineering
Bahauddin Zakariya University
Multan, Pakistan

Fangchun Dong
School of Information and
    Communication Engineering
Hainan University
Haikou, China

MengXing Huang
School of Information and
  Communication Engineering
Hainan University
Haikou, China

Dekai Li
School of Information and
  Communication Engineering
Hainan University
Haikou, China

Jingbing Li
School of Information and
  Communication Engineering
and
State Key Laboratory of Marine Resource
  Utilization in the South China Sea
Hainan University
Haikou, China

Hassaan Malik
Department of Computer Science
National College of Business
  Administration & Economics
Lahore, Pakistan

Shah Marjan
Balochistan University of Information
  Technology, Engineering, and
  Management Sciences (BUITEMS)
Quetta, Pakistan

Saqib Ali Nawaz
School of Information and
  Communication Engineering
Hainan University
Haikou, China

Shah Noor
Department of Computer Engineering
Balochistan University of Information
  Technology, Engineering, and
  Management Sciences (BUITEMS)
Quetta, Pakistan

Muhammad Usman Shoukat
School of Automotive Engineering
Wuhan University of Technology
Wuhan, China

Iqra Tabassum
Balochistan University of Information
  Technology, Engineering, and
  Management Sciences (BUITEMS)
Quetta, Pakistan

Amna Tahir
Department of Computer Science
National College of Business
  Administration & Economics
Lahore, Pakistan

Saima Tareen
Department of Computer Engineering
Balochistan University of Information
  Technology, Engineering, and
  Management Sciences (BUITEMS)
Quetta, Pakistan

Shafi Ullah
Department of Computer Science
Balochistan University of Information
  Technology, Engineering, and
  Management Sciences (BUITEMS)
Quetta, Pakistan

Meng Yang
School of Information and
  Communication Engineering
Hainan University
Haikou, China

Yiyi Yuan
School of Information and
  Communication Engineering
Hainan University
Haikou, China

Pengju Zhang
School of Information and
 Communication Engineering
Hainan University
Haikou, China

Qinqing Zhang
School of Information and
 Communication Engineering
Hainan University
Haikou, China

A Novel Robust Watermarking Algorithm for Encrypted Medical Images Based on Non-Subsampled Shearlet Transform and Schur Decomposition

Meng Yang[1], Jingbing Li[1,2], Uzair Aslam Bhatti[1,2], Yiyi Yuan[1], and Qinqing Zhang[1]

[1]*School of Information and Communication Engineering, Hainan University, Haikou, China*
[2]*State Key Laboratory of Marine Resource Utilization in the South China Sea, Hainan University, Haikou, China*

1.1 INTRODUCTION

As digitalization in the sphere of modern medicine continues to advance, digital medical imaging is increasingly used in the diagnosis and treatment of clinical diseases and other related scientific research. While digitization provides great ease of access to medical information, it also poses a number of data security issues that significantly weaken the protection of patients' private information. The emergence of digital watermarking technology [1,2] is an excellent solution to the issue of how to secure patients' personal information more effectively in the modern medical environment. By embedding patients' personal information or doctors' diagnostic information into medical images in the form of digital watermarks, and encrypting the watermarked information before embedding [3,4], the information security is improved. However, in view of the special characteristics of medical images, zero watermark technology is often used for embedding. The conflict between the imperceptibility and robustness of digital watermarking is effectively resolved by zero watermark technology [5,6], a new digital watermarking

DOI: 10.1201/9781003427674-1

technique that builds copyright information from key image features without modifying the carrier image data. The majority of current research that related to digital watermarking methods for medical images are concentrated in plaintext domain, meaning that the watermark information is embedded and extracted from carrier medical images that are not encrypted [7–9]. Although the plaintext domain algorithm achieves the invisibility of watermark and improves the robustness of the algorithm. However, there are still certain drawbacks, ignoring the security guarantee of the carrier medical image itself. Given that carrier medical images often carry patient pathology information, once intercepted during transmission, this private information can be compromised. Consequently, image encryption technology is crucial. The watermark information will be added to the encrypted medical image after that initial medical image has been encrypted, and invoking third-party concepts to embed and extract the watermark, we can effectively get the embedded watermark information even without using the original medical image and with the image encrypted. With the help of this encrypted domain algorithm [10–12], it is possible to guarantee the secrecy and integrity of both the watermark data and the medical image itself. In recent years, digital watermarking algorithms concerning image encryption have emerged. Reference [13] proposed encryption of original medical images under DWT-DCT transform domain combined with logistic chaotic mapping to increase the carrier medical image's security. The dual-tree complex wavelet transform and discrete cosine transform (DTCWT-DCT) were introduced in reference [14] for achieving a robust watermarking algorithm in the encrypted domain. Firstly, the DTCWT-DCT transform acts on the medical image, and then the transformed coefficient matrix is dotted with the logistic chaos matrix to get the encrypted medical image by inverting the algorithm. Reference [15] proposed a DWT-DCT-tent image encryption method. Firstly, a coefficient matrix is generated in the DWT-DCT transform domain. Next, a pseudo-random sequence is generated using the tent chaos system, and an encryption matrix is generated by a symbolic function. Finally, this encryption matrix is dotted multiplied with the transform domain coefficient matrix to achieve encryption of the image. The original image and the watermark may both be effectively protected by using image encryption techniques, which also improves the watermarking algorithm's security as well.

Feature extraction as a crucial step in zero watermarking technology, the current algorithm for extracting features from images is relatively mature [16–18]. Reference [19] suggested extracting the depth features from medical images by pre-training the network. Then a DCT transform is applied to them and the feature sequence is created by combining the perceptual hash function. According to reference [20], it was suggested that in order to acquire precise edge points on the original medical images, edge detection and Zernike moment processing be applied to the images first, followed by a discrete cosine transform (DCT) to produce feature vectors. Reference [21] showed extracting the corner points of medical images first using the Harris corner point detection technique, and then describing the retrieved corner points using the speeded-up robust features (SURF) approach to build a feature descriptor matrix. After that, the perceptual hashing approach is used to process the feature descriptor matrix in order to produce the feature sequence for the medical image. In recent years, Schur decomposition has the property of stability due to the value of the diagonal elements of the upper triangular matrix generated after its decomposition. It is often used in the feature extraction step of images, combined with perceptual hashing, to extract image feature sequences, and is widely used in digital watermarking algorithms [22–24]. A quick and reliable zero watermarking technique for double encryption was described in references [25,26] and is based on matrix Schur decomposition. The algorithm first performs matrix Schur decomposition of the original image of low-frequency block to obtain stable values, and constructs the transition matrix, combines the perceptual hash, constructs the feature matrix, and selects feature regions to generate feature vectors.

In this chapter, we propose a robust watermarking algorithm for encrypted medical images based on DWT-DCT-tent encryption and NSST-Schur feature extraction by combining the advantages of the above algorithms. The algorithm guarantees that embedded watermark information's accuracy, invisibility, and robustness. Additionally, it strengthens the security of the carrier medical image and the watermark during transmission, which has good application prospects.

1.2 BASIC THEORY

1.2.1 Discrete Wavelet Transform (DWT)

A transform technique with variable resolution that integrates the time and frequency domains is the wavelet transform, which is a novel transform approach to analysis. Its time window's dimensions are automatically modified on a regular basis, which is more in accordance with how people see. Its primary strength is the ability to emphasize particular features through transformation, which effectively allows it to extract crucial information from the signal. It also does multi-scale refinement analysis of the function or signal through scaling, translating, and other operational functions. This makes the wavelet transform an excellent tool. One may think of an image as a two-dimensional signal; as a result, it is frequently used in algorithms for image analysis and processing. It can successfully fend off attacks from noise and compression, and even improve algorithm robustness. Four sub-bands of the image are separated, each with a different direction and frequency following the one-level wavelet decomposition. If multiple levels of DWT decomposition are to be done on the image, this transform is repeated on its upper level low-frequency sub-band. As shown in Figure 1.1, a two-level discrete wavelet

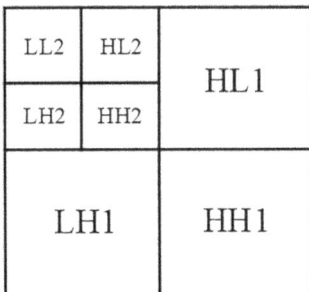

FIGURE 1.1   DWT two-level decomposition diagram.

transform decomposition is shown. As the image is decomposed by DWT, the high-frequency sub-band partly responds to the image detail features and the low-frequency sub-band partly responds to the image approximate features. In addition, the low-frequency sub-band is where the majority of its energy is focused, so it is usually called the approximation subgraph.

1.2.2 Non-Subsampled Shearlet Transform (NSST)

NSST is a new transformation method based on the shearlet transform with translation invariance, which depends on the non-subsampled pyramid (NSP) and a modified shearlet filter (SF) to achieve multiscale decomposition and orientation decomposition of an image. Firstly, the low-frequency and high-frequency sub-band coefficients are first obtained by performing a k-level decomposition of the original image using NSP by NSST; then, by means of an improved SF, the high-frequency sub-band coefficients are decomposed in multiple directions. Finally, the decomposition generates a low-frequency sub-band coefficient and high-frequency sub-band k coefficients.

In this experiment, to acquire the low-frequency sub-band image of the encrypted brain medical image, we conduct NSST on it and perform further feature extraction. Figure 1.2 displays the low-frequency sub-band image of the encrypted brain image following NSST three-level decomposition.

1.2.3 Matrix Schur Decomposition

A matrix, $A \in R^{n \times n}$, is broken down into $U \in R^{n \times n}$ and $T \in R^{n \times n}$, where U is a unitary matrix and T is an upper triangular matrix. See equation (1.1).

$$A = U \times T \times U^{T} \tag{1.1}$$

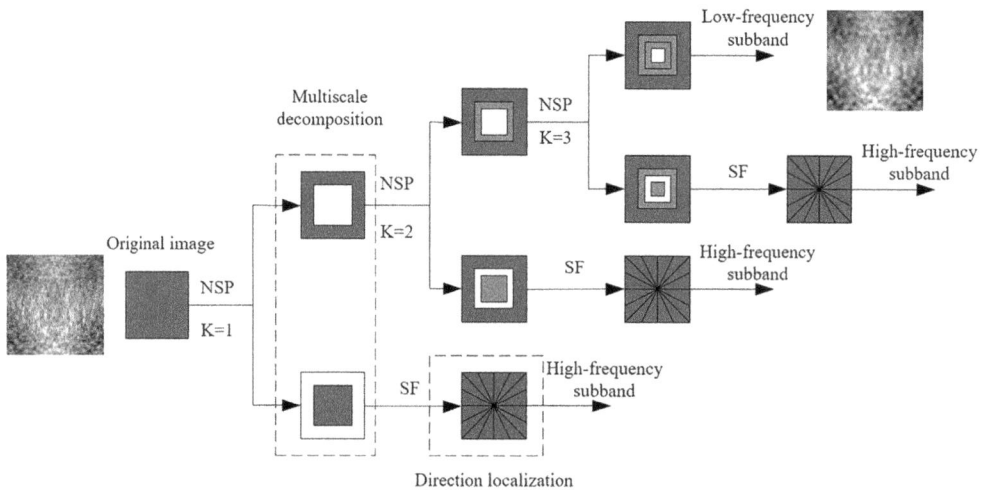

FIGURE 1.2   NSST three-level decomposition chart.

For example:

$$\begin{bmatrix} 2 & 4 & 3 \\ 5 & 3 & 3 \\ 6 & 1 & 2 \end{bmatrix} = U \times \begin{bmatrix} 9.5947 & 1.8791 & -2.0049 \\ 0 & -2.6343 & 2.5396 \\ 0 & 0 & 0.0396 \end{bmatrix} \times U^T$$

A and T are similar, so they have the same eigenvalues. The diagonal elements of the upper triangular matrix make up its eigenvalues, and T is an upper triangular matrix with the same eigenvalues as A. Therefore, T is usually called the Schur standard type of A. The energy of the matrix after the Schur decomposition is mainly concentrated on its eigenvalues, and they are very stable. This property of Schur decomposition can be used to extract stable feature points of images in image processing. Moreover, Schur decomposition is a fast, low false alarm and scaling invariant, which makes its application to digital watermarking techniques effective in improving algorithm performance.

1.2.4 Chaos Encryption System

A chaotic system is a deterministic system that exhibits irregular, seemingly random motion and whose behavior is uncertain, unrepeatable, and unpredictable, which is the phenomenon of chaos. Common chaotic systems include logistic map, tent map, etc.

1.2.4.1 Tent

The tent chaotic system is a segmented linear one-dimensional mapping, so named since its function image is similar to that of a tent. The tent mapping is a chaotic mapping in its parameter range. It is frequently employed in chaotic encryption techniques, such as image encryption. Its system equation is given in equation (1.2):

$$\begin{cases} x(n+1) = \mu x(n), \ 0 \le x(n) \le 0.5 \\ x(n+1) = \mu[1 - x(n)], \ 0.5 < x(n) \le 1 \end{cases} \quad (1.2)$$

where $x(n) \in (0, 1)$ and $\mu \in (0, 2)$. When $\mu > 1$, the system belongs to the chaotic state and has a uniform distribution function and good correlation.

1.2.4.2 Logistic Map

The most common, well-known, and widely used chaotic system is the logistic map. Logistic chaotic mapping is a very simple, yet significant nonlinear iterative equation that has a deterministic form and the system does not contain any random factors, yet the system can produce seemingly completely random results. Based on the above characteristics, the random sequences generated by chaotic systems can be used as encryption sequences. Chaos encryption has become an emerging encryption technique. The composition of the logistic map chaotic system is formulated as follows. See equation (1.3).

$$X_{K+1} = \mu \cdot X_K \cdot (1 - X_K) \quad (1.3)$$

when $3.5699456 < \mu \leq 4$, the situation of this system is utter chaos, μ is the branching parameter, K indicates how many iterations there were, and X_0 is the initial value.

1.3 PROPOSED ALGORITHM

1.3.1 Medical Image Encryption

The confidentiality of the carrier's medical image must be preserved on the basis of achieving zero watermark. The medical image is encrypted before the image's features are extracted to achieve the protection of the medical image itself. Firstly, the wavelet coefficients of each sub-band (LL, HL, LH, HH) are obtained by applying DWT to the original medical image of the brain $I(i, j)$; see equation (1.4). Then, each sub-band wavelet coefficient is subjected to DCT in order to get the coefficient matrix $D(i, j)$; see equation (1.5). Next, set parameter values for Tent chaotic system, generate chaotic sequence $X(j)$, and the binary encryption matrix $C(i, j)$ is generated by the symbolic function $sgn(x)$, $D(i, j)$ is then dotted with $C(i, j)$ to get the encryption coefficient matrix $ED'(i, j)$; see equation (1.6). Lastly, the matrix $ED'(i, j)$ is subjected to the Inverse Discrete Cosine Transform (IDCT) to produce the encrypted sub-band wavelet coefficient matrix $ED(i, j)$, see equation (1.7), and then ultimately produce the encrypted medical image $E(i, j)$ by applying the Inverse Discrete Cosine Transform (IDWT) to the matrix $ED(i, j)$. See equation (1.8). Figure 1.3 illustrates the original medical image encryption procedure.

$$\{LL, HL, LH, HH\} = DWT2(I(i, j)) \tag{1.4}$$

$$D(i, j) = DCT2(LL, HL, LH, HH) \tag{1.5}$$

$$ED'(i, j) = D(i, j)C(i, j) \tag{1.6}$$

$$ED(i, j) = IDCT2(ED'(i, j)) \tag{1.7}$$

$$E(i, j) = IDWT2(ED(i, j)) \tag{1.8}$$

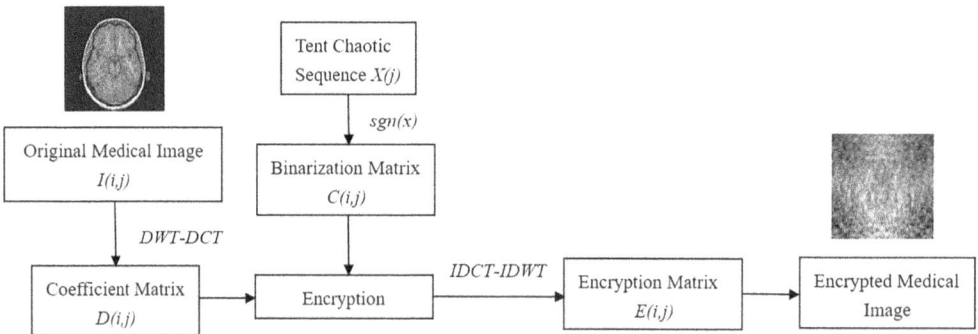

FIGURE 1.3   The process of original brain medical image encryption.

1.3.2 Feature Extraction

In this chapter, we randomly select a 512×512 size medical image of the brain, represented by $I(i, j)$. Firstly, we encrypt the medical image $I(i, j)$ using DWT-DCT transform and Tent chaos system to get the encrypted medical image $E(i, j)$. Secondly, the encrypted medical image is NSST transformed, its low-frequency sub-band is taken and 8×8 chunks are processed, and then Schur decomposition is performed on the chunks to obtain the stable values and construct the transition matrix. Finally, the element values of this matrix are compared with the mean value of the matrix and combined with the perceptual hash binarization. To generate the feature sequence, any data that exceeds the mean value or is equal to it are immediately replaced with 1, as well as, all other data are replaced with 0. Considering both the robustness and the capacity of the embedded watermark, we choose a 4×8 feature region, 32-bit feature sequence $V(j)$. Figure 1.4 depicts the feature extraction procedure of the encrypted brain medical image. Table 1.1 lists the extracted feature sequences of encrypted medical image under different attacks. We can see that the image feature vectors extracted in this algorithm do not change with the different attacks, which demonstrates the stability of this feature extraction approach; see Table 1.1.

To confirm that the derived feature sequences are capable of clearly differentiating between various medical images, we selected six different medical images, see Figure 1.5, and their encrypted images, see Figure 1.6. We extracted the feature sequences of these six encrypted medical images, see Table 1.2, respectively. The feature sequences taken from the six encrypted medical images were put through a correlation test, see Table 1.3, and we discovered that across distinct images the NC values are smaller than 0.5. In particular, since image (c) and image (e) have some similarity in contours, NC = 0.59, the same image's NC value is 1. The results shown are in line with the properties of human

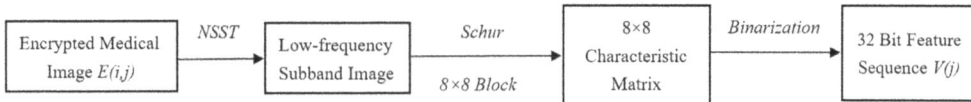

FIGURE 1.4   Feature sequence extraction process.

TABLE 1.1   Feature Sequence Coefficients Under Different Attacks (32-Bit)

Image manipulation	Sequence of coefficient signs	NC
Original encrypted image	00011000001110011111111101111110	1.00
Gaussian noise (20%)	00011000001110011111111101111110	1.00
JPEG compression (2%)	00011000001110011111111101111110	1.00
Median filter [5 × 5] (10 times)	00011000001110011111111101111110	1.00
Rotation (clockwise, 15°)	00011000001110011111111101111110	1.00
Scaling (2.0 times)	00011000001110011111111101111110	1.00
Translation (5%, left)	00011000001110011111111101111110	1.00
Translation (20%, down)	00011000001110011111111101111110	1.00

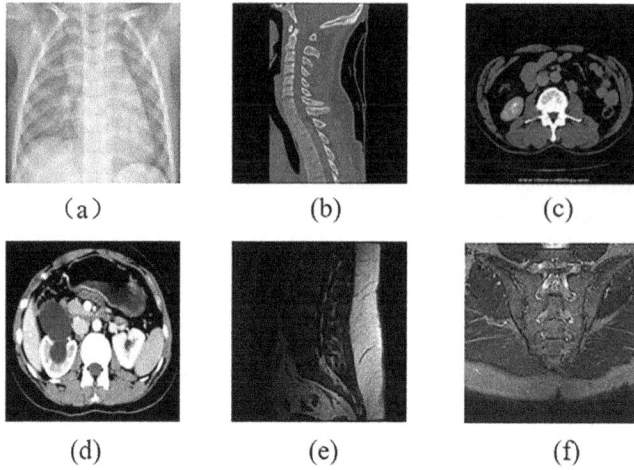

FIGURE 1.5 Six different medical images: (a) pleural, (b) cervical vertebra, (c) abdomen, (d) pancreas, (e) lumbar spine, and (f) sacrum.

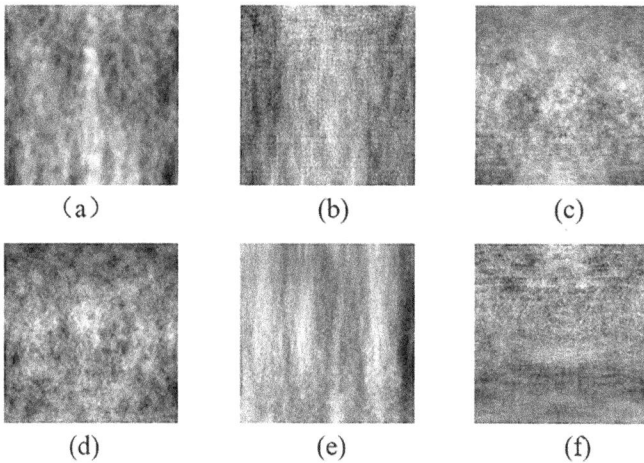

FIGURE 1.6 Six different encrypted medical images.

TABLE 1.2 Extracted Feature Sequences of Six Different Encrypted Medical Images (32-Bit)

Image	Sequence of coefficient signs
(a)	00001000010110000101100001111000
(b)	00111100001111000011110000111100
(c)	00111000001111011111111111011011
(d)	00000000011111101111111101011111
(e)	11000010111000101110010111110010
(f)	10111100101111111111111101111111

TABLE 1.3   NC Values Between Different Images

NC	(a)	(b)	(c)	(d)	(e)	(f)
(a)	**1.0000**	0.3290	0.2041	0.2888	0.0206	0.3102
(b)	0.3290	**1.0000**	0.2697	0.1291	−0.4384	0.4303
(c)	0.2041	0.2697	**1.0000**	0.5919	−0.0929	0.4526
(d)	0.2888	0.1291	0.5919	**1.0000**	0.0485	0.3778
(e)	0.0206	−0.4384	−0.0929	0.0485	**1.0000**	−0.2318
(f)	0.3102	0.4303	0.4526	0.3778	−0.2318	**1.0000**

vision. Therefore, the 32-bit binary sequence extracted by this algorithm can be used as a feature sequence for encrypted medical images.

1.3.3 Embed Watermark

For the purpose of making sure that the inserted watermark information is accurate and secure, the watermark is encrypted before embedding the watermark information. Firstly, starting with the initial value X_0 of the logistic chaos system, a one-dimensional chaotic sequence $X(j)$ is created, and then the binary encryption matrix $B(i, j)$ is generated by the dimensional formula. Finally, the encrypted watermark $BW(i, j)$ is created by XORing, the binary encryption matrix with the binary watermark $W(i, j)$. See equation (1.9). The extracted encrypted medical image feature sequence $V(j)$ and the encrypted watermark $BW(i, j)$ are put through an XOR logic operation to embed the watermark information into the encrypted medical image; additionally, the logical key $K(i, j)$ is acquired and kept by a third party. See equation (1.10). The process of embedding watermark is seen in Figure 1.7.

$$BW(i, j) = B(i, j) \oplus W(i, j) \tag{1.9}$$

$$K(i, j) = V(j) \oplus BW(i, j) \tag{1.10}$$

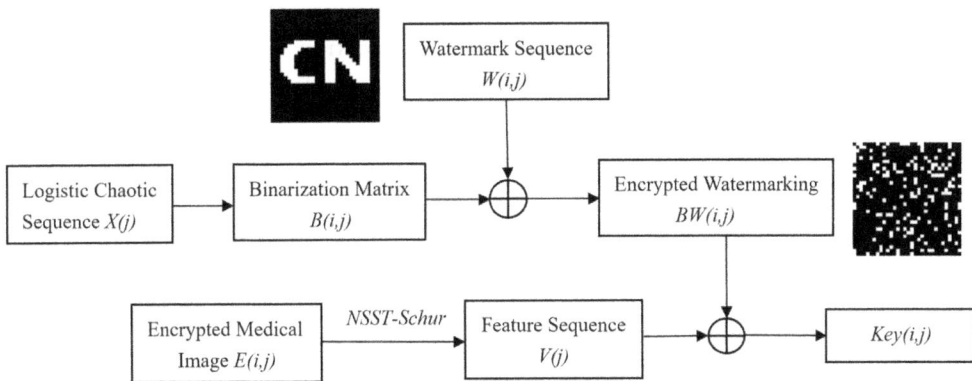

FIGURE 1.7   The embedding process of watermark.

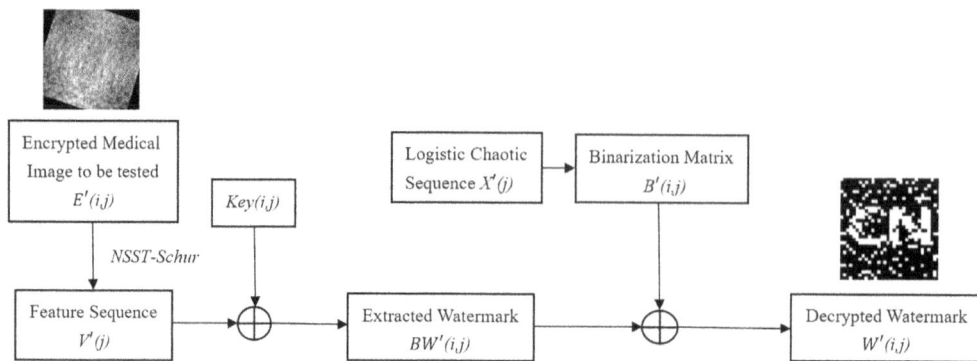

FIGURE 1.8   The process of extracting the watermark.

1.3.4 Extraction of Watermark

After attacks, extract the feature sequence from the tested encrypted medical image, and the encrypted watermark $BW'(i, j)$ can be extracted by performing a XOR logic operation between the feature sequence $V'(j)$ of the tested encrypted medical image and the logical key $K(i, j)$; see equation (1.11). And then uses logistic to generate a chaotic sequence with the same parameters to decrypt the watermarked image $W'(i, j)$, see equation (1.12), and finally realizes the extraction of the watermark; see Figure 1.8.

$$BW'(i, j) = V'(j) \oplus K(i, j) \tag{1.11}$$

$$W'(i, j) = BW'(j) \oplus B'(i, j) \tag{1.12}$$

1.4 EXPERIMENTS AND ANALYSIS OF RESULTS

1.4.1 Simulation Experiment

In this chapter, MATLAB 2019b was chosen as the experimental simulation platform. A medical image of the brain was used as the carrier medical image; pixel size is 512 by 512, 32 × 32 pixels size binary image containing 'CN' information was used as the watermark image. See Figure 1.9. With regards to the algorithm's encryption process's beginning value and growth parameter settings, Tent chaotic sequence parameters for medical image encryption are set: $X_0 = 0.34$, $\mu = 0.8$, and logistic chaotic sequence parameters for watermark encryption are set: $X_0 = 0.2$, $\mu = 4$. The experimental effect evaluation criteria used peak signal-to-noise ratio (PSNR), see equation (1.13), and normalized correlation (NC), see equation (1.14). For the purpose of evaluating the objectivity of medical picture quality, PSNR is employed. NC can more accurately use the data to objectively evaluate the similarity between images, and in this chapter, using NC as a reference, it is possible to assess how closely the extracted watermark resembles the original watermark that is embedded. When NC ≥ 0.5, this means the two have a good correlation. In our

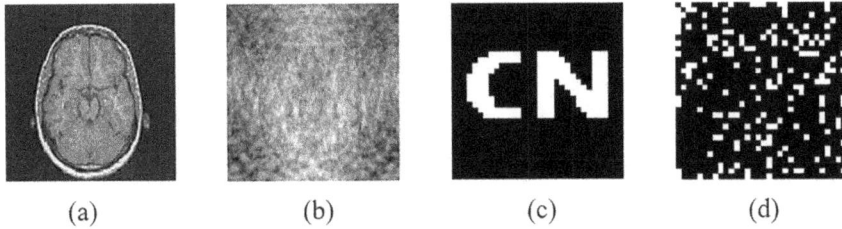

FIGURE 1.9 Medical pictures and watermarks: (a) original brain medical image, (b) encrypted brain medical image, (c) original binary watermark, and (d) encrypted watermark

experiments, we perform conventional and geometric attacks on an encrypted medical image embedded with watermarked information, respectively, to check the robustness of this algorithm.

$$PSNR = 10log_{10} \frac{MNmax_{i,j}(I_{(i,j)})^2}{\Sigma_i \Sigma_j (I_{(i,j)} - I'_{(i,j)})^2} \tag{1.13}$$

$$NC = \frac{\Sigma_i \Sigma_j W_{(i,j)} W'_{(i,j)}}{\Sigma_i \Sigma_j W_{(i,j)}^2} \tag{1.14}$$

1.4.2 Attacks Results

To evaluate how robust the method utilized in this study is, we test several common attacks on encrypted medical image of the brain embedded with watermarked information. Table 1.4 presents the test findings. Figure 1.10 displays a few of the encrypted medical images that were created following the attacks, together with the accompanying extracted watermarks. From Table 1.4, the experimental result data show that after 10% of the Gaussian noise intensity, JPEG compression 1%, and the

TABLE 1.4   PSNR and NC Values Under Test Attacks

Attack types	Intensity of attacks	PSNR/dB	NC
Gaussian noise	5%	13.42	1.00
	10%	11.16	1.00
	20%	9.44	1.00
JPEG compression	1%	24.83	1.00
	5%	27.95	1.00
	20%	34.56	1.00
Median filter (10 times)	[3 × 3]	36.17	1.00
	[5 × 5]	28.56	1.00
	[7 × 7]	25.73	1.00

(Continued)

TABLE 1.4 (Continued)   PSNR and NC Values Under Test Attacks

Attack types	Intensity of attacks	PSNR/dB	NC
Rotation (clockwise)	5°	17.52	0.93
	18°	14.73	0.68
	25°	14.11	0.52
Scaling	×0.5	-	1.00
	×0.8	-	1.00
	×2.0	-	1.00
Left translation	3%	17.08	0.87
	5%	16.03	0.75
	7%	15.42	0.68
Down translation	5%	16.18	0.87
	15%	13.40	0.61
	25%	11.55	0.51

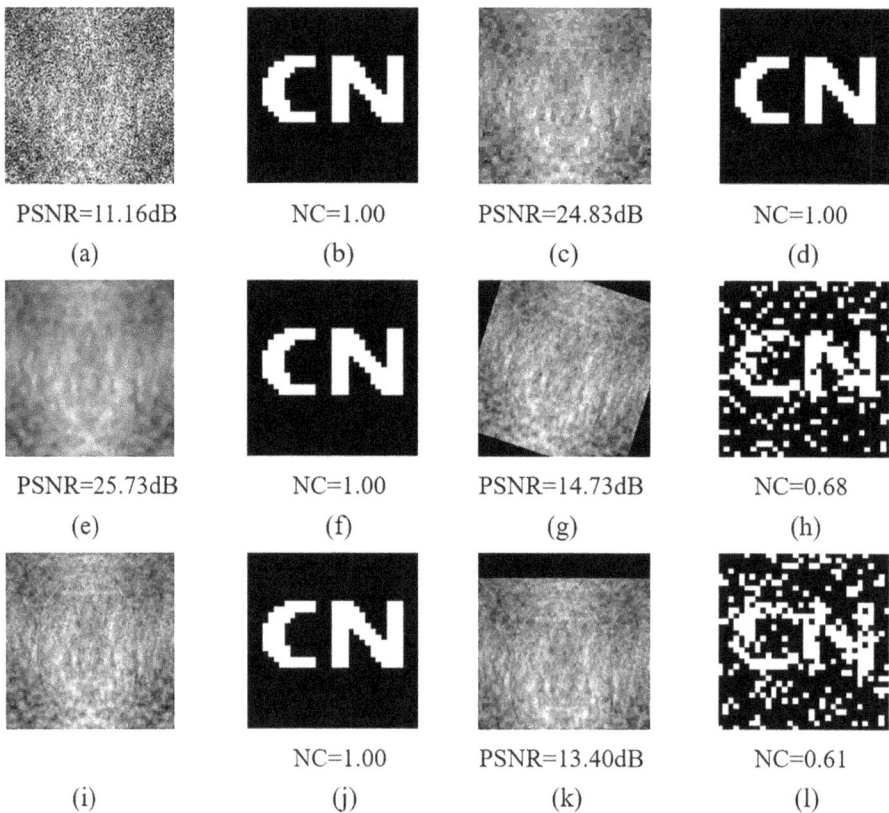

PSNR=11.16dB (a)   NC=1.00 (b)   PSNR=24.83dB (c)   NC=1.00 (d)

PSNR=25.73dB (e)   NC=1.00 (f)   PSNR=14.73dB (g)   NC=0.68 (h)

(i)   NC=1.00 (j)   PSNR=13.40dB (k)   NC=0.61 (l)

FIGURE 1.10   Encrypted medical images after attacks and the corresponding extracted watermark images: (a) Gaussian noise 10%, (b) extracted watermarks under attack a, (c) JPEG compression 1%, (d) extracted watermarks under attack c, (e) median filter [7 × 7] (10 times), (f) extracted watermarks under attack e, (g) clockwise rotation 18°, (h) extracted watermarks under attack g, (i) scaling two times, (j) extracted watermarks under attack i, (k) down translation 15%, and (l) extracted watermarks under attack.

median filter with window size [7 × 7] is repeated 10 times, the NC values of the watermarks obtained in the experiments are the same, both being 1. Therefore, it is possible to fully extract the watermark. It shows that this algorithm offers excellent resistance to the conventional attacks. At the same time, for resisting geometric attacks with higher attack strength, such as rotating 18° clockwise, scaling twice and translation down 15%, the NC values are all still more than 0.5, making it possible to precisely extract the watermark information. In conclusion, it can be seen that this encrypted domain approach not only provides security guarantee for watermark and medical image data, but also demonstrates excellent robustness and great attack resistance.

1.4.3 Contrastion to Plaintext Domain Algorithm

We compare and analyze the proposed encrypted domain algorithm with the plaintext domain algorithm. From Table 1.5, we can see that there is little difference between a plaintext domain and an encrypted domain in terms of PSNR and NC values. And the encrypted domain algorithm is better than the plaintext domain in resisting conventional attacks, although it is slightly weaker in resisting geometric attacks, but both PSNR and NC values are not much different from plaintext domain data. Even if the attack intensity is high, the watermark information can still be extracted completely; see Table 1.5.

TABLE 1.5   Comparison With Unencrypted Algorithm

Attack types	Intensity of attacks	PSNR (dB)		NC value	
		Plaintext domain	Encrypted domain	Plaintext domain	Encrypted domain
Gaussian noise	3%	16.22	15.39	1.00	1.00
	10%	11.89	11.16	1.00	1.00
	15%	10.60	10.08	1.00	1.00
JPEG compression	1%	26.28	24.83	1.00	1.00
	8%	30.29	30.43	1.00	1.00
	15%	32.91	33.43	1.00	1.00
Median filter (10 times)	[3 × 3]	34.00	36.17	1.00	1.00
	[5 × 5]	28.56	28.58	0.92	**1.00**
	[7 × 7]	26.40	25.73	0.85	**1.00**
Rotation (clockwise)	8°	16.21	16.44	**0.93**	0.82
	18°	14.68	14.73	**0.88**	0.68
	21°	14.57	14.45	**0.82**	0.63
Scaling	×0.5	-	-	0.92	**1.00**
	×2.0	-	-	1.00	1.00
	×4.0	-	-	1.00	1.00
Left translation	3%	15.13	17.08	**1.00**	0.87
	5%	14.48	16.03	**0.93**	0.75
	7%	14.31	15.42	**0.87**	0.68
Down translation	5%	15.11	16.18	**0.85**	0.87
	10%	14.74	14.52	**0.76**	0.63
	15%	14.28	13.40	**0.63**	0.61

TABLE 1.6   Comparison With Other Encrypted Algorithms

Attacks	NC value		
	DCT	DWT	DWT-DCT (proposed)
Gaussian noise 15%	1.00	0.64	1.00
JPEG compression 1%	0.94	1.00	1.00
Median filter ([7 × 7], 10 times)	1.00	0.93	1.00
Rotation 21° (clockwise)	0.63	0.16	0.63
Scaling ×2.0	1.00	0.92	1.00
Translation 5% (left)	0.68	0.29	0.75

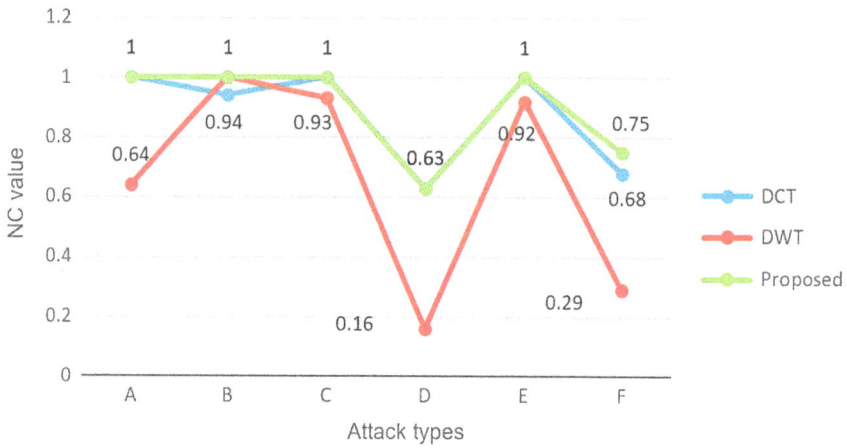

FIGURE 1.11   Comparison with other encrypted algorithms. (A) Gaussian noise 15%, (B) JPEG compression 1%, (C) median filter [7 × 7] (10 times), (D) clockwise rotation 21°, (E) scaling two times, and (F) left translation 5%

1.4.4 Contrastion to Other Encrypted Algorithms

In this chapter, we are based on the DWT-DCT transform to encrypt carrier medical images, and the following will be compared and analyzed with two encryption methods based on DCT and DWT, respectively. See Table 1.6. We can clearly find that the proposed algorithm for encrypting medical images is significantly superior to the other two; in particular, it works well against geometrical attacks.

To show the advantages of this encryption algorithm more clearly, the data of these three different encryption algorithms under the same attack are presented in the form of a line graph; see Figure 1.11. We discover that the means of encryption described in this chapter are substantially more resistant to both conventional and geometric assaults than the other two. Future work will be based on the development of new methods of deep learning [27–36].

1.5 CONCLUSION

In this chapter, we propose a robust watermarking algorithm for medical images based on DWT-DCT-Tent encryption method and NSST-Schur feature extraction. The information included in medical images is protected by medical image encryption technology. It also significantly increases medical image security, which used in online medical diagnosis. The algorithm also combines the translation invariance of NSST and the scaling invariance of Schur decomposition, which makes the algorithm robustness significantly enhanced. The concept of zero watermark technology and third party is used for embedding and extracting processes, which greatly raises the invisibility and security of the watermark. This algorithm offers excellent resistance against geometric and conventional attacks, according to experimental data, notably in geometric attacks. It has a good application prospect.

REFERENCES

1. Arsalan, M., Qureshi, A., Khan, A., & Rajarajan, M. (2017). Protection of medical images and patient-related information in healthcare: using an intelligent and reversible watermarking technique. *Applied Soft Computing*, 51, 168–179.
2. Bhardwaj, R. (2022). Hiding patient information in medical images: an encrypted dual image reversible and secure patient data hiding algorithm for E-healthcare. *Multimedia Tools and Applications*, 81, 1125–1152.
3. Liu, L. F., Wei, Z. X., & Xiang, H. Y. (2022). A novel image encryption algorithm based on compound-coupled logistic chaotic map. *Multimedia Tools and Applications*, 81(14), 19999–20019.
4. Rajendran, S., & Doraipandian, M. (2018). A nonlinear two dimensional logistic-tent map for secure image communication. *International Journal of Information and Computer Security*, 10(2-3), 201–215.
5. Sheng, M. S., Li, J. B., & Liu, J. (2022). Robust zero-watermarking algorithm for medical images based on Hadamard-DWT-DCT. *International Journal of Wireless and Mobile Computing*, 23(2), 183–192.
6. Liu, W. Y., Li, J. B., Shao, C. Y., Ma, J. X., Huang, M. X., & Bhatti, U. A. (2022). Robust zero watermarking algorithm for medical images using local binary pattern and discrete cosine transform. *Communications in Computer and Information Science*, 1588, 350–362.
7. Luo, Y. L., Li, L. J., Liu, J. X., Tang, S. B., Cao, L. C., Zhang, S. S., ... & Cao, Y. (2021). A multi-scale image watermarking based on integer wavelet transform and singular value decomposition. *Expert Systems with Applications*, 168.
8. Fan, D., Li, Y. Y., Gao, S., Chi, W. D., & Lv, C. Z. (2022). A novel zero watermark optimization algorithm based on Gabor transform and discrete cosine transform. *Concurrency and Computation: Practice and Experience*, 34(14).
9. Sanivarapu, P. V. (2022). Adaptive tamper detection watermarking scheme for medical images in transform domain. *Multimedia Tools and Applications*, 81(8), 11605–11619.
10. Xiang, S. J., Ruan, G. Q., Li, H., & He, J. Y. (2022). Robust watermarking of databases in order-preserving encrypted domain. *Frontiers of Computer Science*, 16(2).
11. Liu, J. L., Li, J. B., Chen, Y. W., Zou, X. X., Cheng, J. R., Liu, Y. L. & Bhatti, U. A. (2019). A robust zero-watermarking based on SIFT-DCT for medical images in the encrypted domain. *Computers, Materials and Continua*, 61(1), 135–141.
12. Liu, Y. L., Li, J. B., Liu, J., Cheng, J. R., Liu, J. L., Wang, L. R., & Bai, X. B. (2019). Robust encrypted watermarking for medical images based on DWT-DCT and tent mapping in encrypted domain. *Lecture Notes in Computer Science*, 11633, 584–596.
13. Fang, Y. X., Liu, J., Li, J. B., Yi, D., Cui, W. F., Xiao, X. L., ... & Bhatti, U. A. (2021). A novel robust watermarking algorithm for encrypted medical image based on Bandelet-DCT. *Smart Innovation, Systems and Technologies*, 242, 61–73.

14. Liu, J., Li, J. B., Cheng, J. R., Ma, J. X., Sadiq, N., Han, B. R., ... & Ai, Y. (2019). A novel robust watermarking algorithm for encrypted medical image based on DTCWT-DCT and chaotic map. Computers, Materials & Continua, 61(2), 889–910.

15. Xiao, T., Li, J. B., Liu, J., Cheng, J. R., & Bhatti, U. A. (2018). A robust algorithm of encrypted face recognition based on DWT-DCT and tent map. *Lecture Notes in Computer Science*, 11064, 508–518.

16. Jiang, L. Z., Xu, C. X., Wang, X. F., Luo, B., & Wang, H. Q. (2020). Secure outsourcing SIFT: efficient and privacy-preserving image feature extraction in the encrypted domain. *IEEE Transactions on Dependable and Secure Computing*, 17(1), 179–193.

17. Li, W., Huang, Q., & Srivastava, G. (2021). Contour feature extraction of medical image based on multi-threshold optimization. *Mobile Networks and Applications*, 26(1), 381–389.

18. Han, L., Wang, J. W., Zhang, Y., Sun, X., & Wu, X. H. (2022). Research on adaptive ORB-SURF image matching algorithm based on fusion of edge features. *IEEE Access*, 10, 109488–109497.

19. Sheng, M. S., Li, J. B., Bhatti, U. A., Liu, J., Huang, M. X., & Chen, Y. W. (2023). Zero watermarking algorithm for medical image based on Resnet50-DCT. *Computers, Materials and Continua*, 75(1), 293–309.

20. Yang, C. S., Li, J. B., Bhatti, U. A., Liu, J., Ma, J. X., & Huang, M. X. (2021). Robust zero watermarking algorithm for medical images based on Zernike-DCT. *Security and Communication Networks*, 2021.

21. Gong, C., Li, J. B., Bhatti, U. A., Gong, M., Ma, J. X., & Huang, M. X. (2021). Robust and secure zero-watermarking algorithm for medical images based on Harris-SURF-DCT and chaotic map. *Security and Communication Networks*, 2021.

22. Su, Q. T., Su, L., Wang, G., Li, L. D., & Ning, J. T. (2020). A novel colour image watermarking scheme based on Schur decomposition. *International Journal of Embedded Systems*, 12(1), 31–38.

23. Sripradha, R., & Deepa, K. (2021). Robust and imperceptible digital image watermarking based on DWT-DCT-Schur. *Communications in Computer and Information Science*, 1365, 337–351.

24. Liu, P., Wu, H., Luo, L., & Wang, D. S. (2022). DT CWT and Schur decomposition based robust watermarking algorithm to geometric attacks. *Multimedia Tools and Applications*, 81(2), 2637–2679.

25. Liu, W. J., Sun, S. Y., & Qu H. C. (2019). Fast zero watermarking algorithm for Schur decomposition. *Computer Science and Exploration*, 13(3), 494–504.

26. Li, D., Li, J., Bhatti, U. A., Nawaz, S. A., Liu, J., Chen, Y. W., & Cao, L. (2023). Hybrid encrypted watermarking algorithm for medical images based on DCT and improved DarkNet53. *Electronics*, 12(7), 1554.

27. Bhatti, U. A., Tang, H., Wu, G., Marjan, S., & Hussain, A. (2023). Deep learning with graph convolutional networks: an overview and latest applications in computational intelligence. *International Journal of Intelligent Systems*, 2023, 1–28.

28. Sheng, M., Li, J., Bhatti, U. A., Liu, J., Huang, M., & Chen, Y. W. (2023). Zero watermarking algorithm for medical image based on Resnet50-DCT. *CMC-Computers Materials & Continua*, 75(1), 293–309.

29. Liu, J., Li, J., Ma, J., Sadiq, N., Bhatti, U. A., & Ai, Y. (2019). A robust multi-watermarking algorithm for medical images based on DTCWT-DCT and Henon map. *Applied Sciences*, 9(4), 700.

30. Fan, Y., Li, J., Bhatti, U. A., Shao, C., Gong, C., Cheng, J., & Chen, Y. (2023). A multi-watermarking algorithm for medical images using inception V3 and DCT. *CMC-Computers Materials & Continua*, 74(1), 1279–1302.

31. Li, T., Li, J., Liu, J., Huang, M., Chen, Y. W., & Bhatti, U. A. (2022). Robust watermarking algorithm for medical images based on log-polar transform. *EURASIP Journal on Wireless Communications and Networking*, 2022(1), 1–11.
32. Bhatti, U. A., Huang, M., Wu, D., Zhang, Y., Mehmood, A., & Han, H. (2019). Recommendation system using feature extraction and pattern recognition in clinical care systems. *Enterprise Information Systems*, 13(3), 329–351.
33. Bhatti, U. A., Yu, Z., Chanussot, J., Zeeshan, Z., Yuan, L., Luo, W., … & Mehmood, A. (2021). Local similarity-based spatial–spectral fusion hyperspectral image classification with deep CNN and Gabor filtering. *IEEE Transactions on Geoscience and Remote Sensing*, 60, 1–15.
34. Liu, J., Li, J., Ma, J., Sadiq, N., Bhatti, U. A., & Ai, Y. (2019). A robust multi-watermarking algorithm for medical images based on DTCWT-DCT and Henon map. *Applied Sciences*, 9(4), 700.
35. Bhatti, U. A., Yu, Z., Li, J., Nawaz, S. A., Mehmood, A., Zhang, K., & Yuan, L. (2020). Hybrid watermarking algorithm using Clifford algebra with Arnold scrambling and chaotic encryption. *IEEE Access*, 8, 76386–76398.
36. Liu, J., Li, J., Zhang, K., Bhatti, U. A., & Ai, Y. (2019). Zero-watermarking algorithm for medical images based on dual-tree complex wavelet transform and discrete cosine transform. *Journal of Medical Imaging and Health Informatics*, 9(1), 188–194.

CHAPTER 2

Robust Zero Watermarking Algorithm for Encrypted Medical Images Based on SUSAN-DCT

Jingbing Li[1,2], Qinqing Zhang[1], Meng Yang[1], and Yiyi Yuan[1]

[1]*School of Information and Communication Engineering, Hainan University, Haikou, China*
[2]*State Key Laboratory of Marine Resource Utilization in the South China Sea,*
 Hainan University, Haikou, China

2.1 INTRODUCTION

There has always been interest in the development of the medical industry. It is inevitable to combine it with network science and technology. The powerful transmission capability of web technology can give the medical business the advantage of being faster. As medical information goes digital, more and more medical data and personal information will be transferred online. In this process of transmission, it will inevitably be attacked by nature or humans, which will damage or leak the transmitted data information and cause great social harm. Therefore, it is worth studying to encrypt medical information data. The research of digital watermarking [1] technology is essentially information hiding, but it can be used to improve the security of information by using medical data or personal privacy information as watermarking. The watermarking technology in the field of image is to combine the watermarking information with the image and make the image transport the watermarking information as the information carrier. However, the traditional watermarking technology will cause a certain degree of distortion to the carrier image. In the medical field, doctors carry out pathological diagnosis through medical images [2], and medical images must maintain a high degree of originality, so the embedding of watermarking cannot cause any change to the medical image data. To meet this demand and to make the watermark simultaneously invisible and robust, zero-watermark [3] technology is introduced, which will extract the features from the carrier image. The advantage of zero watermarking over traditional watermarking technology is that the embedding of the

DOI: 10.1201/9781003427674-2

watermark does not cause any degree of change to the medical image, simply by embedding the features of the medical image, and the more stable the extracted image feature information is, the more complete the watermark information will be. Constructing a zero-watermark using image features ensures the originality of the medical image while successfully embedding the watermark information into the carrier image.

2.2 LITERATURE REVIEW

Literature [4] proposes a watermarking algorithm based on DWT-DCT, which makes use of the anti-noise and anti-compression advantages of DWT and DCT to embed watermarking in color images. While this algorithm has good robustness against Gaussian noise attacks, it has poor robustness against geometrical attacks. Literature [5] combines DWT and SVD. Experimental results demonstrate that this algorithm has good robustness and certain shear resistance in the face of compression attacks. Literature [6] combines RHFM on the basis of DT-DCT to effectively improve the security of the algorithm, but it still does not have good robustness against geometric attacks. In reference [7], in order to avoid the time cost caused by repeated operations, QGFD is used to calculate image features, and this algorithm can effectively resist geometric attacks. Document [8] uses image edge information to embed watermark, which has a good ability to resist compression attacks. Literature [9] proposes an efficient watermarking scheme, but it has poor robustness against shear attacks. Literature [10] proposes a zero-watermarking algorithm that is resistant to shear attack and can still extract effective watermarking when image information is lost. Literature [11] combines Gabor transform on the basis of DCT and takes advantage of the information concentration of both to improve the anti-attack ability of the algorithm and has good robustness in anti-shear attack. Literature [12] uses contourlet transform to transform the three channels of color images and Gould transform to transform the bandpass direction coefficient. This algorithm has good resistance to geometric attacks, but its efficiency is mediocre because it involves gradient calculation. Literature [13] involves sine and cosine algorithms, and uses logical chaotic mapping to improve algorithm security, which can resist geometric attacks. In literature [14], the ring template is used to extract features from different channels of color images. The algorithm has a good ability to fight against common geometric attacks; especially in the aspect of anti-rotation attacks, it has good robustness.

In order to give consideration to the robustness and security of the algorithm and improve the operation efficiency, a zero-watermarking algorithm based on SUSAN DCT is proposed. By using SUSAN transform to extract image contour, and combining DCT and Hu invariant moments to generate feature matrix, the algorithm has good performance against attacks, in particular rotation attacks and shear attacks. Because SUSAN transform has no complex operation and gradient calculation, the efficiency of this algorithm is better.

2.3 BASIC THEORY AND PROPOSED ALGORITHM

2.3.1 SUSAN Edge Detection

The smallest univalue segment assimilating nucleus (SUSAN) [15] algorithm can be used for edge extraction. In the first step, an approximate circular template is used to move on

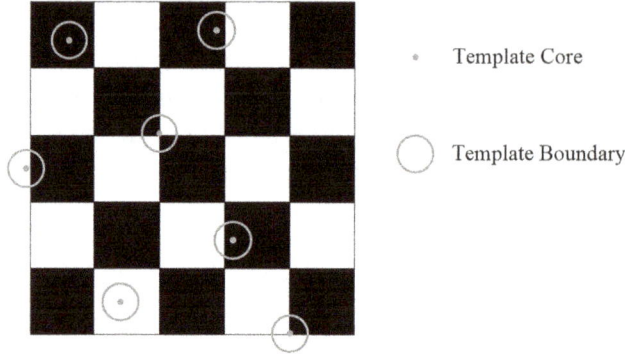

FIGURE 2.1 SUSAN edge extraction template.

the target image. Image edges extracted from circular template are shown in Figure 2.1. If the difference between the gray level of the inner pixel of the template and the gray level of the central pixel of the template is less than a certain threshold, it is considered that the point and the kernel have the same or similar gray level.

The area composed of the pixels meeting such conditions is called USAN. The discriminant function is shown in equation (2.1).

$$c(r, r_0) = \begin{cases} 1, & |I(r) - I(r_0)| \le t \\ 0, & |I(r) - I(r_0)| > t \end{cases} \tag{2.1}$$

where r_0 represents the position of the core point in the two-dimensional image, r represents the position of other points in the template, and t represents a threshold value of brightness interpolation, which determines the maximum brightness difference between points in the USAN region.

In the second step, the number of pixel values $n(r_0)$ with similar brightness values between the circular template and the core point was counted. The statistical formula is shown in equation (2.2).

$$n(r_0) = \sum c(r, r_0) \tag{2.2}$$

The third step is to use the corner response function, as shown in equation (2.3). If the USAN value of a certain pixel is less than a certain threshold, it is considered the initial corner point. And g is generally set as about 0.75 time of the maximum number of points similar to the center.

$$R(r_0) = \begin{cases} g - n(r_0), & n(r_0) < g \\ 0, & n(r_0) \ge g \end{cases} \tag{2.3}$$

The fourth step is to obtain the final edge by non-extreme suppression of the initial edge.

2.3.2 Hu Moments

Momentum is a concept in probability and statistics, which is a numeric feature of random variables. The Hu invariant moment [16] is calculated using the central moment invariant to the image transformation, which has some translational invariance, scale invariance, and rotational invariance. The formulas for calculating the seven Hu invariant moments are shown in equations (2.4) to (2.10).

$$hu[0] = \mu_{20} + \mu_{02} \tag{2.4}$$

$$hu[1] = (\mu_{20} - \mu_{02})^2 + 4\mu_{11}^2 \tag{2.5}$$

$$hu[2] = (\mu_{30} - 3\mu_{12})^2 + 3(\mu_{21} - \mu_{03})^2 \tag{2.6}$$

$$hu[3] = (\mu_{30} + \mu_{12})^2 + (\mu_{21} + \mu_{03})^2 \tag{2.7}$$

$$hu[4] = (\mu_{30} - 3\mu_{12})(\mu_{30} + \mu_{12})[(\mu_{30} + \mu_{12})^2 - 3(\mu_{21} + \mu_{03})^2] \\ + 3(\mu_{21} - \mu_{03})(\mu_{21} + \mu_{03})[3(\mu_{30} + \mu_{12})^2 - (\mu_{21} + \mu_{03})^2] \tag{2.8}$$

$$hu[5] = (\mu_{20} - \mu_{02})[(\mu_{30} + \mu_{12})^2 - (\mu_{21} + \mu_{03})^2] \\ + 4\mu_{11}(\mu_{30} + \mu_{12})(\mu_{21} + \mu_{03}) \tag{2.9}$$

$$hu[6] = (3\mu_{21} - \mu_{03})(\mu_{30} + \mu_{12})[(\mu_{30} + \mu_{12})^2 - 3(\mu_{21} + \mu_{03})^2] \\ - (\mu_{30} - 3\mu_{12})(\mu_{21} + \mu_{03})[3(\mu_{30} + \mu_{12})^2 - (\mu_{21} + \mu_{03})^2] \tag{2.10}$$

Hu invariant moments can reflect the characteristic information of the image, especially the invariant moments based on the second-order moments are more stable for the representation of the image and have better ability to describe the shape of the subject in the image.

2.3.3 Logical Mapping

The main idea of chaotic mapping is to generate chaotic sequences by some iterative way, with randomness and extreme sensitivity to the selection of initial values. Image encryption based on logistic mapping, on the other hand, changes the image pixel values by the generated random sequence, and small changes in the initial values can lead to large differences in the sequence. The iterative formula is shown in equation (2.11).

$$X_{k+1} = \mu X_k(1 - X_k), \quad k = 0, 1, 2, ..., n \tag{2.11}$$

When different parameters μ are changed, the dynamic limit behavior of the equation will be different. The experiment shows that when $0 < X_0 \leq 1$ and $3.5699456 < \mu \leq 4$, the logistic mapping enters the chaotic state [17].

2.3.4 Proposed Algorithm

2.3.4.1 Medical Image Encryption

The medical image encryption process is shown in Figure 2.2.

The first step is to perform discrete wavelet transform on the original medical image $I(i, j)$ to obtain low-frequency and high-frequency signals. The second step is to carry out discrete cosine transform for each signal to obtain the coefficient matrix $D(i, j)$ of discrete wavelet transform. In the third step, the chaotic sequence of tent map is binarized by the step function, $sgn(x)$, and the binary encryption matrix, $T(i, j)$, is obtained. The fourth step is to obtain the encrypted coefficient matrix $E(i, j)$ by XOR operation of $T(i, j)$ and $D(i, j)$. The fifth step is to carry out inverse discrete cosine transform and inverse discrete wavelet transform on E successively to obtain the encrypted medical image $I'(i, j)$.

2.3.4.2 Watermark Encryption

The watermark encryption process is shown in Figure 2.3.

The first step is to obtain the binary encryption matrix, $B(i, j)$, by binarizing the chaotic sequence of the logical map. In the second step, XOR operation is performed between $B(i, j)$ and the original watermark image, $W(i, j)$, to obtain the encrypted watermark image, $EW(i, j)$.

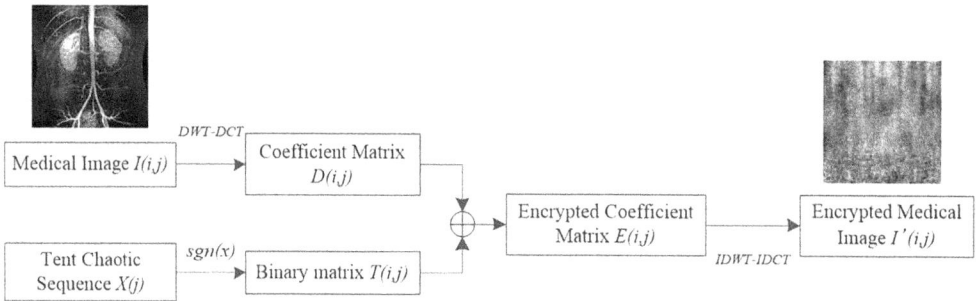

FIGURE 2.2   Medical image encryption.

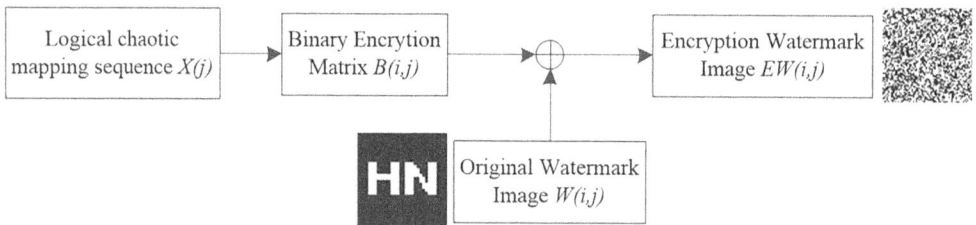

FIGURE 2.3   Watermark encryption.

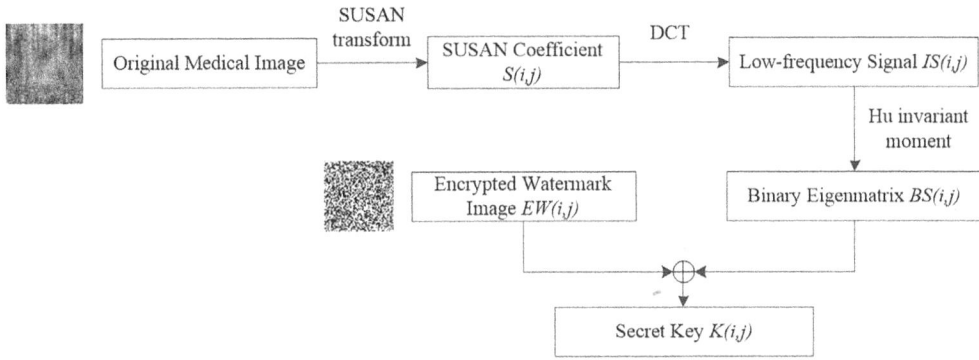

FIGURE 2.4   Watermark embedding.

2.3.4.3 Watermark Embedding

The watermark embedding process is shown in Figure 2.4.

The first step is to obtain the SUSAN coefficient matrix $S(i, j)$ after SUSAN transformation of the pre-encrypted image. In the second step, $S(i, j)$ is transformed by discrete cosine to obtain the low-frequency signal $IS(i, j)$. In the third step, Hu invariant moment is used to calculate and binarize $IS(i, j)$, and the binary eigenmatrix $BS(i, j)$ is obtained. In the fourth step, $BS(i, j)$ is XOR processed with the encrypted watermark image $EW(i, j)$, and logic of the secret key $K(i, j)$ is obtained; the watermark embedded the encryption medical image successfully.

2.3.4.4 Extraction and Decryption of Watermark

The watermark embedding process is shown in Figure 2.5.

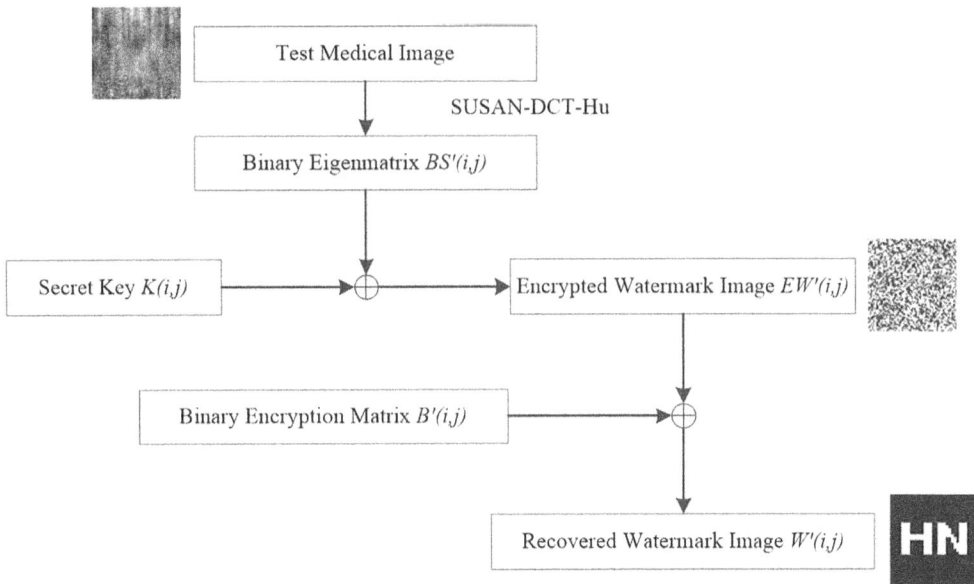

FIGURE 2.5   Watermark extraction and decryption.

The medical image to be tested goes through the first three steps of the watermarking embedding process to obtain the binary eigenmatrix $BS'(i, j)$ of the image to be tested. The encrypted watermark image $EW'(i, j)$ is obtained by XOR of $BS'(i, j)$ and the key $K(i, j)$. Finally, the extracted watermark image $W'(i, j)$ is obtained by XOR of the matrix $B'(i, j)$ and $EW'(i, j)$.

2.4 EXPERIMENT AND RESULTS

2.4.1 Evaluation Parameter

In order to evaluate the watermark quality, the peak signal-to-noise ratio (PSNR) and the normalized correlation coefficient [18] are selected as evaluation indexes. The PSNR is used to quantify the image quality after the attack and may reflect the degree of image distortion [19–29]. To evaluate the quality of the watermark, we use the NC, and measure the similarity between the extracted watermark and the original watermark by computing the number of mutual relationships between the two. PSNR and NC can be expressed respectively by equations (2.12) and (2.13).

$$PSNR = 10 \log_{10} \left[\frac{MNmax\,(I(i, j))^2}{\Sigma_i \Sigma_j\,[I(i, j) - I'(i, j)]^2} \right] \tag{2.12}$$

$$NC = \frac{\Sigma_i \Sigma_j\ W(i, j)\,W'(i, j)}{\Sigma_i \Sigma_j\,[W(i, j)]^2} \tag{2.13}$$

The image quality is directly proportional to the PSNR. If the image is under attack, the quality of the image will be affected, and the PSNR will decrease accordingly. The NC measures the correlation between the images. When it is greater than 0.5, two images can be regarded as having a certain similarity.

2.4.2 Experimental Setup

The experimental environment was Windows 10, and the experimental platform was Matlab R2017b; the size of medical image selected was 512 × 512, and the size of watermark image was selected as 64 × 64. In order to verify whether the features extracted by this algorithm are valid features, that is, whether the algorithm can distinguish different medical images, six medical images of different parts are selected for testing, as shown in Figure 2.6. The experimental data after the test is shown in Table 2.1. The NC between the same images is 1.00, and the correlation between the characteristic data of medical images of different parts is low, indicating that the algorithm can extract different characteristic information from different medical images and distinguish different medical images (Figure 2.7).

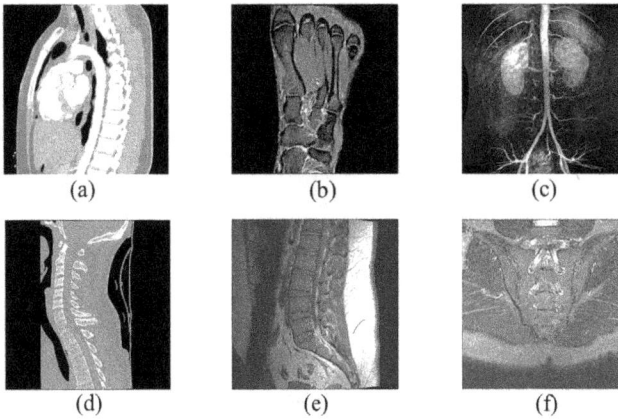

FIGURE 2.6   Different medical images.

TABLE 2.1   Correlation Coefficient Between the Feature Sequences of Six Test Images (64 Bits)

Image	Arm	Foot	Lung	Spine	Knee	Sacroiliac bone
Arm	1	0.21	0.13	0.32	−0.11	0.20
Foot	0.21	1	0.31	0.30	−0.15	0.14
Lung	0.13	0.31	1	0.17	−0.19	0.13
Spine	0.32	0.30	0.17	1	−0.32	0.17
Knee	−0.11	−0.15	−0.19	−0.32	1	−0.01
Sacroiliac bone	0.20	0.14	0.13	0.17	−0.01	1

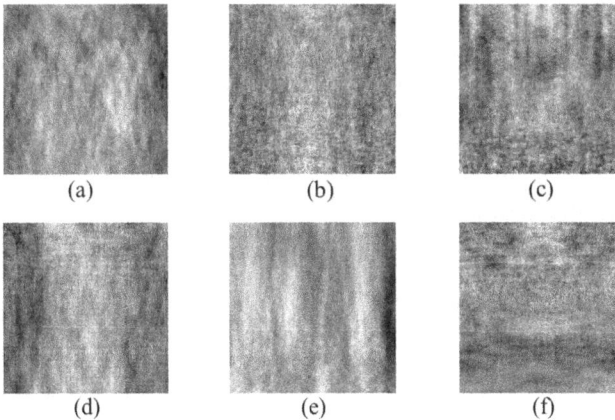

FIGURE 2.7   Encrypted medical images.

2.4.3 Results and Analysis

2.4.3.1 Gaussian Noise Attack

The encrypted lung images were subjected to a Gaussian noise attack, with the intensity ranging from 1% to 20%. Test data and results in this range are shown in Table 2.2. When the attack intensity is increased from 1% to 15%, PSNR has been reduced from 15.94dB

TABLE 2.2   Gaussian Noise Attack Data

Noise intensity (%)	1	5	10	15	20
PSNR	15.94	9.33	6.47	4.87	3.85
NC	0.95	0.79	0.58	0.63	0.53

to 4.87dB. At this time, the extracted watermark has a correlation coefficient of 0.63 with the original watermark. If the intensity is increased to 20%, it is almost impossible to distinguish image information with naked eyes under this attack intensity. However, the NC value of the watermark image extracted from it is still greater than 0.5, which indicates that the embedded watermark is secure and effective, and the algorithm is resistant to a certain level of Gaussian noise attacks.

2.4.3.2 JPEG Compression Attack

The experimental data and results of the JPEG compression attack performed are shown in Table 2.3, which shows that when the compression strength increases from 20% to 2%, although PSNR decreases from 31.50 dB to 22.05 dB, and when the compression strength is 10%, the distortion degree of the image is large, but the watermark correlation coefficient extracted in this attack range is almost stable around 0.98, far greater than 0.5. The results show that the algorithm performs well in the face of JPEG compression attacks.

2.4.3.3 Median Filter Attack

When the median filtering attack experiment was performed on the encrypted lung image, the experiment of five-times filtering and ten-times filtering was selected under the window sizes of 3 × 3, 5 × 5, and 7 × 7, and the test data and results were shown in Table 2.4. It can be seen that after the attacks of different window sizes and filtering times of the encrypted lung images, the NC is all above 0.9 and stable at about 0.96. The image quality of the encrypted lung image is somewhat reduced after ten times of filtering under the window size of 7 × 7, but the NC is still 0.93, indicating that the algorithm is highly resistant to median filter attack.

TABLE 2.3   JPEG Compression Attack Data

Compression quality (%)	2	5	10	15	20
PSNR	22.05	25.21	28.08	29.85	31.50
NC	1.00	1.00	0.98	1.00	1.00

TABLE 2.4   Median Filter Attack Data

Times	5			10		
Parameter	[3 × 3]	[5 × 5]	[7 × 7]	[3 × 3]	[5 × 5]	[7 × 7]
PSNR	28.68	23.98	21.96	28.33	22.47	20.58
NC	0.97	0.96	0.96	0.97	0.96	0.93

TABLE 2.5   Rotation Attack Data

Rotate clockwise (°)	2	10	13	20	26	30	33	40	45
PSNR	16.20	12.87	12.77	12.52	12.21	12.05	11.92	11.46	11.30
NC	0.92	0.72	0.74	0.72	0.74	0.60	0.60	0.54	0.55

2.4.3.4 Rotation Attack

During the rotation attack experiment, the test image is subjected to clockwise rotation attack. As can be seen through Table 2.5, when the rotation angle increases from 2° to 10°, the NC value decreases significantly, but when the rotation angle continues to increase, the NC value keeps decreasing steadily. Overall, the PSNR of the picture decreases when the rotation angle increases from 2° to 45°. Even though the quality of images rotated by 30° is somewhat reduced, the NC is still greater than 0.5, indicating that correct watermark images can be extracted from medical images after rotation. It also shows that the algorithm is able to resist strong rotation attacks and has good robustness in the face of rotation attacks.

2.4.3.5 Scaling Attack

Scale attack experiments were conducted on the encrypted lung images, and experimental data and results obtained under different attack intensifies were shown in Table 2.6. The size of the image changes after the scaling attack, so the corresponding PSNR cannot be calculated. Data in the table shows that when the scaling attack is performed on the encrypted lung image, the narrowing ratio ranges from 0.1 to 0.5, and the watermark correlation coefficient extracted under the attack intensity within this interval is stable above 0.6. When the magnification attack is performed on the encrypted lung image, the magnification ratio ranges from 1.5 to 3.0, and the watermark correlation coefficient extracted under the attack intensity within this interval is still greater than 0.5. It can be considered that the algorithm is able to resist scaling attacks of a certain strength.

2.4.3.6 Translation Attack

When performing translation attack, the attack method of horizontal shift to the left is selected. The experimental data and results are shown in Table 2.7. When the encrypted

TABLE 2.6   Scaling Attack Data

Zoom (x)	0.1	0.3	0.5	1.5	2.0	3.0
NC	0.76	0.66	0.61	0.61	0.69	0.52

TABLE 2.7   Translation Attack Data

Translation to left (%)	1	5	10	15	20
PSNR	21.64	17.65	16.90	16.64	16.55
NC	0.86	0.75	0.67	0.65	0.65

TABLE 2.8   Cropping Attack Data

Cropping along the x-axis (%)	1	5	10	20	30	35
NC	1.00	0.84	0.82	0.57	0.64	0.69

lung image is shifted 20% to the left, the area filled after deletion appears in the right part of the image. As shown in Table 2.7, when the attack intensity increases from 1% to 20%, although the PSNR value continues to decline, the extracted watermark NC value is greater than 0.6, indicating that the watermark image has been successfully extracted. The experimental results show the good performance of the algorithm on the translation attack.

2.4.3.7 Cropping Attack
When performing cropping attack, the mode of shear is selected on the x-axis. The experimental data is shown in Table 2.8. When 30% of the contents of the encrypted lung image were cut off, there was a missing area in the image and the size of the image changed, so the PSNR after the attack could not be calculated. As can be seen from the data shown in Table 2.8, as the clipping area increases from 1% to 35%, more and more image areas and details are lost, and the extracted watermark correlation coefficient can still reach about 0.6. This shows that the algorithm has good performance in the face of shear attacks and has good shear resistance and robustness.

2.5 CONCLUSION
In this chapter, a robust zero-watermarking algorithm based on SUSAN-DCT is proposed for encrypted medical images. Firstly, DWT-DCT and the tent chaos map are used for encryption of the original medical image. SUSAN transformation was then carried out on the encrypted medical image and DCT transform was performed after contour information was obtained. After that, the Hu invariant moment was used to calculate the obtained low-frequency signal, and the eigenmatrix of the encrypted medical image was obtained after binarization of the results. Finally, the feature matrix and the watermarking image after logical chaos encryption are XOR operation, and the watermarking information is successfully embedded into the encrypted medical image, and the construction of zero watermarking is realized. The experimental results show that the algorithm has good robustness against conventional attacks and good performance against geometric attacks, especially against rotation attacks and shearing, and can resist higher attack strength. The algorithm has some resistance to scaling; however, this aspect can still be optimized to improve the robustness in scaling. The algorithm not only uses cryptography theory, but also uses zero-watermarking technology to embed and extract watermarks, which can guarantee the safety of the algorithm and balance the invisibility and robustness of the watermark.

REFERENCES

1. Zainab, F. M., & Abdulamir, A. K. (2021). Digital watermark technique: A review. *Journal of Physics: Conference Series*.

2. Zhang, X. B., Zhao, X. L., Wu, Y. Y., Zheng, H. L., & Li, Y. (2023). Federated learning for medical image classification: Advances, challenges and opportunities. *SSRN*.
3. Wen, Q., Sun, T. F., & Wang, S. X. (2003). Concept and application of zero-watermark. *Acta Electronica Sinica*, 31(2), 214–216.
4. Abadi, R. Y., & Moallem, P. (2022). Robust and optimum color image watermarking method based on a combination of DWT and DCT. *Optik*, 261.
5. Yao, J. C., & Shen, J. (2022) Image watermark combining with discrete wavelet transform and singular value decomposition. *Communications in Computer and Information Science*, 1563, 115–124.
6. Li, Z. Y., Zhang, H., Liu, X. L., Wang, C. P., & Wang, X. Y. (2021). Blind and safety-enhanced dual watermarking algorithm with chaotic system encryption based on RHFM and DWT-DCT. *Digital Signal Processing*, 115, 103062.
7. Wang, B. W., Wang, W. S., & Zhao, P. (2021). Multiple color medical images zero-watermark scheme based on quaternion generalized Fourier descriptor and QR code. *ACM International Conference Proceeding Series*, 248–253.
8. Xu, H. Y. (2021). Digital media zero watermark copyright protection algorithm based on embedded intelligent edge computing detection. *Mathematical Biosciences and Engineering*, 18(5), 6771–6789.
9. Wang, B. W., Wang, W. S., Zhao, P., & Xiong, N. X. (2022). A zero-watermark scheme based on quaternion generalized Fourier descriptor for multiple images. *Computers, Materials and Continua*, 71(2), 2633–2652.
10. Wang, J., Palaniappan, S., & He, B. (2022). A zero-watermarking against large-scale cropping attack. *Lecture Notes in Electrical Engineering*, 813, 365–375.
11. Fan, D., Li, Y. Y., Gao, S., Chi, W., & Lv, C. Z. (2022). A novel zero watermark optimization algorithm based on Gabor transform and discrete cosine transform. *Concurrency and Computation: Practice and Experience*, 34(14).
12. Thomas, R., & Sucharitha, M. (2022). Contourlet and Gould transforms for hybrid image watermarking in RGB color images. *Intelligent Automation and Soft Computing*, 33(2), 879–889.
13. Daoui, A., Karmouni, H., Elogri, O., Sayyouri, M., & Qjidaa, H. (2022). Robust image encryption and zero-watermarking scheme using SCA and modified logistic map. *Expert Systems with Applications*, 190.
14. Yang, J. H., Hu, K., Wang, X. C., Wang, H. F., Liu, Q., & Mao, Y. (2022). An efficient and robust zero watermarking algorithm. *Multimedia Tools and Applications*, 81(14), 20127–20145.
15. Smith, S. M., & Brady, J. M. (1997). SUSAN – a new approach to low level image processing. *International Journal of Computer Vision*, 23(1), 45–78.
16. Balakrishnan, S., & Joseph, P. K. (2022). Stratification of risk of atherosclerotic plaque using Hu's moment invariants of segmented ultrasonic images. *Biomedizinische Technik*, 67(5), 391–402.
17. Zhang, Y., He, Y., Zhang, J., & Liu, X. B. (2022). Multiple digital image encryption algorithm based on chaos algorithm. *Mobile Networks and Applications*, 27(4), 1349–1358.
18. Zhu, T., Qu, W., & Cao, W. L. (2022). An optimized image watermarking algorithm based on SVD and IWT. *Journal of Supercomputing*, 78(1), 222–237.
19. Li, D., Li, J., Bhatti, U. A., Nawaz, S. A., Liu, J., Chen, Y. W., & Cao, L. (2023). Hybrid encrypted watermarking algorithm for medical images based on DCT and improved DarkNet53. *Electronics*, 12(7), 1554.
20. Bhatti, U. A., Tang, H., Wu, G., Marjan, S., & Hussain, A. (2023). Deep learning with graph convolutional networks: An overview and latest applications in computational intelligence. *International Journal of Intelligent Systems*, 2023, 1–28.

21. Sheng, M., Li, J., Bhatti, U. A., Liu, J., Huang, M., & Chen, Y. W. (2023). Zero watermarking algorithm for medical image based on Resnet50-DCT. *CMC-Computers Materials & Continua*, 75(1), 293–309.

22. Liu, J., Li, J., Ma, J., Sadiq, N., Bhatti, U. A., & Ai, Y. (2019). A robust multi-watermarking algorithm for medical images based on DTCWT-DCT and Henon map. *Applied Sciences*, 9(4), 700.

23. Fan, Y., Li, J., Bhatti, U. A., Shao, C., Gong, C., Cheng, J., & Chen, Y. (2023). A multi-watermarking algorithm for medical images using inception V3 and DCT. *CMC-Computers Materials & Continua*, 74(1), 1279–1302.

24. Li, T., Li, J., Liu, J., Huang, M., Chen, Y. W., & Bhatti, U. A. (2022). Robust watermarking algorithm for medical images based on log-polar transform. *EURASIP Journal on Wireless Communications and Networking*, 2022(1), 1–11.

25. Bhatti, U. A., Huang, M., Wu, D., Zhang, Y., Mehmood, A., & Han, H. (2019). Recommendation system using feature extraction and pattern recognition in clinical care systems. *Enterprise Information Systems*, 13(3), 329–351.

26. Bhatti, U. A., Yu, Z., Chanussot, J., Zeeshan, Z., Yuan, L., Luo, W., … & Mehmood, A. (2021). Local similarity-based spatial–spectral fusion hyperspectral image classification with deep CNN and Gabor filtering. *IEEE Transactions on Geoscience and Remote Sensing*, 60, 1–15.

27. Bhatti, U. A., Yu, Z., Li, J., Nawaz, S. A., Mehmood, A., Zhang, K., & Yuan, L. (2020). Hybrid watermarking algorithm using Clifford algebra with Arnold scrambling and chaotic encryption. *IEEE Access*, 8, 76386–76398.

28. Liu, J., Li, J., Zhang, K., Bhatti, U. A., & Ai, Y. (2019). Zero-watermarking algorithm for medical images based on dual-tree complex wavelet transform and discrete cosine transform. *Journal of Medical Imaging and Health Informatics*, 9(1), 188–194.

29. Zhang, Y., Chen, J., Ma, X., Wang, G., Bhatti, U. A., & Huang, M. (2024). Interactive medical image annotation using improved Attention U-net with compound geodesic distance. *Expert Systems with Applications*, 237, 121282.

CHAPTER 3

Robust Zero Watermarking Algorithm for Encrypted Medical Volume Data Based on PJFM and 3D-DCT

Lei Cao[1], Jingbing Li[1,2], and Uzair Aslam Bhatti[1,2]

[1]*School of Information and Communication Engineering, Hainan University, Haikou, China*
[2]*State Key Laboratory of Marine Resource Utilization in the South China Sea,*
Hainan University, Haikou, China

3.1 INTRODUCTION

The adoption of digital healthcare, smart cities, and intelligent healthcare has accelerated significantly as the field of medical big data develops quickly. The fusion, sharing, and open application of these medical big data have brought unprecedented opportunities for modern medical services and industries, but have also caused a series of data security and privacy issues [1]. For instance, in 2020, one of the most significant healthcare data breaches occurred when a ransomware attack targeted Blankbaud, a cloud server provider. The German vulnerability analysis and management company Greenbone Networks discovered that more than 737 million radiographic images were leaked, involving basic personal identity information and medical details of over 20 million individuals [2]. Attackers can use this information to deploy and implement scams and phishing attacks, thereby gaining high returns and causing social unrest [3]. The most important issue is the leakage and tampering of medical images. Medical images promoting a thorough comprehension of severe conditions, and reducing the incidence of misdiagnosis. They serve as an essential medium for retaining medical information [4]. In contrast to ordinary images, medical images encompass a significant volume of crucial information, upon which doctors rely for making accurate assessments and informed judgments. Therefore, the accuracy and integrity of medical images are highly demanded [5,6]. However, medical images transmitted over networks or stored in the cloud without processing are vulnerable to privacy leakage and are often subject to various malicious attacks. Consequently, securing medical imaging data is a hot research area right now.

DOI: 10.1201/9781003427674-3

Currently, the main technique for ensuring the confidentiality of medical photographs is encryption technology. The protection of information security and confidentiality within the medical picture domain can be ensured by encrypting authentic medical images [7]. The encrypted image assumes a chaotic and disordered state, rendering its content incomprehensible to unauthorized individuals. However, authorized users possess the key required to decrypt the image and restore it to its original form. Spatial domain picture encryption and frequency domain image encryption are the two basic methods used in image encryption. Scrambling encryption and grayscale encryption are additional subcategories of spatial domain picture encryption. On the other hand, the process of encrypting images in the frequency domain entails converting them to the time-frequency domain, such as DCT, DFT, DWT, etc., the frequency domain data is then encrypted after being converted from the picture matrix [8]. Frequency domain encryption is regarded to be more secure than spatial domain encryption. Therefore, this research paper adopts DWT-DCT frequency domain image encryption. Furthermore, since medical images contain sensitive patient information and diagnosis records, it becomes imperative to guarantee image clarity, data integrity, and security during network transmission while effectively extracting patient information. Consequently, by integrating digital watermark technology with confidentiality measures, it becomes feasible to safeguard the security of patient information in addition to the actual medical images.

Digital watermarking is widely regarded as an effective approach for addressing information leakage and safeguarding the security of medical images [9]. By embedding medical information as watermark information within medical images, the algorithm ensures the secure transmission of information, effectively safeguarding patient privacy. Given the unique characteristics of medical images, the conventional embedding of general watermarks can compromise image integrity and potentially mislead doctors during diagnosis. This method makes use of zero-watermark technology to make sure that the embedding of watermark data has no detrimental effects on the visual quality of the original medical image [10]. Currently, the majority of research on digital watermarking algorithms primarily focuses on the plaintext domain [11,12]. In the current state of research, most digital watermarking algorithms work in the plaintext space, where watermark embedding and extraction take place on unencrypted carrier images. However, this approach poses a security risk as intercepted carrier images during network transmission and storage can potentially expose sensitive information. We suggest using digital watermarking techniques in the ciphertext domain to reduce this risk. In order to do this, the watermark data must be inserted within the encrypted domain and the original medical image must be encrypted. A trusted third party can get the encrypted watermark and medical image by utilizing the homomorphic characteristics of the encryption technique, ensuring comprehensive protection for both the medical image and watermark information [13,14]. Moreover, the existing digital watermarking technologies predominantly focus on two-dimensional medical images, neglecting the abundance of three-dimensional images in the medical domain, such as CT and MRI images that comprise multiple slices. Three-dimensional medical data offers a more comprehensive and precise representation of patient information, facilitating easier and more accurate diagnoses for doctors. Consequently, there is a significant need to explore methodologies for embedding watermarks within encrypted medical volumetric data, enabling enhanced protection and security in this context.

Feature extraction constitutes a vital component of digital watermarking technology, and several widely utilized methods currently exist. The most commonly employed techniques encompass the DCT, DWT, and SIFT, as well as feature point extraction methods like SUSAN feature points and Harris-Laplace feature points. These methods play a crucial role in accurately identifying and extracting distinctive features within the watermarking process. Previous research has achieved many results by using these methods in various fields. In recent years, researchers have started studying moment invariants and found that they are highly condensed image features with translation, scale, rotation, and grayscale invariance. Introducing moment invariants into digital watermarking technology can enhance the algorithm's robustness [15], such as ZMs (Zernike Moments), PZMs (Pseudo-Zernike Moments), Zernike (Zernike polynomials), JFMs (Jacobi-Fourier Moments), and PJFMs (Pseudo Jacobi-Fourier Moments). In literature [16], Zernike moments are employed for sub-pixel edge detection and feature extraction in medical volume data. The algorithm exhibits robust anti-noise capabilities. However, it does not provide encryption for the original medical volume data, leaving it susceptible to tampering attacks. In literature [17], a digital watermarking algorithm is proposed based on JFM. This algorithm exploits the rotational invariance of JFM moment amplitudes to embed binary watermark sequences into the amplitude of JFM moments through quantization [18–28]. The algorithm demonstrates effectiveness in countering geometric and conventional attacks. However, it is limited to the application on two-dimensional images and has not yet been extended to medical images or medical volume data. PJFM is a variation of JFM proposed by Amu G L, with fewer moments and lower time consumption but better rotational invariance. Combining the good anti-geometric attack performance of PJFM moments with the good anti-conventional attack performance of 3D-DCT transformation, to increase the reliability of digital watermarking for medical applications, watermarks are incorporated into feature vectors generated from medical volume data.

This chapter presents the robust zero-watermarking technique for encrypted medical volume data. The algorithm incorporates methods for zero watermarking, chaotic encryption technology, PJFMs (Pseudo Jacobi-Fourier Moments), and the 3D-DCT (3D Discrete Cosine Transform) algorithm.

3.2 THE FUNDAMENTAL THEORY

3.2.1 Pseudo Jacobi-Fourier Moment

In 2003, Amu G.L. et al. proposed the theory of Pseudo Jacobi-Fourier moments, which exhibits multi-distortion invariance, including gray level, scale, translation, and rotation. These moments are particularly suitable for image feature extraction [13]. The radial Jacobi polynomial in the interval [0,1] can be defined as:

$$G_n(p, q, r) = \frac{\Gamma(q + n)}{\Gamma(p + 2n)} \sum_{m=0}^{n} (-1)^m \binom{n}{m} \frac{\Gamma(p + 2n - m)}{\Gamma(q + n - m)} r^{n-m} \tag{3.1}$$

Suppose $s = n - m$, and when $p = 4$ and $q = 3$, the Jacobi polynomial can be transformed into:

$$J_n(r) = a_n \sqrt{(1 - r)r} \sum_{s=0}^{n} b_{ns} r^s \tag{3.2}$$

In this way,

$$a_n = (-1)^n \sqrt{\frac{(2n + 4)}{(n + 3)(n + 1)}} \tag{3.3}$$

$$b_{ns} = (-1)^s \frac{(n + s + 3)!}{(n - s)! s! (s + 2)!} \tag{3.4}$$

where n signifies a non-negative integer, m has a value range of $- n \leq m \leq n$, and n and m are the order of the polynomials. The range of s is $0 \leq s \leq n$.

The polynomial satisfies the condition of orthogonal normalization:

$$\int_0^1 J_n(r) J_k(r) r dr = \delta_{nk} \tag{3.5}$$

As a result, we can define a system of functions, $P_{nm}(r, \theta)$, in polar coordinates (r, θ). It is made up of an angular function, $\exp(jm\theta)$, and a radial function, $J_n(r)$.

$$P_{nm}(r, \theta) = J_n(r) \exp(jm\theta) \tag{3.6}$$

Of course, the function system, $P_{nm}(r, \theta)$, also meets the requirement for orthogonal normalization.

$$\int_0^{2\pi} \int_0^1 P_{nm}(r, \theta) P_{kl}(r, \theta) r dr d\theta = \delta_{nmkl} \tag{3.7}$$

The orthogonal normalized polynomial $\{p_n(r)\}$ of the weight function $p(r)$ is said to be able to expand the function into a generalized Fourier series when it meets the Dirichlet condition, in accordance with the orthogonality theory. Then $f(r, \theta)$ can be expanded as:

$$f(r, \theta) = \sum_{n=0}^{+\infty} \sum_{m=-\infty}^{+\infty} \Phi_{nm} J_n(r) \exp(jm\theta) \tag{3.8}$$

Therefore, the deformed Jacobian ($p = 4$, $q = 3$)-Fourier moment can be obtained:

$$\Phi_{nm} = \frac{1}{2\pi} \int_0^{2\pi} \int_0^1 f(r, \theta) J_n(r) \exp(-jm\theta) r dr d\theta \qquad (3.9)$$

where $f(r, \theta)$ denotes the image function and the value range of r is $[0,1]$ (which displays the target image's scale in the polar coordinate system). Essentially, the deformed Jacobi moment refers to the distance from the centroid of the target image to its center, serving as the origin of the coordinate system.

3.2.2 D-DCT and 3D-IDCT

For data blocks of the size $n_x \times n_y \times n_z$, the 3D-DCT transform calculation is performed using equation (3.3):

$$F(u, v, w) = C(u)C(v)C(w)\left[\sum_{x=0}^{n_x-1}\sum_{y=0}^{n_y-1}\sum_{z=0}^{n_z-1} f(x, y, z) \cdot \frac{(2x+1)u\pi}{2n_x}\frac{(2y+1)v\pi}{2n_y}\frac{(2z+1)w\pi}{2n_z}\right] \qquad (3.10)$$

Here,

$$C(u) = \begin{cases} \sqrt{1/n_x}\ u = 0 \\ \sqrt{2/n_x}\ u \neq 0 \end{cases}, \quad C(v) = \begin{cases} \sqrt{1/n_y}\ v = 0 \\ \sqrt{2/n_y}\ v \neq 0 \end{cases}, \quad C(w) = \begin{cases} \sqrt{1/n_z}\ w = 0 \\ \sqrt{2/n_z}\ w \neq 0 \end{cases} \qquad (3.11)$$

where $f(x, y, z)$ is the volume metadata value of the volume data, V, located at position (x, y, z), and $F(u, v, w)$ represents the volume metadata's equivalent 3D-DCT transformation coefficient [14].

In addition, the size of the block is $n_x \times n_y \times n_z$, calculated by the 3D-IDCT transform:

$$f(x, y, z) = \sum_{u=0}^{n_x-1}\sum_{v=0}^{n_y-1}\sum_{w=0}^{n_z-1}\left[C(u)C(v)C(w)\frac{(2x+1)u\pi}{2n_x} \cdot F(u, v, w)\frac{(2y+1)v\pi}{2n_y}\frac{(2z+1)w\pi}{2n_z}\right] \qquad (3.12)$$

where $C(u)$, $C(v)$, and $C(w)$ are the same as those in the 3D-DCT transform; equation (3.3).

3.2.3 Logistic Mapping

A one-dimensional discrete chaotic system known as logistic mapping operates quickly. By iteratively applying specific equations, it can generate improved chaotic sequences [15]. Both the initial state and the system characteristics have a significant impact on the chaotic sequence that results. The following definitions apply to the logistic mapping:

$$x_{k+1} = \mu x_k (1 - x_k) \qquad (3.13)$$

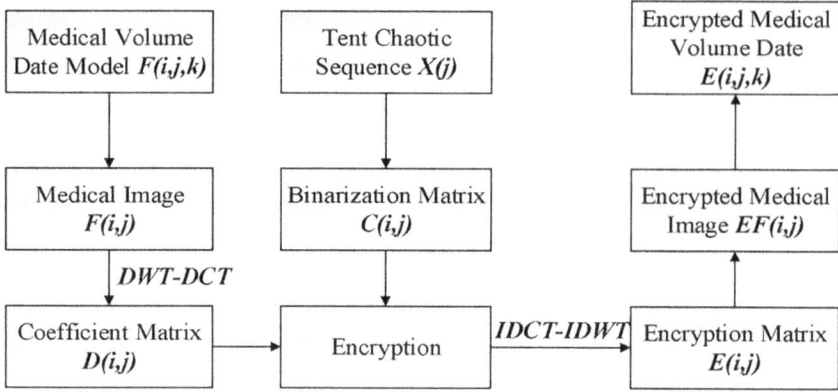

FIGURE 3.1 Medical volume data model encryption process.

where k is the number of iterations, $x_k \in (0, 1)$ is the system variable, and $0 < \mu \leq 4$ stands for the branch parameter. Numerous studies on logistic mapping have demonstrated that as the parameter u approaches its limiting value, that is, $u = 3.5699456$, the steady-state solution period of the system is ∞. As shown in Figure 3.1, when $3.5699456 < \mu \leq 4$, the state of logistic mapping is chaotic. Therefore, to attain chaotic behavior, the value range of μ should be set as follows:

$$3.5699456 < \mu \leq 4 \tag{3.14}$$

3.3 THE PROPOSED METHOD

3.3.1 Medical Volume Data Encryption

Figure 3.1 shows the flowchart of the encryption process for medical volume data. These are the precise steps: To begin with, k medical images $F(i,j)$ are created by slicing the original medical volume data $F(i,j,k)$. Each medical image is then subjected to the DWT in order to extract the wavelet coefficients for the various sub-bands, including cA, cH, cV, and cD. Each wavelet coefficient matrix, $D(i,j)$, is then subjected to the DCT. The binary encryption matrix, $C(i,j)$, is then created using the tent chaotic sequence and sgn (x) function. The encrypted coefficient matrix, $ED'(i,j)$, is created by computing the dot product of the coefficient matrix, $D(i,j)$, with the encryption matrix, $C(i,j)$. The encrypted coefficient matrix, $ED'(i,j)$, is then transformed using the IDCT to produce the encrypted sub-band wavelet coefficient matrix, $ED(i,j)$. The IDWT transform is then applied to $ED(i,j)$ to produce the encrypted medical picture slice, $EF(i,j)$. The reconstruction procedure is then used to incorporate the encrypted medical picture slices into the encrypted medical volume data, $E(i,j,k)$.

$$\mathrm{sgn}(x) = \begin{cases} 1, & x(n) \geq 0 \\ -1, & x(n) < 0 \end{cases} \tag{3.15}$$

$$X'(j) = \text{sgn}(X(j)) \tag{3.16}$$

$$C(i, j) = reshapeX'(j) \tag{3.17}$$

$$ED'(i, j) = D(i, j)C(i, j) \tag{3.18}$$

$$ED(i, j) = IDCT2(ED'(i, j)) \tag{3.19}$$

$$EF(i, j) = IDWT2(ED(i, j)) \tag{3.20}$$

$$E(i, j, k) = reshapeEF(i, j) \tag{3.21}$$

As shown in Figure 3.2, medical images are obtained by slicing the original medical volume data. These medical images are then subjected to DWT-DCT and chaotic tent mapping operations to obtain encrypted medical images, as shown in Figure 3.3. It can be seen that the encrypted medical images cannot be observed by the naked eye and no information can be obtained from them.

3.3.2 Feature Extraction

PJFM and 3D-DCT are combined in the feature extraction procedure. These techniques possess the ability to accurately describe image features with excellent rotational

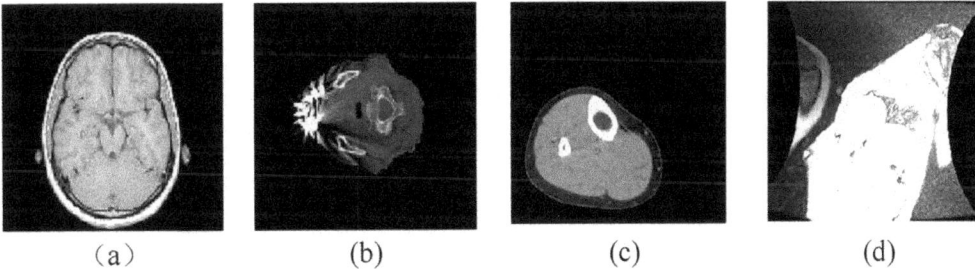

| (a) | (b) | (c) | (d) |

FIGURE 3.2 The original medical images: (a) brain 1, (b) brain 2, (c) leg, and (d) ventricle.

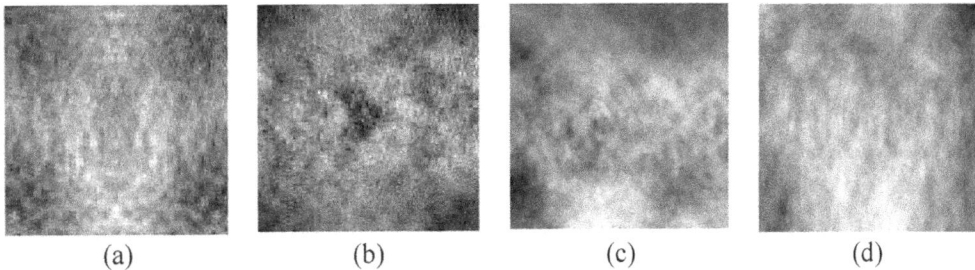

| (a) | (b) | (c) | (d) |

FIGURE 3.3 Encrypted medical images.

invariance and orthogonality. As a result, they can be effectively utilized for extracting feature vectors from medical volume data. In this algorithm, PJFM and 3D-DCT are employed to extract the perceptual hash value of medical volume data features. The target object for testing was chosen as a conventional MRI brain volume data with dimensions of 128 pixels × 128 pixels × 27 pixels, encrypted its sliced images with DWT-DCT, performed PJFM and 3D-DCT transformations, and then subjected it to various attack tests. As for visual feature vectors, we selected a 4 × 8 coefficient matrix from the low-frequency coefficients, and the perceptual hash threshold value is the average of 32-bit coefficients. The threshold value for the perceptual hash is used to compare each coefficient. A coefficient is substituted with "1" if it is larger than or equal to the threshold. Conversely, if a coefficient is less than the threshold, it is replaced with "0". This process results in a 32-bit binary sequence.

As depicted in Figure 3.4, an additional step is conducted to validate the effectiveness of the feature vector obtained through this method, we randomly selected five different volume data to extract visual feature vectors. Table 3.1 data shows that the NC value of all volume data with themselves is 1. Furthermore, the obtained feature vectors exhibit small NC (normalized cross-correlation) values when comparing different volume data. This indicates the ability to differentiate between various volume data, indicating that the encrypted medical volume data's low-frequency coefficients, obtained using this algorithm, can serve as effective visual feature vectors.

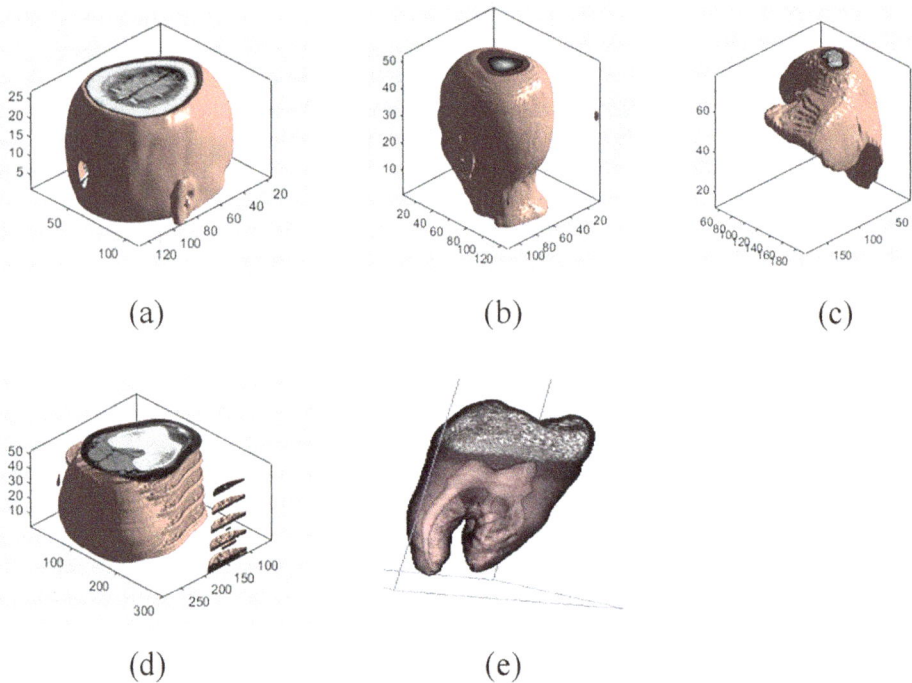

(a)   (b)   (c)

(d)   (e)

FIGURE 3.4   Different volume data: (a) brain 1, (b) brain 2, (c) liver, (d) leg bone, and (e) teeth.

TABLE 3.1 Correlation Coefficients are Calculated Between the Perceptual Hash Values of Different Volume Data

	(a)	(b)	(c)	(d)	(e)
(a)	1	0.3928	0.0976	0.5096	0.3415
(b)	0.3928	1	0.4472	0.2809	0.2809
(c)	0.0976	0.4472	1	−0.0897	0.3410
(d)	0.5096	0.2809	−0.0897	1	0.2271
(e)	0.3415	0.2809	0.3410	0.2271	1

3.3.3 Watermark Encryption and Embedding

The watermark's encryption and embedding processes are shown in Figure 3.5. The watermark encryption generates a chaotic sequence $X(j)$ with a starting value of ×0 using the logistic map. The values of the chaotic sequence are then arranged in ascending order. The positional changes of each value in $X(j)$ before and after sorting are then used to scramble the watermark's pixel space. The watermark for chaotic encryption, $BW(i,j)$, is then obtained.

The encrypted medical volume data is first chopped to produce a two-dimensional medical image for the watermark. The medical image is then transformed using the PJFM algorithm, which yields approximative coefficients. Following that, a 3D-DCT transformation is performed to acquire the coefficient matrix. The low-frequency 4×8 matrix from the coefficient matrix is perceptually hashed to yield a 32-bit visual feature vector, $V(i,j)$, for the medical image. The resultant feature vector, $V(i,j)$, is then bit by bit XORed

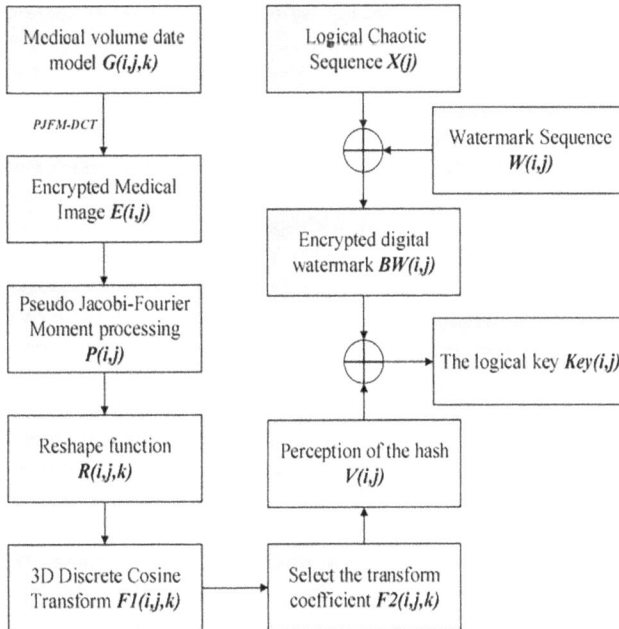

FIGURE 3.5 The flowchart of watermark encryption.

with the encrypted watermark, *BW(i,j)*. This technique creates the logical key, *Key(i,j)*, while also embedding the watermark into the encrypted volume contents. In order to extract the watermark in the future, the logical key is saved in a separate location.

3.3.4 Watermark Extraction and Decryption

Following the reverse process of embedding and encryption, watermarks are extracted and decrypted. Firstly, the encrypted medical image data, which has undergone tampering attacks, is sliced to obtain two-dimensional medical images. Subsequently, PJFM and 3D-DCT transformations are applied to these images, extracting transformed coefficients. These coefficients result in the feature sequence, *V'(i,j)*. The feature sequence, *V'(i,j)*, is then bitwise XORed with the logical key, *Key(i,j)*. Using this technique, the encrypted watermark, *BW'(i,j)*, can be extracted.

$$BW'(i, j) = Key(i, j) \oplus V'(i, j) \tag{3.22}$$

The binary feature sequence is obtained using the same method as the encryption procedure. The chaotic-scrambled watermark, *BW'(i,j)*, position space is then put through XOR and restoration processes based on this binary sequence. By sorting the values of the logistic chaotic sequence, *X(j)*, in ascending order and using the positional changes made before and after sorting for each value in *X(j)*, the positions of the pixels in the watermark *BW'(i,j)* are restored. This restoration procedure results in the watermark, *W'(i,j)*, which has been decrypted. In order to identify the true owner of the medical volume data and watermark, the correlation coefficient NC value between the original watermark, *BW(i,i)*, and the decrypted watermark, *W'(i,j)*, is calculated.

3.4 EXPERIMENTAL RESULTS AND PERFORMANCE EVALUATION

3.4.1 Simulation Experiment

We conducted simulation experiments using MATLAB 2022a as the experimental platform. The experiment utilized medical body data (as shown in Figure 3.2(b)) for testing, with a watermark image consisting of a 32 × 32 pixel "HN" letter image. An initial value of ×0 = 0.34 and a growth parameter of = 0.8 were used in the encryption process of medical photos. An initial value of ×0 = 0.2 and a growth parameter of = 4 were chosen for the watermark image's encryption.

Peak signal-to-noise ratio (PSNR) was utilized to formally evaluate the quality of the medical images. Additionally, the correlation between the original watermark and the recovered watermark was examined using the normalized correlation coefficient (NC). The following is the mathematical formula for PSNR and NC:

$$PSNR = 10 \lg \left[\frac{MN \max_{i,j}(\mathrm{I}(i, j))^2}{\Sigma_i \Sigma_j (\mathrm{I}(i, j) - \mathrm{I}'(i, j))^2} \right] \tag{3.23}$$

$$NC = \frac{\Sigma_i \Sigma_j W_{(i,j)} W'_{(i,j)}}{\Sigma_i \Sigma_j W^2_{(i,j)}} \qquad (3.24)$$

3.4.2 Attacks Results

The medical body data from Table 3.2 was put through geometric and conventional attacks, and the test results are displayed in Table 3.2. According to the findings in Table 3.2, the quality of the derived watermark gradually deteriorates as the intensity of conventional attacks on the medical body data rises. For instance, the watermark may still be clearly retrieved when the Gaussian attack hits 20% and the median filter with a 77 kernel is performed ten times. In this case, the NC value is around 0.8. Additionally, even after 10% JPEG compression, the NC value stays at 1, demonstrating the algorithm's great resistance to common attacks.

Digital watermarking technology is challenged by geometric attacks, such as rotation. Rotation and other geometric assaults were regarded as important components in this investigation. When the clockwise rotation angle reached 40°, the resulting NC value was 0.72, indicating that discernible watermark information could still be extracted. Similarly, for other geometric attacks such as a downward translation of 12%, Z-axis shearing of 40%, and scaling by a factor of 0.65, the extracted watermark's NC values stayed above 0.7.

TABLE 3.2   The Experimental Results Under Attacks

Attack types	Intensity of attacks	PSNR (dB)	NC
Gaussian noise	3%	15.44	1.00
	10%	11.20	0.89
	20%	9.45	0.80
JPEG compression	1%	22.36	0.89
	5%	24.51	1.00
	10%	26.64	1.00
Median filter (10 times)	[3 × 3]	26.71	0.89
	[5 × 5]	22.96	0.79
	[7 × 7]	21.06	0.79
Scaling	× 0.65	-	0.80
	× 0.75	-	1.00
	× 4	-	1.00
Rotation (clock wise)	3°	20.63	1.00
	20°	14.84	0.80
	40°	13.56	0.72
Movement (down)	2%	21.16	1.00
	6%	16.70	0.89
	12%	14.40	0.71
Cropping (Z direction)	5%	-	1.00
	10%	-	0.89
	40%	-	0.71

TABLE 3.3 The PSNR and NC Values in PJFM-DCT Plaintext and Encrypted Domain

Attacks	PSNR (dB)		NC value	
	Plaintext domain	Encrypted domain	Plaintext domain	Encrypted domain
Gaussian noise 10%	3.32	11.20	0.80	0.89
JPEG compression 10%	20.21	26.00	1.00	1.00
Median filter[5 × 5]/10 times	18.69	22.96	1.00	0.79
Scaling ×0.75	-	-	0.89	1.00
Rotation 40° (clock wise)°	13.97	16.63	1.00	0.80
Movement (down) 12%	10.51	14.40	0.80	0.71
Cropping Z direction 10%	-	-	1.00	0.89

This result indicates that the watermark information was still discernible, demonstrating the algorithm's resistance against geometric attacks.

3.4.3 Comparison with Unencrypted Algorithm

Table 3.3 displays the findings of our comparison between the proposed algorithm and the algorithm for unencrypted medical body data. The table demonstrates that there are no appreciable differences in the PSNR and NC values between the ciphertext domain (encrypted) and plaintext domain (unencrypted) data. This finding indicates that the suggested algorithm demonstrates strong homomorphic features, meaning that the encryption process does not significantly degrade the quality of the medical body data.

3.5 CONCLUSION

In this research, a reliable zero-watermarking approach for encrypting body-related medical data is presented. The algorithm utilizes feature extraction techniques such as PJFM and 3D-DCT, along with DWT-DCT for enhanced security and protection by using the rotation invariance and strong feature descriptor of PJFM matrix transformation algorithm, and combining with the 3D-DCT algorithm to extract features with energy concentrated in the low-frequency region. The validity and integrity of medical body data are further ensured by this algorithm's use of zero-watermarking technology, which enables invisible watermark insertion. The watermark can be retrieved even in the absence of the original medical body data or the encrypted body data by taking advantage of the homomorphic property of the encryption technique. The network medical diagnosis's data security is improved by the encryption technology for medical body data, which protects the data inside. The simulation results clearly demonstrate the algorithm's excellent numerical stability and resistance to a variety of attacks.

REFERENCES

1. OuYang, T., Yang, Y. F., & Shu, J. H. (2020). Ethical thinking on patients' awareness of privacy protection in the context of health big data[J]. *Journal of Chaohu University* 22(6), 91–97.

2. Wei, C. X., & Li, W. J. (2022). Study on patient privacy protection and countermeasures in the application of health and medical big data. *Network Security Technology & Application*, pp. 64–67.

3. Memon, N. A., & Alzahrani, A. (2020). Prediction-based reversible watermarking of CT scan images for content authentication and copyright protection. *IEEE Access*. pp. 1–1.

4. Liu, Y., Li, J., Liu, J., Cheng, J., Wang, L., & Bai, X. (2019). *Robust encrypted watermarking for medical images based on DWT-DCT and tent mapping in encrypted domain.* Springer International Publishing, pp. 584–596.

5. Qasim, A. F., Meziane, F., & Aspin, R. (2017). Digital watermarking: Applicability for developing trust in medical imaging workflows state of the art review. *Computer Science Review*, pp. 45–60.

6. Guo, J., Zheng, P., & Huang, J. (2015). Secure watermarking scheme against watermark attacks in the encrypted domain. *Journal of Visual Communication and Image Representation*. pp. 125–135.

7. Avudaiappan, T., Balasubramanian, R., Pandiyan, S. S., Saravanan, M., Lakshmanaprabu, S.k., & Shankar, K. (2018). Medical image security using dual encryption with oppositional based optimization algorithm. *Journal of Medical Systems* 42, 208.

8. Fares, K., Khaldi, A., Redouane, K., & Salah, E. (2021). DCT&DWT based watermarking scheme for medical information security. *Biomedical Signal Processing and Control* 66, 102403.

9. Liu, J., Li, J., Zhang, K., Bhatti, U. A., & Ai, Y. (2019). Zero-watermarking algorithm for medical images based on dual-tree complex wavelet transform and discrete cosine transform. *Journal of Medical Imaging and Health Informatics*, 9, 188–194.

10. Yang, Y., Liu, X., Deng, R. H., & Li, Y. (2017). Lightweight sharable and traceable secure mobile health system. *IEEE Transactions on Dependable and Secure Computing*. pp. 1.

11. Kexin, L. E., & Liangliang, W. (2020). Image encryption scheme based on chaos in cloud environment. *Journal of Shanghai University of Electric Power* 36, 500–504.

12. Gao, H., Zhang, Y., Liang, S., & Li, D. (2006). A new chaotic algorithm for image encryption. *Chaos, Solitons&Fractals* 29, 393–399.

13. Vengadapurvaja, A. M., Nisha, G., Aarthy, R., & Sasikaladevi, N. (2017). An efficient homomorphic medical image encryption algorithm for cloud storage security. *Procedia Computer Science* 115, 643–650.

14. Dai, Y., Wang, H., Yu, Y., & Zhou, Z. (2016). Research on medical image encryption in telemedicine systems. *Technology Health Care* 24 Suppl 2, S435.

15. Ke, T. T., & Chen, Q. (2017). A robust image watermarking to geometric attacks based on Pseudo-Zernike moments. *ShuJu TongXin*. pp. 43–46.

16. Yang, C. S., & Li, J. B.(2021). Robust zero watermarking algorithm for medical images based on Zernike-DCT. *Security and Communication Networks*, 2021, Article ID 4944797, 8.

17. Sun, Y. X., Wang, X. W., Li, L. D., & Li, S. S. (2014). Image watermarking based on Jacobi-Fourier moment. *Computer Engineering and Applications* 50(4), 94–97.

18. Amu, G. L., Yang, X. Y., & Ping, Z. L.(2003). Describing image with pseudo-Jacobi (p=4,q=3)-Fourier moments. *Journal of Optoelectronics Laser* 14(9), 981–985.

19. Li, D., Li, J., Bhatti, U. A., Nawaz, S. A., Liu, J., Chen, Y. W., & Cao, L. (2023). Hybrid encrypted watermarking algorithm for medical images based on DCT and improved DarkNet53. *Electronics* 12(7), 1554.

20. Bhatti, U. A., Tang, H., Wu, G., Marjan, S., & Hussain, A. (2023). Deep learning with graph convolutional networks: An overview and latest applications in computational intelligence. *International Journal of Intelligent Systems* 2023, 1–28.

21. Sheng, M., Li, J., Bhatti, U. A., Liu, J., Huang, M., & Chen, Y. W. (2023). Zero watermarking algorithm for medical image based on Resnet50-DCT. *CMC-Computers Materials & Continua* 75(1), 293–309.

22. Liu, J., Li, J., Ma, J., Sadiq, N., Bhatti, U. A., & Ai, Y. (2019). A robust multi-watermarking algorithm for medical images based on DTCWT-DCT and Henon map. *Applied Sciences* 9(4), 700.

23. Fan, Y., Li, J., Bhatti, U. A., Shao, C., Gong, C., Cheng, J., & Chen, Y. (2023). A multi-watermarking algorithm for medical images using inception V3 and DCT. *CMC-Computers Materials & Continua* 74(1), 1279–1302.

24. Li, T., Li, J., Liu, J., Huang, M., Chen, Y. W., & Bhatti, U. A. (2022). Robust watermarking algorithm for medical images based on log-polar transform. *EURASIP Journal on Wireless Communications and Networking* 2022(1), 1–11.

25. Bhatti, U. A., Huang, M., Wu, D., Zhang, Y., Mehmood, A., & Han, H. (2019). Recommendation system using feature extraction and pattern recognition in clinical care systems. *Enterprise Information Systems* 13(3), 329–351.

26. Bhatti, U. A., Yu, Z., Chanussot, J., Zeeshan, Z., Yuan, L., Luo, W., ... & Mehmood, A. (2021). Local similarity-based spatial–spectral fusion hyperspectral image classification with deep CNN and Gabor filtering. *IEEE Transactions on Geoscience and Remote Sensing* 60, 1–15.

27. Bhatti, U. A., Yu, Z., Li, J., Nawaz, S. A., Mehmood, A., Zhang, K., & Yuan, L. (2020). Hybrid watermarking algorithm using Clifford algebra with Arnold scrambling and chaotic encryption. *IEEE Access* 8, 76386–76398.

28. Liu, J., Li, J., Zhang, K., Bhatti, U. A., & Ai, Y. (2019). Zero-watermarking algorithm for medical images based on dual-tree complex wavelet transform and discrete cosine transform. *Journal of Medical Imaging and Health Informatics* 9(1), 188–194.

CHAPTER 4

Robust Zero Watermarking Algorithm for Medical Images Based on BRISK and DCT

Fangchun Dong[1], Jingbing Li[1,2], and Uzair Aslam Bhatti[1,2]

[1]*School of Information and Communication Engineering, Hainan University, Haikou, China*
[2]*State Key Laboratory of Marine Resource Utilization in the South China Sea,*
 Hainan University, Haikou, China

4.1 INTRODUCTION

As network communication technology continues to advance, the volume of data transmitted over networks is steadily rising. In order to prevent user information from leaking, digital watermarking technology is also constantly updated and iterated. It has gradually become a trend to use digital watermarking technology to protect user information privacy and provide strong protection for personal information security [1]. Medical images play a crucial role in facilitating accurate medical diagnosis, treatment, and scientific research. It provides a wealth of clinical information and helps doctors make more accurate diagnoses and more effective treatment plans [2]. Science and technology advancements have permitted the merging of current information technology with medical treatment, and more and more doctors and patients use telemedicine for diagnosis [3]. However, with the widespread dissemination of medical image information on the Internet, the security and integrity of patient information are facing severe challenges [4]. In this context, medical image watermarking technology emerged as the times require to provide technical support for protecting patient information privacy. Medical image watermarking technology realizes copyright protection, integrity verification, and content authentication of medical images by embedding invisible or imperceptible watermark data in medical images [5]. This technology not only needs to embed the watermark without affecting the image quality and diagnostic accuracy, but also needs to have good robustness and invisibility to resist common image attack processing [6]. At the current stage, most medical image algorithms still have not fully alleviated the problem

DOI: 10.1201/9781003427674-4

of ownership protection, especially in resisting geometric attacks, and some algorithms can only show strong robustness to specific partial geometric attacks. As such, it is critical to create an innovative and strong watermarking algorithm that can survive a wider variety of assaults.

There are two types of digital watermarking techniques: spatial domain watermarking and frequency domain watermarking. Spatial domain watermarking involves directly embedding the watermark information into the original data, which includes techniques such as least significant bit (LSB) replacement, pixel value mapping, and other methods. Such methods are relatively simple, but are vulnerable to image attacks. Wang, Huanying et al. [7] devised a color picture watermarking system that combines QR decomposition with spatial domain. By calculating QR decomposition elements in the spatial domain and performing watermark embedding and extraction in the frequency domain, the proposed method demonstrates a certain level of robustness. Basha, Shaik Hedayath et al. [8] used the ESP technique to construct Euclidean space points for watermark embedding and the Diffie-Hellman key to improve resistance to different assaults such as JPEG compression, cropping, and rotation, and demonstrated great robustness against these attacks. Cao, H. et al. [9] used quantization approaches to obtain significant resilience in watermark embedding and extraction by exploiting the link between the DFT-DC component and domain pixel value. Unlike spatial domain watermarking, frequency domain watermarking transforms the original data into the frequency domain before inserting the watermark information [10]. This approach is notable for its increased resilience and capacity to withstand many forms of attacks. Tang, Ming et al. [11] employed the FRFT transform to acquire the picture magnitude, followed by the DWT transformation and SVD to embed the watermark information in the low-frequency sub-band of the second-level DWT. To increase algorithm security, FRFT was utilized to encrypt the watermark information. The suggested approach is resistant to a variety of assaults, including rotation, chopping, Gaussian filtering, and median filtering. Jing Liu et al. [12] combined DTCWT-DCT transform with perceptual hashing technology and utilized zero-watermarking technology to achieve embedding and extraction of watermarks. The proposed algorithm shows strong resilience against conventional as well as geometric assaults, particularly impressive performance in the face of geometric attacks. Asmeen, Fauzia et al. [13] modified the original picture using the DWT technique, then used SVD to the fourth lower coefficient of the changed image. The watermark data was then placed into the chosen image. The suggested technique is very resistant to different assaults in sub-bands.

In recent years, Local feature extraction methods have grown in popularity in the realm of watermarking technology, and researchers have used these algorithms to increase the resilience of watermarking systems. Traditional local feature extraction algorithms, such as SIFT, SURF, and KAZE [14–16], have good rotation and scale invariance in image feature extraction and matching. In addition, binary feature extraction algorithms, such as BRIEF, ORB, and BRISK [17–19], have faster running speeds. Hamidi A. et al. [20] proposed a hybrid watermarking method that combines DWT-DCT and SIFT, exploiting the geometric invariance of SIFT to improve watermarking against geometric attacks with good robustness. Similarly, A. Soualmi et al. [21] suggested an invisible watermarking approach for medical picture tampering detection by integrating SURF and Weber descriptor (WD) with the Arnold algorithm. They used Weber descriptor embedding and extraction techniques to put the watermark around the SURF point in the region of interest (ROI) of medical pictures. Furthermore, zero-watermarking technology has been used in the field of medical pictures due to its capacity to protect the integrity of the original data while avoiding any negative impacts on the original data [22]. Cheng Zeng et al. [23] extracted characteristics from the original medical pictures using the KAZE feature extraction technique, which were then transformed using DCT to obtain a feature sequence. The algorithm utilized perceptual hashing and zero-watermarking techniques to embed and extract the watermark, demonstrating robustness against geometric attacks. Based on the above research, it can be found that there is relatively less research on binary feature extraction algorithms for medical image watermarking [24–34]. Additionally, most of the existing watermarking algorithms suffer from drawbacks such as slow computational speed and low robustness, especially resisting geometric attacks. Aiming at these problems, this paper proposes a BRISK-DCT watermarking algorithm.

4.2 FUNDAMENTAL THEORY

4.2.1 BRISK Feature Extraction Algorithm

The BRISK (Binary Robust Invariant Scalable Kepoints) algorithm is an image local feature extraction algorithm proposed by Stefan Leutenegge et al. [19]. The BRISK algorithm is characterized by its ability to maintain its performance and accuracy even when the image undergoes transformations such as scaling and rotation, indicating that it possesses scale invariance and rotation invariance. This algorithm reduces the computing cost and runs faster than SIFT and SURF algorithms. It mainly uses FAST for feature point detection and constructs an image pyramid for multi-scale expression.

4.2.1.1 Scale Space Keypoints Detection

1. Construction of scale space pyramid. Construct N octave layers (denoted by c_i) and N intra-octave layers (denoted by d_i), octave layer: the c_0 layer represents the original image with a scale of 1, the c_1 layer is two times the down sampling of the c_0 layer, and the c_2 layer is two times the down sampling of the c_1 layer, and so on; intra-octave layer: The d_0 layer is obtained by down sampling the c_0 layer by 1.5 times, the d_1 layer is the down sampling of the d_0 layer by two times, the d_2 layer is the down sampling of the d_1 layer by two times, and so on. See equation (4.1) for the scale calculation formula of each layer.

$$\begin{cases} t_{c_i} = 2^i \\ t_{d_i} = 1.5 \times 2^i \end{cases} \tag{4.1}$$

In the equation, t_{c_i} represents the scale relationship between the octave layer and the original image, and t_{d_i} indicates the scale relations between the intra-octave layer and the original image.

2. Feature point detection. FAST corner detection is performed at each level of the scale-space pyramid.

3. Non-maximum suppression. By comparing the detected candidate feature points with their location space and their neighbor points in the upper and lower scale spaces, and eliminating candidate points with smaller response values than other points, the purpose of non-maximum value suppression is achieved.

4. Obtain feature point information. The poles and their coordinate positions are obtained by interpolating the FAST poles with two-dimensional quadratic functions at the positions corresponding to the layer where the poles are located and the upper and lower layers; then the scales corresponding to the poles are obtained. The scale space key point detection is shown in Figure 4.1.

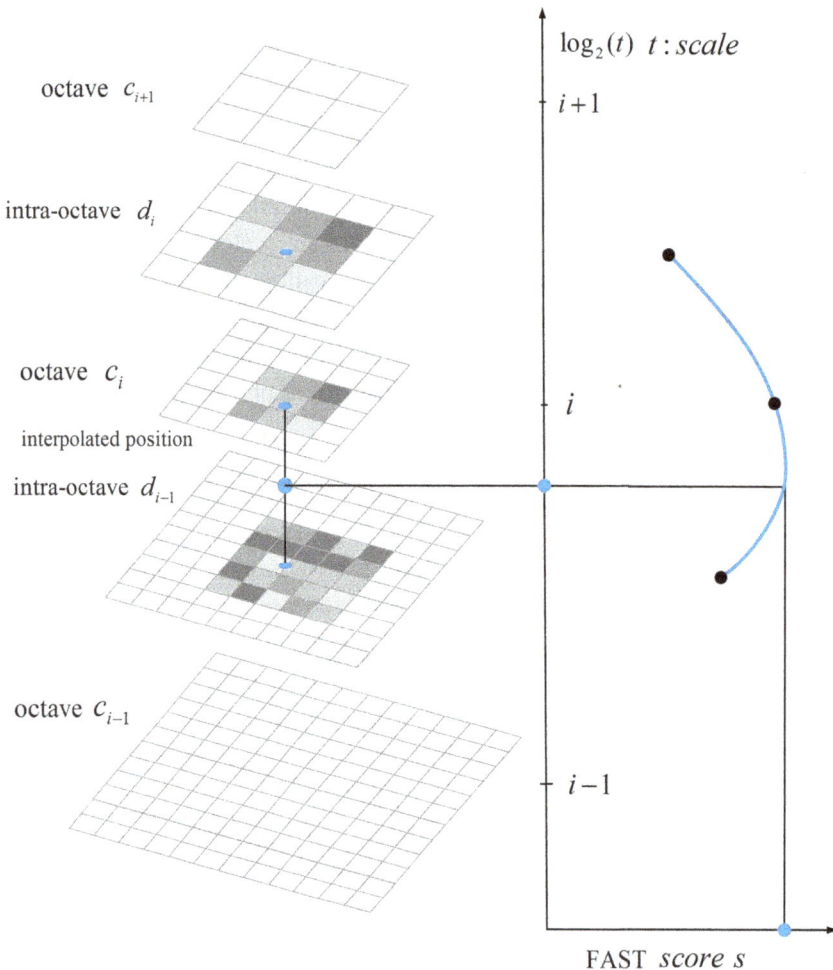

FIGURE 4.1 Scale-space keypoints detection.

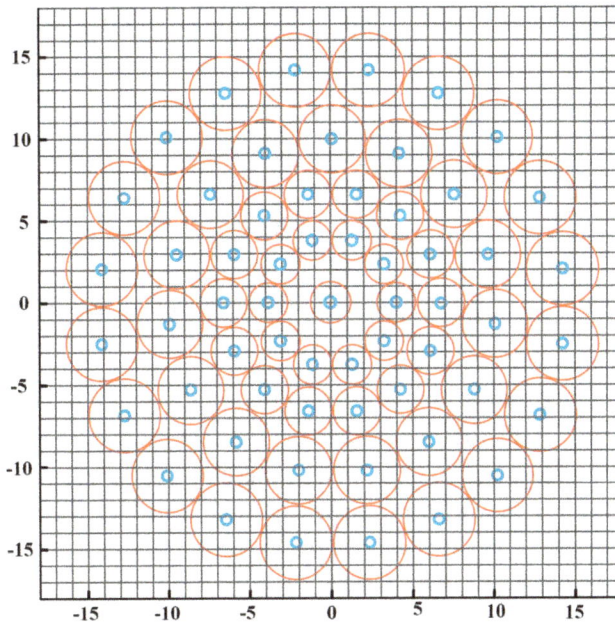

FIGURE 4.2   BRISK sampling mode.

4.2.1.2 Keypoints Description

After the keypoints are obtained, the feature description is carried out. The BRISK descriptor is made up of a binary string that is linked to the results of the brightness comparison test, identifying the direction feature of each keypoint for direction normalization, strengthening the keypoint with strong robustness; when the scale is 1, the BRISK sampling mode is shown in Figure 4.2. The sampling mode is to use $N = 60$ points for sampling. The blue circle in Figure 4.2 indicates the position of the sampling point, and the red circle indicates the Gaussian kernel standard deviation of the intensity value of the smoothed sampling point.

1. BRISK sampling method and rotation evaluation. The BRISK descriptor adopts the keypoints field mode, as shown in Figure 4.2, which defines the equidistant distribution of N keypoints positions. In order to avoid the overlapping effect of sampling points in this mode, Gaussian smoothing is used to pass through the image. The keypoint k of is localized and scaled in such a way that it takes into account points in $N(N − 1)/2$ of sampled point pairs. The smoothed values at these points are used to estimate the local gradient, as shown in equation (4.2).

$$g(P_i, P_j) = (P_j, P_i) \cdot \frac{I(P_j, \sigma_j) − I(P_i, \sigma_i)}{||P_j − P_i||^2} \tag{4.2}$$

In the equation, $g(P_i, P_j)$ is the local gradient value, (P_j, P_i) is the sampling point pair, σ_i is the standard deviation, and $I(P_j, \sigma_j)$, $I(P_i, \sigma_i)$ is the smoothed intensity value.

The maximum distance setting of the threshold is 9.75t, and the minimum is 13.67t. t is the scale of keypoint k, through iteration, evaluate the direction of the entire keypoint k as:

$$g = g(P_x, P_y) = \frac{1}{L} \cdot \sum_{\substack{(P_i,P_j)\in M}}^{n} g(P_i, P_j) \tag{4.3}$$

In the equation, M is the defined long-distance pairing subset, and is the local gradient value.

2. Build BRISK descriptor. The sampling method of BRISK is to rotate $\alpha = \arctan 2(g_y, g_x)$ at the keypoint k. The rotation and scale normalization descriptors are obtained, and the vector descriptors consist of all short-distance intensity point pairs. Each bit is shown in equation (4.4). In the equation, S is the short-distance pairing subset. According to the sampling method and the distance threshold, a 512-bit byte is obtained.

$$b = \begin{cases} 1, & I(P_j^\alpha, \sigma_j) > I(P_j^\alpha, \sigma_j) \\ 0, & \text{otherwise} \end{cases} \quad \forall \ (P_i^\alpha, P_j^\alpha) \in S \tag{4.4}$$

4.2.2 Discrete Cosine Transform (DCT)

DCT transform is often used for lossy data compression of images, which has separability and energy concentration. 2D-DCT is a technique for converting 2D images to the frequency domain. It is based on the idea of discrete cosine transform (DCT), which divides 2D images into many blocks and applies a one-dimensional DCT transform to each block. As shown in equation (4.5):

$$F(u, v) = C(u)C(v) \sum_{x=0}^{M-1} \sum_{y=0}^{N-1} f(x, y) \cos\left[\frac{(x + 0.5)u\pi}{M}\right] \cos\left[\frac{(y + 0.5)v\pi}{N}\right] \tag{4.5}$$
$$u = 0, 1, \dots, M - 1; \ v = 0, 1, \dots, N - 1$$

$$C(u) = \begin{cases} \sqrt{\dfrac{1}{M}}, & u = 0 \\ \sqrt{\dfrac{2}{M}}, & u = 1, 2, \dots, M - 1 \end{cases}, \quad C(v) = \begin{cases} \sqrt{\dfrac{1}{N}}, & v = 0 \\ \sqrt{\dfrac{2}{N}}, & v = 1, 2, \dots, N - 1 \end{cases} \tag{4.6}$$

4.2.3 Logistic Mapping

Logistic mapping is a chaotic mapping. Logistic mapping can generate very complex dynamic behaviors with a high degree of randomness and unpredictability. Therefore, it is widely used in the field of cryptography and communication. The mathematical expression of logistic mapping is shown in equation (4.7):

$$X_{K+1} = \mu \times X_K \times (1 - X_K)\, \mu \in [0, 4],\ X_K \in (0, 1) \tag{4.7}$$

4.3 PROPOSED ALGORITHM

The present study proposes a new digital watermarking system for medical pictures, which is based on the integration of the BRISK feature extraction algorithm, the discrete cosine transform (DCT) domain transformation, and the perceptual hashing function.

4.3.1 Medical Image Feature Extraction

This chapter randomly selects a brain CT image with a pixel size of 512 pixels × 512 pixels, using the BRISK feature extraction algorithm to extract feature vectors $B(i, j)$ of the medical image, and analyze the extracted features. After DCT transformation of the vector, a characteristic matrix $D(i, j)$ is obtained, and the top-left corner of the characteristic matrix is chosen as obtain a 2 pixels × 16 pixels low-frequency coefficient matrix, $V(i, j)$, and the sign conversion is performed on the low-frequency coefficient matrix. The elements greater than 0 in the matrix are set to 1, and other elements are set to 0. Get the hash value $H(i, j)$, which is a sequence of binary features. The image feature extraction flowchart are shown in Figure 4.3.

4.3.2 Watermark Encryption

To enhance the watermarking algorithm's anti-interference and security, a chaotic logistic mapping system is used to encrypt the watermarked image by performing a permutation operation. First, branching parameters are taken into the chaotic state, and then the watermarked image is perturbed to get the encrypted watermark $L(i, j)$.

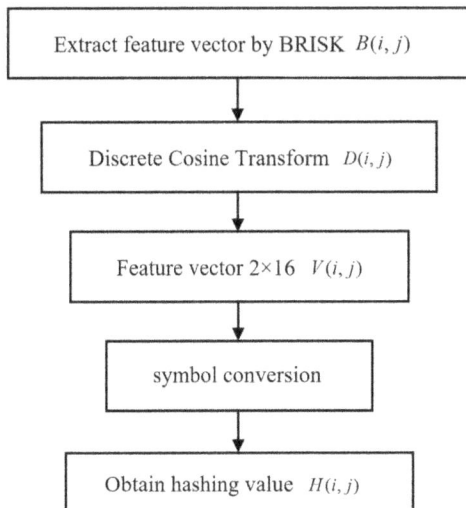

FIGURE 4.3   Image feature extraction flowchart.

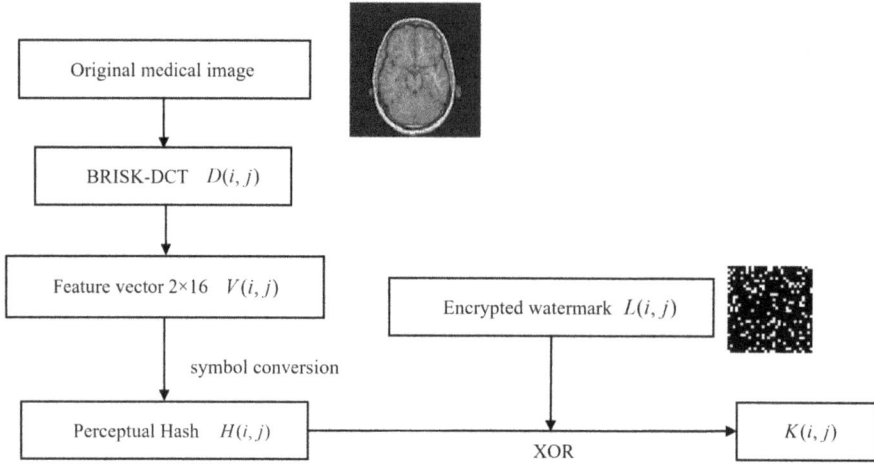

FIGURE 4.4 Embed watermark flowchart.

4.3.3 Embed Watermark

After applying the BRISK-DCT algorithm to extract the characteristics of the original medical picture, a binary feature sequence, $H(i, j)$, is generated. We then use an XOR operation to combine this binary feature sequence, $H(i, j)$, with the encrypted watermark, $L(i, j)$, creating a logical key, $K(i, j)$, for extracting the watermark. The proposed watermark embedding technique employs zero-watermark embedding technology, which does not alter the original medical picture. The embedding process is illustrated in Figure 4.4.

4.3.4 Watermark Extraction and Decryption

The watermark extraction process in this study employs the BRISK-DCT algorithm for image feature extraction, resulting in a binary feature sequence, $H'(i, j)$. An XOR operation is performed between the binary feature sequence, $H'(i, j)$, and the logical key, $K(i, j)$, to extract the encrypted watermark, $L'(i, j)$. The encrypted watermark, $L'(i, j)$, is restored to its original value through the use of the initial value of the logistic map. The procedure is demonstrated in Figure 4.5.

The watermark extraction procedure utilizes the BRISK-DCT algorithm for feature extraction of the image; to obtain a binary feature sequence, $H'(i, j)$; $H'(i, j)$ to perform an XOR operation with the logical key, $K(i, j)$, to obtain the extracted encrypted watermark $L'(i, j)$; and restore the encrypted watermark, $L'(i, j)$, to the original value through the initial value of the logistic map. The specific watermark extraction steps are shown in Figure 4.5.

4.4 EXPERIMENTS AND RESULTS

This chapter employs the MATLAB 2022b simulation platform to conduct experiments on medical images, with the medical image dimensions being 512 × 512 pixels and

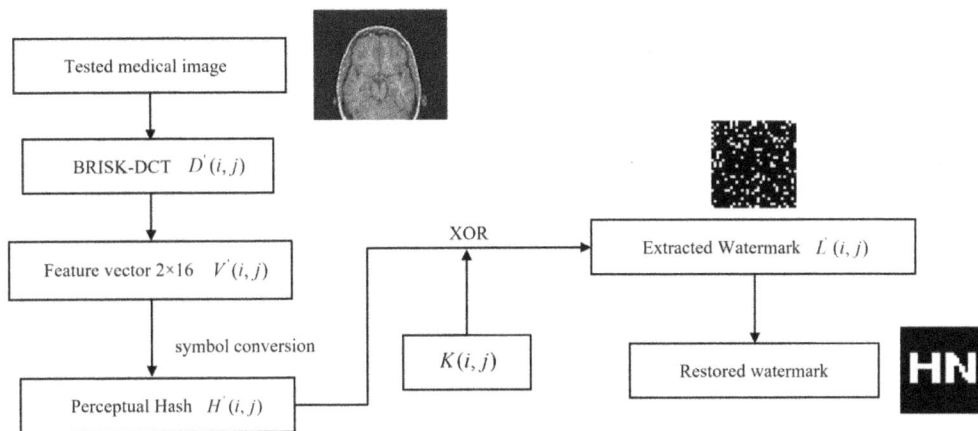

FIGURE 4.5   Flowchart of watermark extraction and decryption.

watermark image dimensions being 32 × 32 pixels. To evaluate the robustness of the proposed algorithm, the normalized correlation coefficient (NC) was used as a metric. The NC measures the similarity between the extracted watermark and the original watermark after applying various attacks to the medical image. In addition, the peak signal-to-noise ratio (PSNR) was used to assess image quality. The NC value is 1 in the absence of any attacks on medical images.

4.4.1 Test Different Images

Before evaluating the attack-resistance performance of medical images, we first apply the algorithm to test ten distinct medical images, as illustrated in Figure 4.6. We then calculate their correlation coefficients to verify whether the algorithm can differentiate between various medical images. Table 4.1 displays the outcomes of the tests, which demonstrate that the algorithm is indeed capable of distinguishing different medical images.

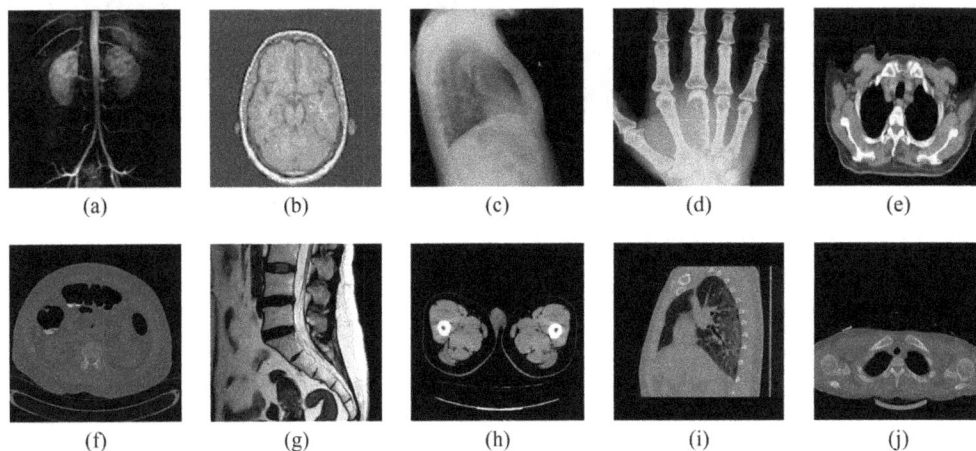

FIGURE 4.6   Different medical images.

TABLE 4.1   Correlation Coefficients of Different Medical Images (32 Bits)

Image	(a)	(b)	(c)	(d)	(e)	(f)	(g)	(h)	(i)	(j)
(a)	1.00	0.29	0.06	0.09	0.22	0.35	0.13	0.39	0.33	−0.04
(b)	0.29	1.00	0.11	0.30	0.26	0.31	0.03	0.18	0.26	0.36
(c)	0.06	0.11	1.00	0.13	0.18	0.06	0.19	0.22	0.09	0.19
(d)	0.09	0.3	0.13	1.00	0.33	0.31	0.32	0.24	0.34	0.38
(e)	0.22	0.26	0.18	0.33	1.00	0.37	0.36	0.22	0.39	0.39
(f)	0.35	0.31	0.06	0.31	0.37	1.00	0.30	0.44	0.28	0.18
(g)	0.13	0.03	0.19	0.32	0.36	0.30	1.00	0.25	0.23	0.19
(h)	0.39	0.18	0.22	0.24	0.22	0.44	0.25	1.00	0.14	−0.05
(i)	0.33	0.26	0.09	0.34	0.39	0.28	0.23	0.14	1.00	0.20
(j)	0.04	0.36	0.19	0.38	0.39	0.18	0.19	0.05	0.20	1.00

4.4.2 Conventional Attacks

The watermark is extracted after different degrees of JPEG compression and median filter attack on the medical image that already contains the encrypted watermark, as shown in Figure 4.7. The outcomes of the experiments are demonstrated in Tables 4.2 and 4.3, indicating that the proposed algorithm exhibits robustness against attacks such as JPEG compression and median filtering.

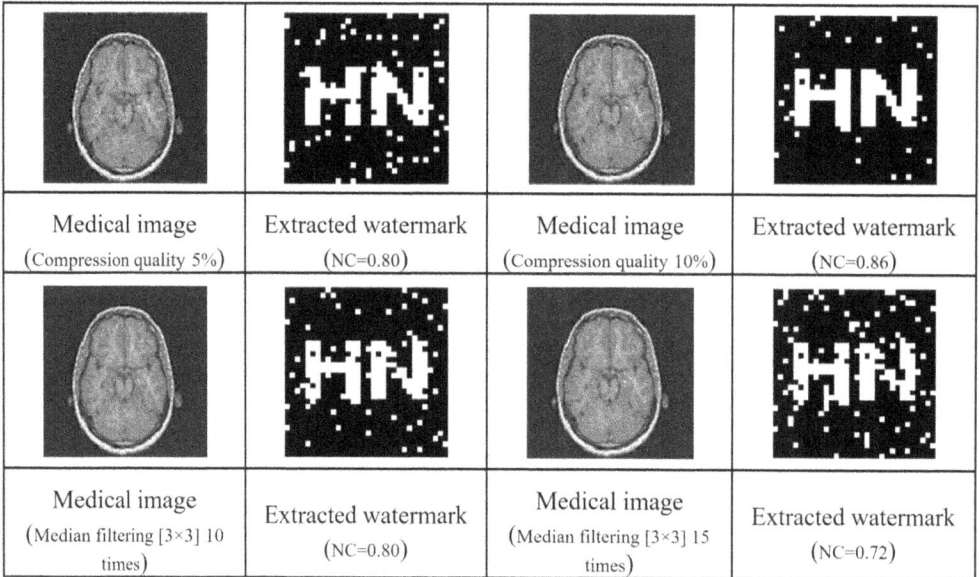

Medical image (Compression quality 5%)	Extracted watermark (NC=0.80)	Medical image (Compression quality 10%)	Extracted watermark (NC=0.86)
Medical image (Median filtering [3×3] 10 times)	Extracted watermark (NC=0.80)	Medical image (Median filtering [3×3] 15 times)	Extracted watermark (NC=0.72)

FIGURE 4.7   Medical images under conventional attacks.

TABLE 4.2   Experimental Results Under JPEG Compression Attacks

Compression quality (%)	5	15	25	35	45
PSNR (dB)	28.44	32.91	34.39	35.19	35.74
NC	0.80	0.89	0.89	1	1

TABLE 4.3 Experimental Results Under Median Filtering Attacks

Median filtering [3 × 3] (times)	5	10	15	20
PSNR (dB)	34.49	34.00	33.86	33.82
NC	0.79	0.79	0.71	0.63

4.4.3 Geometric Attacks

After applying various geometric attacks to the medical images containing encrypted watermarks, as shown in Figure 4.8. The proposed algorithm's robustness was assessed. The results indicate that the suggested technique is very resistant against rotation

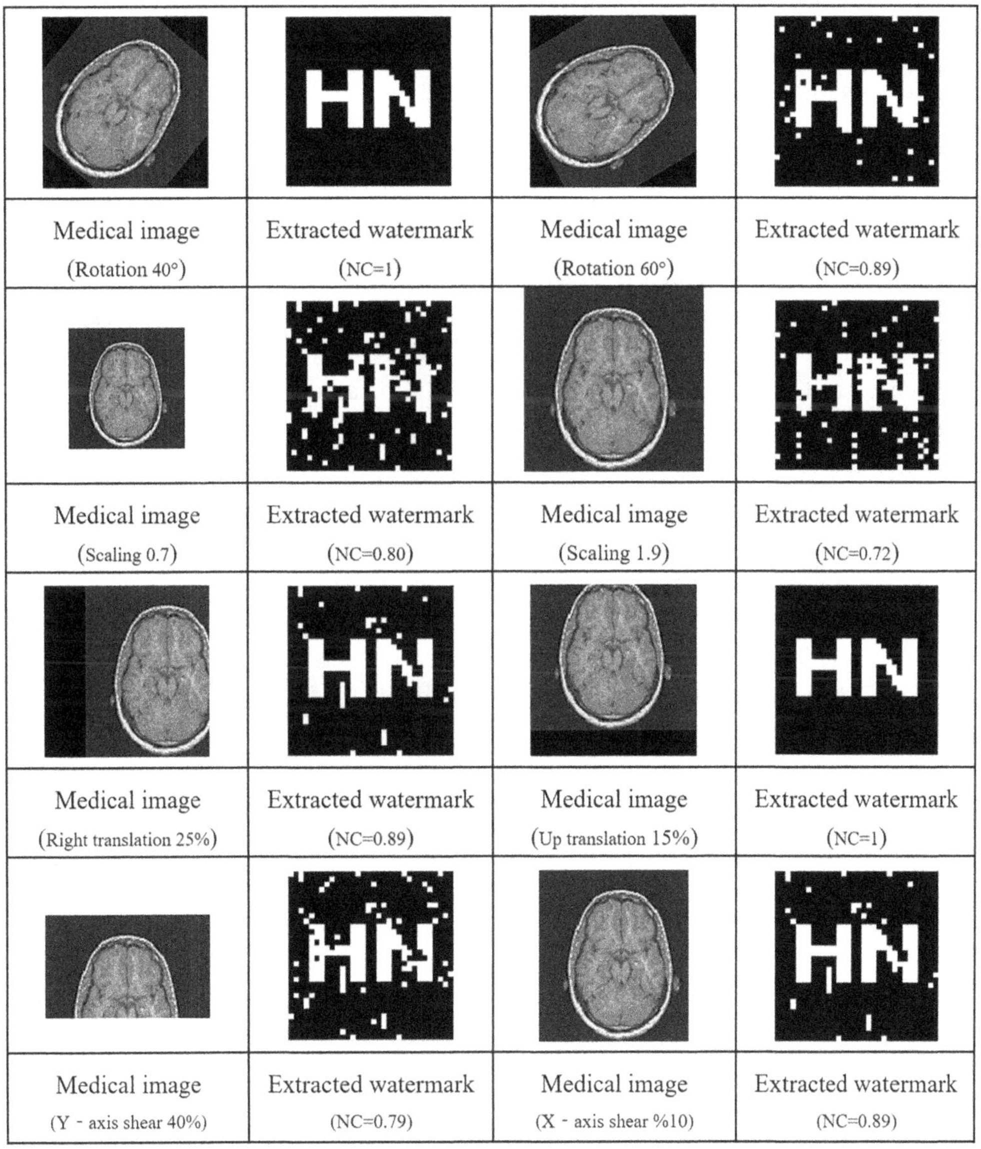

FIGURE 4.8 Medical images under geometric attacks.

TABLE 4.4 Experimental Results Under Rotation Attacks

Angle (°)	10	20	30	40	50	60
PSNR (dB)	15.60	14.60	14.40	14.05	13.71	13.55
NC	0.79	0.89	0.89	1	0.80	0.89

assaults, as evidenced by a PSNR value of 13.55 and an NC value of 0.89 even at a rotation angle of 60 degrees. For scaling ratios between 0.7 and 1.9, most NC values were found to be greater than 0.7. Furthermore, even when the encrypted medical image was shifted to the right by 40% or shifted upward by 35%, the NC values were 0.72 and 0.79, respectively. The algorithm also demonstrated good resistance against shear attacks, with an NC value of 0.79 even when the encrypted medical image was sheared by 40%. Finally, when the x-axis was cropped by 40%, the NC value was 0.79, indicating the algorithm's robustness against cropping attacks (Tables 4.4–4.7).

4.4.4 Compare with Other Algorithms

In order to evaluate the robustness of the proposed BRISK-DCT algorithm, it was compared with other existing algorithms. Figure 4.9 illustrates the comparison results, where DCT [35] is represented in green, DWT-DCT [36] is represented in black, SIFT-DCT [24] is represented in blue, and the proposed BRISK-DCT algorithm is represented in red. The experimental data suggests that BRISK-DCT algorithm exhibits superior resistance against geometric attacks in comparison to the other algorithms.

TABLE 4.5 Experimental Results Under Scaling Attacks

Scaling	0.7	0.9	1.1	1.3	1.5	1.7	1.9
NC	0.69	0.80	0.79	0.89	0.79	0.71	0.79

TABLE 4.6 Experimental Results Under Translation Attacks

Right translation				Up translation			
Scale (%)	5	15	25	40	Scale (%)	5	15	25	35
PSNR (dB)	14.51	12.90	11.30	10.08	PSNR (dB)	14.64	13.13	11.93	11.05
NC	0.89	0.80	0.89	0.72	NC	0.80	1	0.89	0.79

TABLE 4.7 Experimental Results Under Cropping Attacks

Y-axis shear				X-axis shear			
Ratio (%)	10	20	30	40	Ratio (%)	10	20	30	40
NC	0.79	0.71	0.70	0.79	NC	0.89	0.80	0.71	0.79

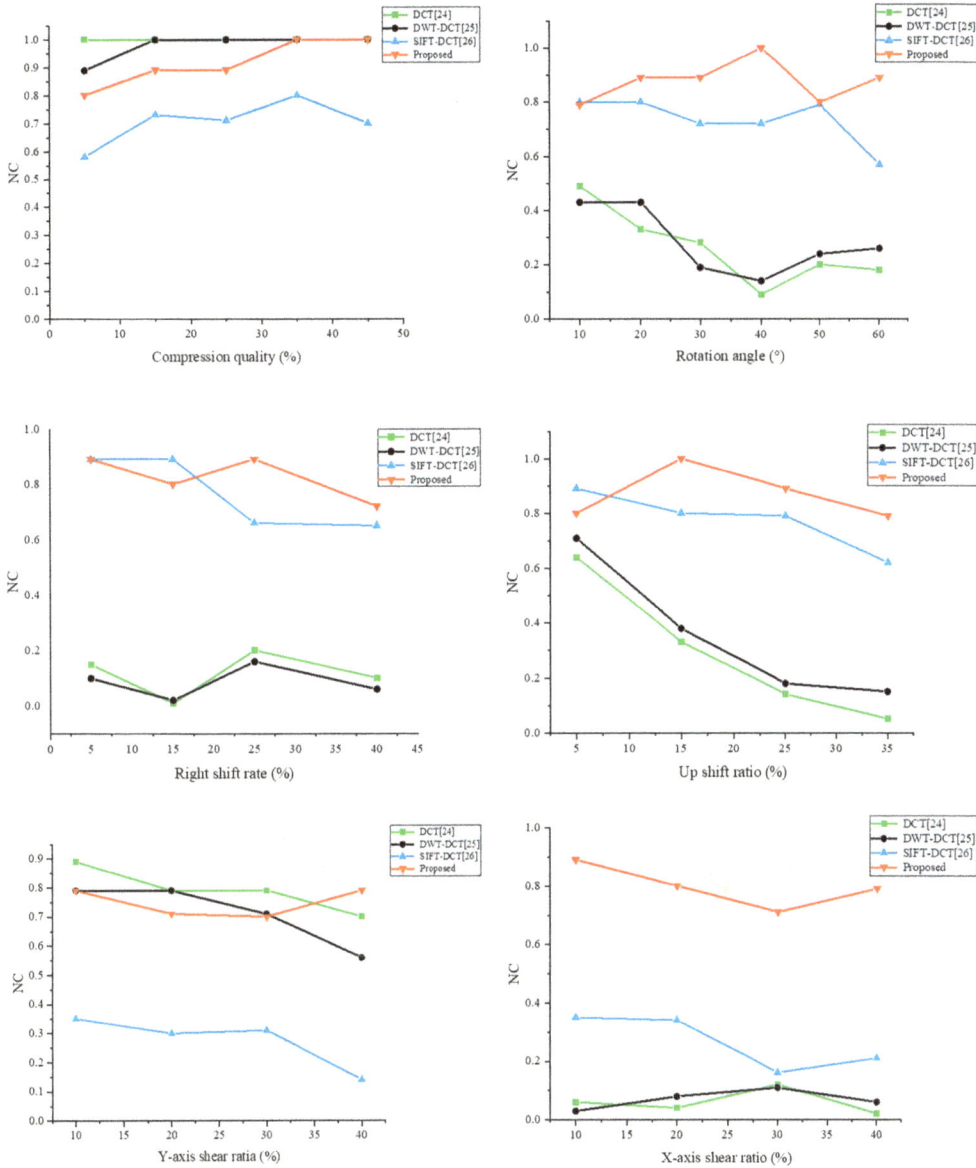

FIGURE 4.9 Comparison of the results of different algorithms.

4.5 CONCLUSION

This study describes a zero-watermarking system for medical photographs that integrates chaotic encryption technology, a perceptual hashing algorithm, and zero-watermarking technology to assure the security of the watermark information. The experimental findings show that the proposed approach is resistant to conventional and geometric assaults, rendering it suitable for medical image applications. However, to address the issue of poor robustness against certain attacks, future research aims to enhance the algorithm while also improving its ability to extract image features effectively.

REFERENCES

1. Evsutin, O., Melman, A., & Meshcheryakov, R. (2020). Digital steganography and water-marking for digital images: A review of current research directions. *IEEE Access*, 8, 166589–166611.
2. Tian, Y., & Fu, S. (2020). A descriptive framework for the field of deep learning applications in medical images. *Knowledge-Based Systems*, 210, 106445.
3. Amine, K., Fares, K., Redouane, K. M., & Salah, E. (2022). Medical image watermarking for telemedicine application security. *Journal of Circuits, Systems and Computers*, 31(5).
4. Venkateswarlu, I. B. (2020). Fast medical image security using color channel encryption. *Brazilian Archives of Biology and Technology*, 63, e20180473.
5. Thabit, R. (2021). Review of medical image authentication techniques and their recent trends. *Multimedia Tools and Applications*, 80(9), 13439–13473.
6. Raj, N. R. N., & Shreelekshmi, R. (2021). A survey on fragile watermarking based image authentication schemes. *Multimedia Tools and Applications*, 80(13), 19307–19333.
7. Wang, H., & Su, Q. (2022). A color image watermarking method combined QR decomposition and spatial domain. *Multimedia Tools and Applications*, 81(26), 37895–37916.
8. Basha, S. H., & Jaison, B. (2022). A novel secured Euclidean space points algorithm for blind spatial image watermarking. *EURASIP Journal on Image and Video Processing*, 2022(1), 1–15.
9. Cao, H., Hu, F., Sun, Y., Chen, S., & Su, Q. (2022). Robust and reversible color image watermarking based on DFT in the spatial domain. *Optik*, 262, 1–15.
10. Chopra, A., Gupta, S., & Dhall, S. (2020). Analysis of frequency domain watermarking techniques in presence of geometric and simple attacks. *Multimedia Tools and Applications*, 79(1–2), 501–554.
11. Tang, M., & Zhou, F. (2022). A robust and secure watermarking algorithm based on DWT and SVD in the fractional order Fourier transform domain. *Array*, 15, 100230.
12. Liu, J., Li, J., Cheng, J., Ma, J., Sadiq, N., Han, B., ... Ai, Y. (2019). A novel robust watermarking algorithm for encrypted medical image based on DTCWT-DCT and chaotic map. *Computers, Materials and Continua*, 61(2), 889–910.
13. Yasmeen, F., & Uddin, M. S. (2022). A novel watermarking scheme based on discrete wavelet transform-singular value decomposition. *Security and Privacy*, 5(3), 1–15.
14. Lowe, D. G. (2004). Distinctive image features from scale-invariant keypoints. *International Journal of Computer Vision*, 60(2), 91–110.
15. Bay, H., Ess, A., Tuytelaars, T., & Van Gool, L. (2008). Speeded-up robust features (SURF). *Computer Vision and Image Understanding*, 110(3), 346–359.
16. Alcantarilla, P. F., Bartoli, A., & Davison, A. J. (2012). KAZE Features. In A. Fitzgibbon, S. Lazebnik, P. Perona, Y. Sato, & C. Schmid (Eds.), *Computer Vision – ECCV 2012* (pp. 214–227). Berlin, Heidelberg: Springer.
17. Sjöstrand, T., Mrenna, S., & Skands, P. (2008). A brief introduction to PYTHIA 8.1. *Computer Physics Communications*, 178(11), 852–867.
18. Rublee, E., Rabaud, V., Konolige, K., & Bradski, G. (2011). ORB: An efficient alternative to SIFT or SURF. 2011 International Conference on Computer Vision, 2564–2571.
19. Leutenegger, S., Chli, M., & Siegwart, R. Y. (2011). BRISK: Binary robust invariant scalable keypoints. 2011 International Conference on Computer Vision, 2548–2555.
20. Hamidi, M., El Haziti, M., Cherifi, H., & El Hassouni, M. (2021). A hybrid robust image watermarking method based on dwt-dct and sift for copyright protection. *Journal of Imaging*, 7(10), 1–15.

21. Soualmi, A., Alti, A., & Laouamer, L. (2022). An Imperceptible watermarking scheme for medical image tamper detection. *International Journal of Information Security and Privacy*, 16(1), 1–15.
22. Wang, B., Wang, W., Zhao, P., & Xiong, N. (2022). A zero-watermark scheme based on quaternion generalized Fourier descriptor for multiple images. *Computers, Materials and Continua*, 71(2), 2633–2652.
23. Zeng, C., Liu, J., Li, J., Cheng, J., Zhou, J., Nawaz, S. A., ... Bhatti, U. A. (2022). Multi-watermarking algorithm for medical image based on KAZE-DCT. *Journal of Ambient Intelligence and Humanized Computing*, 1–9.
24. Fang, Y., Liu, J., Li, J., Cheng, J., Hu, J., Yi, D., Xiao, X., & Bhatti, U. A. (2022). Robust zero-watermarking algorithm for medical images based on SIFT and Bandelet-DCT. *Multimedia Tools and Applications*, 81(12), 16863–16879.
25. Li, D., Li, J., Bhatti, U. A., Nawaz, S. A., Liu, J., Chen, Y. W., & Cao, L. (2023). Hybrid encrypted watermarking algorithm for medical images based on DCT and improved DarkNet53. *Electronics*, 12(7), 1554.
26. Bhatti, U. A., Tang, H., Wu, G., Marjan, S., & Hussain, A. (2023). Deep learning with graph convolutional networks: An overview and latest applications in computational intelligence. *International Journal of Intelligent Systems*, 2023, 1–28.
27. Sheng, M., Li, J., Bhatti, U. A., Liu, J., Huang, M., & Chen, Y. W. (2023). Zero watermarking algorithm for medical image based on Resnet50-DCT. *CMC-Computers Materials & ContinuA*, 75(1), 293–309.
28. Liu, J., Li, J., Ma, J., Sadiq, N., Bhatti, U. A., & Ai, Y. (2019). A robust multi-watermarking algorithm for medical images based on DTCWT-DCT and Henon map. *Applied Sciences*, 9(4), 700.
29. Fan, Y., Li, J., Bhatti, U. A., Shao, C., Gong, C., Cheng, J., & Chen, Y. (2023). A multi-watermarking algorithm for medical images using inception V3 and DCT. *CMC-Computers Materials & Continua*, 74(1), 1279–1302.
30. Li, T., Li, J., Liu, J., Huang, M., Chen, Y. W., & Bhatti, U. A. (2022). Robust watermarking algorithm for medical images based on log-polar transform. *EURASIP Journal on Wireless Communications and Networking*, 2022(1), 1–11.
31. Bhatti, U. A., Huang, M., Wu, D., Zhang, Y., Mehmood, A., & Han, H. (2019). Recommendation system using feature extraction and pattern recognition in clinical care systems. *Enterprise Information Systems*, 13(3), 329–351.
32. Bhatti, U. A., Yu, Z., Chanussot, J., Zeeshan, Z., Yuan, L., Luo, W., ... Mehmood, A. (2021). Local similarity-based spatial–spectral fusion hyperspectral image classification with deep CNN and Gabor filtering. *IEEE Transactions on Geoscience and Remote Sensing*, 60, 1–15.
33. Bhatti, U. A., Yu, Z., Li, J., Nawaz, S. A., Mehmood, A., Zhang, K., & Yuan, L. (2020). Hybrid watermarking algorithm using Clifford algebra with Arnold scrambling and chaotic encryption. *IEEE Access*, 8, 76386–76398.
34. Liu, J., Li, J., Zhang, K., Bhatti, U. A., & Ai, Y. (2019). Zero-watermarking algorithm for medical images based on dual-tree complex wavelet transform and discrete cosine transform. *Journal of Medical Imaging and Health Informatics*, 9(1), 188–194.
35. Liu, Y. L., & Li, J. B. (2013). DCT and logistic map based multiple robust watermarks for medical image. *Applied Research of Computers*, 30(11), 3430–3433.
36. Liu, Y., & Li, J. (2012). The medical image watermarking algorithm using DWT-DCT and logistic. 2012 7th International Conference on Computing and Convergence Technology (ICCCT), 599–603.

CHAPTER 5

Robust Color Images Zero-Watermarking Algorithm Based on Stationary Wavelet Transform and Daisy Descriptor

Yiyi Yuan[1], Jingbing Li[1,2], Uzair Aslam Bhatti[1,2], Meng Yang[1], and Qinqing Zhang[1]

[1]School of Information and Communication Engineering, Hainan University, Haikou, China
[2]State Key Laboratory of Marine Resource Utilization in the South China Sea, Hainan University, Haikou, China

5.1 INTRODUCTION

With the rapid growth of Internet technology, it has become easier and easier to store, copy, and disseminate digital multimedia contents. However, the information age provides people with great convenience but also brings a series of security problems of multimedia information [1]. In particular, illegal theft, tampering, and dissemination of multimedia information in the process of public network transmission are very common [2]. The copyright protection of multimedia data is crucial for this reason.

The technical means to protect multimedia information usually include the following: digital encryption, digital signature, digital watermark, etc. [2,3]. Among the many protection methods, digital watermarking technology is one of the effective means to achieve copyright protection of multimedia information [4,5]. Digital watermarking technology mainly proves the copyright ownership by embedding some invisible information and proving the copyright ownership by extracting the degree of association between the watermarked data and the original data when needed.

DOI: 10.1201/9781003427674-5

In different application environments, watermarking has its own specific requirements; thus, it is obviously unrealistic to expect to design a watermarking algorithm that can be applied to any application environment. In general, digital watermarking techniques have the following basic requirements [6]: (1) imperceptibility, (2) robustness, (3) security, and (4) capacity. The aim of this study is to design a watermarking scheme to achieve imperceptibility, robustness, and security of watermarking. To this end, we use a combination of local image feature descriptors, transform domain and perceptual hashing techniques to extract robust feature vectors of the image, and use zero-watermarking techniques to embed and extract watermark information without changing the pixel values of the original image in order to achieve the imperceptibility of the watermark. Meanwhile, to enhance the security of the watermark, we use chaotic encryption system to encrypt the watermark.

The structure of the paper is as follows: Section 5.1 describes the concept and research requirements of watermarking techniques. Section 5.2 introduces the latest literature and related works. Section 5.3 introduces the basic concepts of smooth wavelet transform, Daisy descriptor, and Tent chaos mapping, respectively, and the algorithm design process (mainly including embedding and extraction of watermark). Section 5.4 verifies the feasibility of the algorithm and the simulation. Section 5.4 verifies the feasibility of the algorithm and the analysis of the simulation results. Section 5.5 gives a brief summary of the proposed algorithm and provides directions for future research on the topic.

5.2 LITERATURE REVIEW

In this section, the latest research advances in digital image watermarking are reviewed. Digital image watermarking techniques mainly include null-domain watermarking technique, frequency-domain watermarking technique, and chaotic watermarking technique [7]. (1) Space-domain watermarking technique refers to embedding the watermark into the pixel values of the original image, and the commonly used algorithms include least significant bit (LSB) [8] and direct sequence spread spectrum (DSSS) [9], etc. [8] proposed a hybrid watermarking technique based on LSB and discrete wavelet transform (DWT), which divides the watermarked image into two parts and embeds them in the null and frequency domains, respectively. In [9], a DSSS sequence of length 15 is generated for each polar coordinate of the watermarked image and added to the mid-frequency band of the image DCT to produce an image with watermark. (2) Frequency-domain watermarking technique refers to embedding the watermark into the frequency-domain coefficients of the original image, and the commonly used algorithms include discrete cosine transform (DCT) and discrete wavelet transform (DWT), etc. [10] embedded the watermark by modifying the singular values of the low-frequency sub-bands after the DWT transform of the image. [11] used several variants of the Fourier transform on the image to hide the parity of the transformed coefficients in the mid-frequency band. [12] put up a color image watermarking approach based on singular value decomposition and a smooth wavelet transform. (3) Chaotic watermarking technique refers to the use of chaotic mapping to embed the watermark into the original image, which has the advantage of high security and robustness. In [13], a Tent-henon mapping dual chaotic watermark encryption scheme is designed to embed the encrypted watermark into the tested image.

In addition, local image feature descriptors are now widely used in the design of digital watermarking algorithms. These descriptors are capable to distinguish locally similar regions of an image. Among them, the scale invariant feature transform (SIFT) is one of the most commonly used descriptors [14]. Gradient location and orientation histogram (GLOH) and Daisy descriptors are two extended extension algorithms of SIFT descriptors, which can guarantee a certain invariance to image geometric transformations and linear luminance variations. In addition, there are other commonly used feature descriptors such as speeded-up robust features (SURF), accelerated segmentation test (FAST), and local binary pattern (LBP), etc. In [14], the SIFT algorithm was used for watermarked images and the obtained features were used as a key for watermark extraction. A reversible watermarking technique based on SURF feature points was proposed in [15]. [16] used the position information of matching SIFT key points to recover the watermark synchronization information. [17] used the FAST technique to extract image feature points and embed the watermark sequence into a circular patch centered on the feature points. [18] used LBP to design a dual watermarks scheme using LBP to generate a systematic vector to perform XOR with the watermark.

5.3 MATERIAL AND TECHNIQUES

5.3.1 Daisy Descriptor

Existing local descriptors, such as SIFT and SURF, have been applied to image watermarking algorithms, proving that local descriptors can extract robust features from images. This paper adopts the Daisy image local feature descriptor proposed by Tola et al. [19], which keeps the SIFT and GLOH's robustness while fast calculating the feature descriptor matrix of each pixel on the image. Figure 5.1 shows the regional structure of the Daisy descriptor used in this paper. As can be seen from Figure 5.1, this algorithm

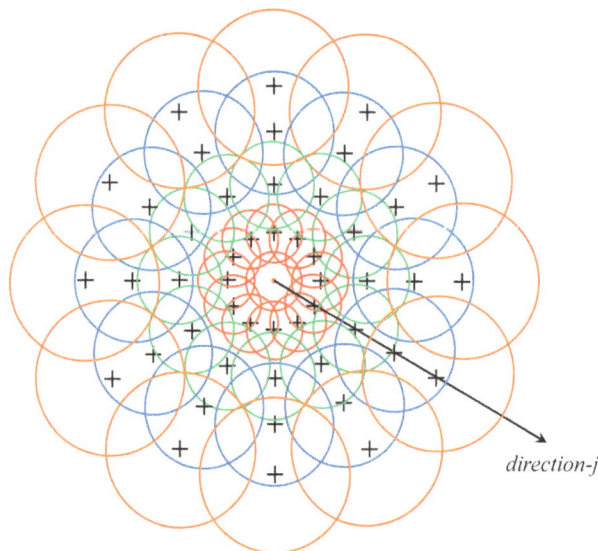

direction-j

FIGURE 5.1 The region structure of the Daisy descriptor used in this chapter.

mainly adopts a circular structure, with each circle representing a region. The characteristic of this structure is that when rotation occurs, the regions around the center point will not change [20], thus achieving a certain degree of rotation robustness.

Assume we select a point p, (u_0, v_0), on the image I as the central pixel for the Daisy descriptor. First, we take four circular rings with different radii centered at p, increasing in size by a multiple factor. The radius of each ring is proportional to the standard deviation of the Gaussian kernel. Additionally, for each ring, we obtain 12 sampling points at intervals of 30 degrees. For each sampling point, (u, v), and direction i, the gradient value is represented by $G_i(u, v)$, and we calculate the gradient for each of the 12 directions.

$$G_i(u, v) = \left(\frac{\partial I}{\partial i}\right)^+, \quad 1 \le i \le 12 \tag{5.1}$$

$$\left(\frac{\partial I}{\partial i}\right)^+ = max\left(\frac{\partial I}{\partial i}, 0\right) \tag{5.2}$$

Then, the gradient map is convolved with a Gaussian kernel using different Σ values. By performing multiple Gaussian convolutions, the Gaussian convolution values of the sampling points on different layers can be obtained. For a point, (u, v), its gradient histogram, $h_\Sigma(u, v)$, can be expressed as follows:

$$G_i^\Sigma = G_\Sigma \times G_i(u, v) \tag{5.3}$$

$$h_\Sigma(u, v) = [G_1^\Sigma(u, v), G_2^\Sigma(u, v), \dots, G_{12}^\Sigma(u, v)]^\top \tag{5.4}$$

where $G_1^\Sigma, G_2^\Sigma, \dots, G_{12}^\Sigma$ represent the gradient histograms of 12 different directions.

Finally, the gradient histograms for the central pixel point p and the 4×12 sampling points on the circular ring are computed. The descriptor for the feature point p located in the center is represented by the vectors of these 49 points. Thus, the Daisy descriptor for point p is as follows:

$$D(u_0, v_0) = \begin{bmatrix} h_{\Sigma_1}^\top(u_0, v_0) \\ h_{\Sigma_1}^\top(l_1(u_0, v_0, r_1)) & \cdots & h_{\Sigma_1}^\top(l_{12}(u_0, v_0, r_1)) \\ \vdots & \cdots & \vdots \\ h_{\Sigma_4}^\top(l_1(u_0, v_0, r_4)) & \cdots & h_{\Sigma_4}^\top(l_{12}(u_0, v_0, r_4)) \end{bmatrix} \tag{5.5}$$

where $l_i(u, v, r_j)$ represents the position of the ith sampling point on the jth concentric circle in the structure centered at the pixel point, (u_0, v_0). The Daisy descriptor, $D(u_0, v_0)$, of the pixel point, (u_0, v_0), is a matrix with a size of 49×12.

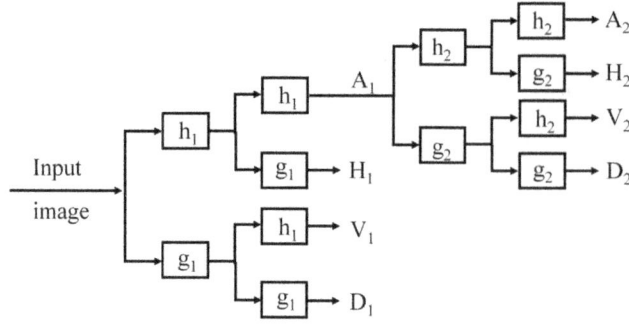

FIGURE 5.2   Schematic diagram of two-layer SWT.

5.3.2 Stationary Wavelet Transform

Stationary wavelet transform (SWT) transform is up-sampled throughout the decomposition [21], which keeps the length of the processed approximation and detail coefficients unchanged. The SWT can extract some abnormal and sudden signals from the original signal, making it possible to denoise the signal.

The flowchart of the two-level SWT is shown in Figure 5.2. $h_{(n)}$ and $g_{(n)}$ represent the low-pass and high-pass filters of the SWT transform, and the filters used in each step should satisfy the condition that the filter used in the $n+1$ step is upsampled from the filter used in the n step. After SWT decomposition, the original image's low-frequency, horizontal, vertical, and diagonal high-frequency components are indicated by the letters $A_{(n)}$, $H_{(n)}$, $V_{(n)}$, and $D_{(n)}$, respectively, as shown in equations (5.6)–(5.9).

$$A_{(n)}(i, j) = \sum_{x,y} h_{(n)}(x)h_{(n)}(y)A_{(n-1)}(x + i, y + j) \tag{5.6}$$

$$H_{(n)}(i, j) = \sum_{x,y} h_{(n)}(x)g_{(n)}(y)A_{(n-1)}(x + i, y + j) \tag{5.7}$$

$$V_{(n)}(i, j) = \sum_{x,y} g_{(n)}(x)h_{(n)}(y)A_{(n-1)}(x + i, y + j) \tag{5.8}$$

$$D_{(n)}(i, j) = \sum_{x,y} g_{(n)}(x)g_{(n)}(y)A_{(n-1)}(x + i, y + j) \tag{5.9}$$

5.3.3 Tent Chaotic Map

This study uses chaotic systems to generate chaotic sequences and scramble watermark images. The Tent map belongs to a classic chaotic system [22,23]. Compared with the Logistic map, the iteration process of the Tent map is more controllable and faster, and therefore, its dynamic performance is better. The Tent map can be expressed as:

$$x_{i+1} = \begin{cases} \dfrac{x_i}{\beta} x_i & \in [0, \beta) \\ \dfrac{1 - x_i}{1 - \beta} x_i & \in [\beta, 1] \end{cases} \tag{5.10}$$

In formula (5.10), the Tent map is chaotic when the parameter $\beta \in (0, 1)$. It is a typical Tent map when β is close to 0.5. The sequence has a uniform distribution and a density of the distribution that is fairly uniform for various parameters. However, it should be noted that the initial value x_i and parameter β of the system cannot be the same, otherwise it will become a periodic system.

5.3.4 Proposed Algorithm

This algorithm is a zero-watermark algorithm based on the SWT-Daisy-DCT and Tent chaotic encryption system. It uses the SWT-Daisy-DCT transformation to extract a set of feature vectors from an image and combines them with the watermark information encrypted by the Tent chaotic encryption system. The resulting information is stored in a third-party organization to protect copyright.

5.3.4.1 Feature Extraction

To obtain a robust set of feature vectors for test images, the specific steps are as follows: first, the RGB color space, $I_r(i, j)$, of the original image is converted to the YCbCr color space, $I_y(i, j)$. Then, a two-level SWT is carried out on the Cb component, $B(i, j)$, using the "db1" wavelet basis. The central point of the low-frequency component, $A_{(n)}$, after transformation is selected, and the Daisy descriptor matrix, $D(i, j)$, is calculated for this point. The descriptor matrix is then subjected to a DCT transform, and the 8×4 low-frequency coefficient matrix, $F(i, j)$, is selected. Finally, a hash function is used to generate a 32-bit feature vector, $V(i, j)$.

$$\{A_2, \quad H_2, \quad V_2, \quad D_2\} = swt2(B(i, j)) \tag{5.11}$$

$$F(i, j) = dct2(D(i, j)) \tag{5.12}$$

$$V(i, j) = sign \begin{cases} F(i, j) \geq 0, & 1 \\ F(i, j) < 0, & 0 \end{cases} \tag{5.13}$$

5.3.4.2 Watermark Encryption and Embedding

We encrypt the watermark before embedding it to guarantee the safety of the watermark image [23–31]. In this paper, we use the Tent chaotic map to preprocess the watermark. First, we set the initial parameters and generate the chaotic sequence, $X(j)$. Then, we use dimensionality augmentation and symbol operation to generate the binary encryption matrix, $E(i, j)$. Finally, we perform the XOR operation between the binary watermark, $W(i, j)$, and the binary encryption matrix, $E(i, j)$, to obtain the encrypted watermark, $EW(i. j)$. The watermark embedding process uses the feature extraction method mentioned above to obtain the 32-bit feature vector, $V(i, j)$, of the color image, and performs the XOR operation between it and the encrypted watermark, $EW(i. j)$, to obtain the key, $Key(i, j)$, stored in a third party. The watermark encryption and embedding process is shown in Figure 5.3.

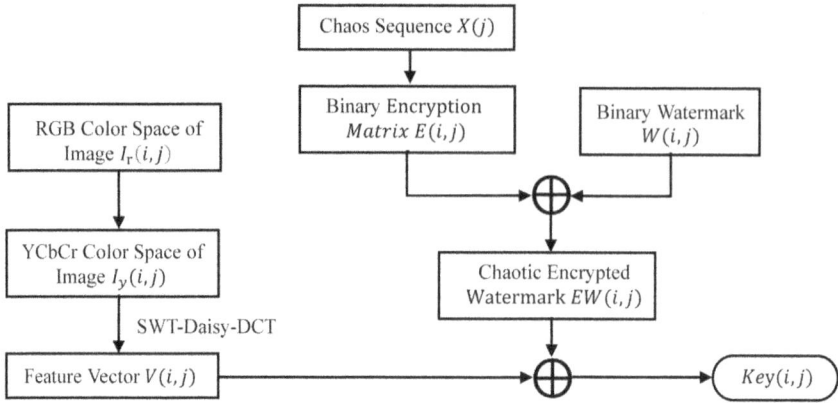

FIGURE 5.3 The process of watermark encryption and embedding.

$$EW(i, j) = W(i, j) \oplus E(i, j) \tag{5.14}$$

$$Key(i, j) = EW(i, j) \oplus V(i, j) \tag{5.15}$$

5.3.4.3 Watermark Extraction and Recovery

Firstly, the image $I'_r(i, j)$ after the attack is subjected to SWT-Daisy-DCT transformation to extract a 32-bit feature vector, $V'(i, j)$, which is then XORed with the logical key, $Key(i, j)$, stored in a third party to extract the encrypted watermark image, $EW'(i, j)$. Then, the binary encryption matrix, $E(i, j)$, is XORed with $EW'(i, j)$ to obtain the restored watermark image, $W'(i, j)$. The watermark extraction and restoration process is shown in Figure 5.4.

$$EW'(i, j) = V'(i, j) \oplus Key(i, j) \tag{5.16}$$

$$W'(i, j) = E(i, j) \oplus EW'(i, j) \tag{5.17}$$

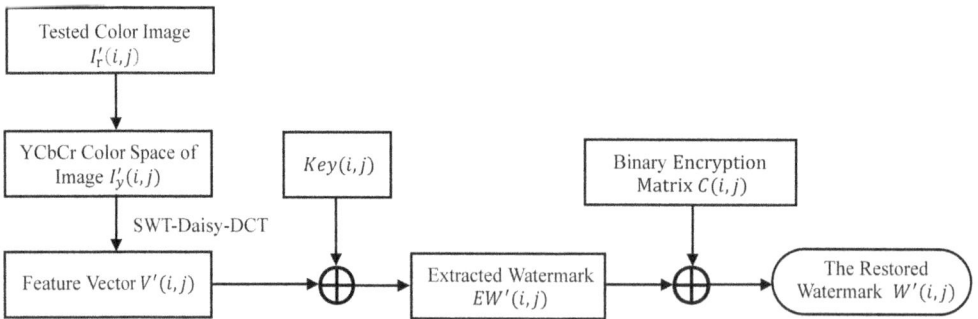

FIGURE 5.4 The process of watermark extraction and recovery.

5.4 EXPERIMENT AND RESULTS

5.4.1 Evaluation Parameter

In the experiment, a "lena" image randomly selected from Figure 1.5 with a pixel size of 512×512 was chosen as the test image. In addition, a binary image with a size of 32×32 pixels and containing "CN" watermark information was selected as the watermark image to verify the invisibility and robustness of the watermarking technique.

1. Robustness: The experiments in this article mainly use the normalized correlation (NC) to verify the robustness of watermarking technology. The NC value is directly proportional to the robustness. When NC $\in$ (0.5,1], the extracted watermark information can be used as a basis for copyright protection.

$$NC = \frac{\sum_i \sum_j W(i, j) W'(i, j)}{\sqrt{\sum_i \sum_j W(i, j)^2} \sqrt{\sum_i \sum_j W'(i, j)^2}} \tag{5.18}$$

where $W(i, j)$ and $W'(i, j)$ are the original watermark and the extracted watermark, respectively.

2. Invisibility: This study extracts the feature binary sequence of the test image and establishes a certain relationship with the binary watermark information to obtain the corresponding key without modifying the original data, thus satisfying the requirement of invisibility of the image watermark [25]. The main experiment in this paper uses peak signal-to-noise ratio (PSNR) to measure the invisibility of the watermark. PSNR is used in the main experiment in this research to gauge the watermark's imperceptibility. The PSNR is defined as follows:

$$PSNR = 10 lg \frac{MN \max_{i,j} (I(i, j))^2}{\sum_i \sum_j (I(i, j) - I'(i, j))^2} \tag{5.19}$$

where max is the maximum possible pixel value in the image, usually 255. M and N represent the size of the tested image $I(i, j)$, where $M = 512$, $N = 512$.

5.4.2 Feasibility Analysis

5.4.2.1 Distinguishability Analysis

In the experiment, we selected ten different color images to verify the distinguishability of the algorithm. Figure 5.5 and Table 5.1 show the NC values between the feature vectors (32-bit) of these ten different color images. The data shows that the NC values between distinct images are all less than 0.5 and are all 1.00 for their individual NC values. Different images can be distinguished, which is in line with human visual characteristics.

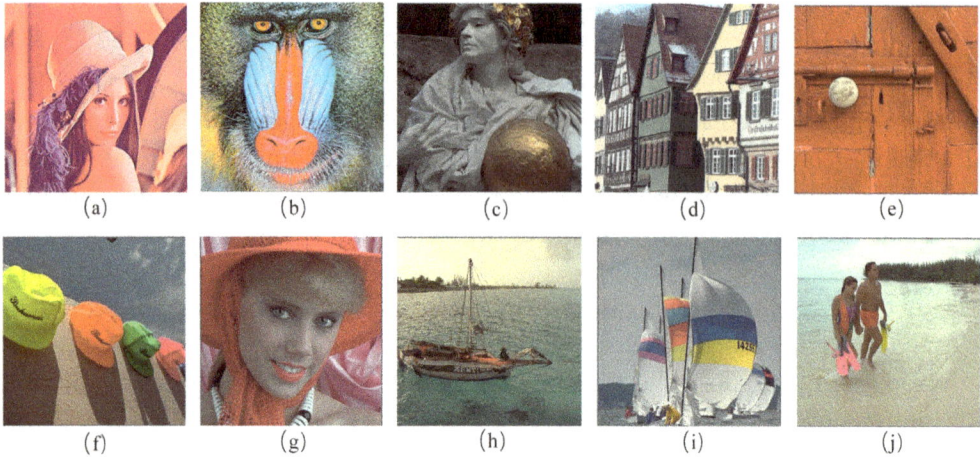

FIGURE 5.5   Different color tested images: (a) Lena, (b) baboon, (c) statue, (d) house, (e) wall, (f) hat, (g) woman, (h) boat, (i) sailing, and (j) sand.

TABLE 5.1   Correlation between Feature Vectors (32-bit) of Various Images

Image	Lena	Baboon	Statue	House	Wall	Hat	Woman	Boat	Sailing	Sand
Lena	**1.00**	0.07	0.15	−0.06	−0.13	0.25	0.26	−0.49	0.21	0.18
Baboon	0.07	**1.00**	−0.10	0.12	−0.16	0.30	−0.21	−0.19	0.09	−0.25
Statue	0.15	−0.10	**1.00**	0.32	0.06	−0.15	−0.33	0.11	0.11	−0.19
House	−0.06	0.12	0.32	**1.00**	0.19	−0.06	−0.44	0.06	−0.12	−0.25
Wall	−0.13	−0.16	0.06	0.19	**1.00**	−0.24	−0.09	0.01	−0.27	0.31
Hat	0.25	0.30	−0.15	−0.06	−0.24	**1.00**	−0.13	−0.25	−0.21	0.06
Woman	0.26	−0.21	−0.33	−0.44	−0.09	−0.13	**1.00**	−0.01	0.14	0.31
Boat	−0.49	−0.19	0.11	0.06	0.01	−0.25	−0.01	**1.00**	0.04	−0.18
Sailing	0.21	0.09	0.11	−0.12	−0.27	−0.21	0.14	0.04	**1.00**	0.12
Sand	0.18	−0.25	−0.19	−0.25	0.31	0.06	0.31	−0.18	0.12	**1.00**

5.4.2.2 Robustness Analysis

In the experiment, we selected a "lena" color image with 512 × 512 in size as the test image. We used the method proposed in section 5.3.4.1 to extract feature vectors and perform SWT-Daisy-DCT transformation on the test image. To illustrate, Table 5.2 shows the first 16 symbol sequences extracted from the "lena" image under different attacks, which are "1001 1010 1110 1110." From Table 5.2, it can be seen that all the attacked images have the same symbol sequence as the original image, and the NC value is 1.00, which meets the requirement of robustness.

5.4.3 Results and Analysis

5.4.3.1 Conventional Attack

Table 5.3 displays experimental results under various Gaussian noise attack strengths. When the noise intensity is under 30%, the NC values of the extracting watermark are all

TABLE 5.2   The First 16-bit Symbol Sequence Extracted from "lena" Image under Different Attacks

Image processing	Symbol sequence	NC
Original image	1001 1010 1110 1110	1.0
Gaussian noise (5%)	1001 1010 1110 1110	1.0
JPEG compression (10%)	1001 1010 1110 1110	1.0
Median filter [5×5]	1001 1010 1110 1110	1.0
Rotation (clockwise, 5°)	1001 1010 1110 1110	1.0
Scaling (× 0.5)	1001 1010 1110 1110	1.0
Clipping (20%, Y direction)	1001 1010 1110 1110	1.0
Clipping (20%, X direction)	1001 1010 1110 1110	1.0
Clipping (1/9, Top-left)	1001 1010 1110 1110	1.0

TABLE 5.3   Experimental Data of Gaussian Noise Attack

Noise intensity (%)	2	5	10	15	20	25	30
PSNR	17.37	13.86	11.52	10.34	9.62	9.11	8.73
NC	0.75	0.68	0.70	0.68	0.68	0.67	0.57

greater than 0.5, a relatively complete watermark information can be identified. The experimental data under different degrees of JPEG compression attacks are shown in Table 5.4. Under the given different compression qualities, the NC values of extracting watermark are all greater than 0.8, and watermark information can be identified well. Under different sizes of filtering attacks, the experimental results of NC values given in the table are all greater than 0.94, and watermark information can be identified well. Under median filtering, when the filter size is [5 × 5], NC = 1.00. Under mean filtering, the NC values of the watermark corresponding to the given several filter sizes attacks are all 1.00 (Table 5.5). The experimental results of several conventional attacks are shown in Figure 5.6.

TABLE 5.4   Experimental Data of JPEG Compression

Compression quality (%)	5%	10%	20%	25%	30%
PSNR	24.60	28.32	31.23	32.05	32.67
NC	1.00	0.94	0.88	0.94	0.94

TABLE 5.5   Experimental Data of Median Filtering and Mean Filtering

Type of filtering	Median filtering				Mean filtering			
Size of filtering	[3 × 3]	[5 × 5]	[7 × 7]	[9 × 9]	[3 × 3]	[5 × 5]	[7 × 7]	[9 × 9]
PSNR	40.20	33.14	29.42	27.32	5.17	5.17	5.17	5.17
NC	1.00	1.00	0.94	0.94	1.00	1.00	1.00	1.00

Type of attack	The attacked image	YCbCr color space	The extracted watermark	NC
Gaussian noise 10%				0.70
JPEG compression 10%				0.94
Median filtering [7×7]				0.94
Mean filtering [7×7]				1.00

FIGURE 5.6 Experimental effect of conventional attack.

5.4.3.2 Geometric Attack

Table 5.6 displays experimental results under various rotation attack strengths. It has been found to be particularly effective against rotation attacks, due to the Daisy descriptor's good rotation invariance. The NC value is not less than 0.6 when the rotation angle of the tested image to be measured is within 50°. In Table 5.7, scaling attacks also have a minimal impact on the watermark image, with NC values consistently below 0.93 when the scaling factor is between 0.5 and 3. Experimental data under different types and proportions of shear attacks are shown in Table 5.8. It can be found that even when the image information is close to half lost when shearing along the coordinate axis reaches 40%, the NC value can still reach about 0.90. Similarly, the performance of the algorithm is equally good for both top-left clipping and random clipping. The experimental results of several geometric attacks are shown in Figure 5.7.

TABLE 5.6 Experimental Data of Rotation

Rotation angle	5°	10°	15°	20°	25°	30°	35°	40°	45°	50°
PSNR	14.38	12.27	11.42	11.04	10.71	10.43	10.27	10.25	10.24	10.22
NC	1.00	0.95	0.89	0.89	0.79	0.80	0.73	0.68	0.68	0.60

TABLE 5.7 Experimental Data of Scaling

Scaling	0.5	1	1.5	2	3
NC	0.94	1.00	0.93	0.93	0.93

TABLE 5.8   Experimental Data of Clipping

Clipping type	Y-axis			X-axis			Top-left			Random
Ratio	10%	25%	40%	10%	30%	40%	1/16	1/9	1/6	-
NC	1.00	1.00	0.89	1.00	1.00	0.93	1.00	1.00	0.95	1.00

Clipping type	The attacked image	YCbCr color space	The extracted watermark	NC
Rotation 20%			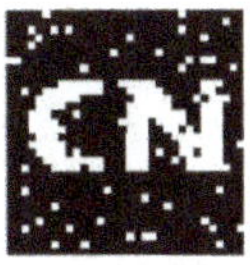	0.89
Scaling (×2)				0.93
Y-axis clipping 25%				1.00
X-axis clipping 30%				1.00
Top-left clipping 1/6				0.95

FIGURE 5.7   Experimental effect of geometric attack.

Overall, the above experimental findings demonstrate the algorithm's better robustness against both conventional and geometric attacks.

5.5 CONCLUSION

This article proposes a robust zero-watermarking algorithm for color images, which combines the local image feature descriptor Daisy with the transformation domain SWT and DCT. The algorithm integrates the advantages of Daisy algorithm's rotational invariance, SWT transformation's noise reduction function, and DCT transformation's robustness to low-frequency components. In the implementation of the algorithm, the SWT-Daisy-DCT transform is first used to extract the 32-bit low-frequency coefficients of the tested image, and then perceptual hash technology is applied to generate a binary feature sequence. Finally, the encrypted watermark is logically related to the feature sequence, and a key is generated and stored with a third-party organization. The main feature of this research is the integration of multiple technical means, which achieves imperceptibility, robustness, and security of the watermark and is suitable for various application environments. Experimental data show that the feature vector extraction algorithm has strong robustness against conventional and geometric attacks. In summary, the comprehensive approach of this research provides an effective solution for watermarking technology and is worthy of further research and application.

REFERENCES

1. Li, D., Li, J., Bhatti, U. A., Nawaz, S. A., Liu, J., Chen, Y. W., & Cao, L. (2023). Hybrid encrypted watermarking algorithm for medical images based on DCT and improved DarkNet53. *Electronics*, 12(7): 1554.
2. Bhatti, U. A., Tang, H., Wu, G., Marjan, S., & Hussain, A. (2023). Deep learning with graph convolutional networks: An overview and latest applications in computational intelligence. *International Journal of Intelligent Systems*, 2023: 1–28.
3. Sheng, M., Li, J., Bhatti, U. A., Liu, J., Huang, M., & Chen, Y. W. (2023). Zero watermarking algorithm for medical image based on Resnet50-DCT. *CMC-Computers Materials & Continua*, 75(1): 293–309.
4. Fan, Y., Li, J., Bhatti, U. A., Shao, C., Gong, C., Cheng, J., & Chen, Y. (2023). A multi-watermarking algorithm for medical images using inception V3 and DCT. *CMC-Computers Materials & Continua*, 74(1): 1279–1302.
5. Fares K., Khaldi A., Redouane K., et al. (2021). DCT & DWT based watermarking scheme for medical information security. *Biomedical Signal Processing and Control*, 66: 102403.
6. Bhatti, U. A., Yu, Z., Chanussot, J., Zeeshan, Z., Yuan, L., Luo, W., … Mehmood, A. (2021). Local similarity-based spatial–spectral fusion hyperspectral image classification with deep CNN and Gabor filtering. *IEEE Transactions on Geoscience and Remote Sensing*, 60: 1–15.
7. Yuan Z., Liu D., Zhang X., et al. (2020). New image blind watermarking method based on two-dimensional discrete cosine transform. *Optik*, 204: 164152.
8. Ghrare S. E., Alamari A. A. M., Emhemed H. A. (2022). Digital Image Watermarking Method Based on LSB and DWT Hybrid Technique. 2022 IEEE 2nd International Maghreb Meeting of the Conference on Sciences and Techniques of Automatic Control and Computer Engineering (MI-STA). IEEE, 465–470.
9. Kumar S., Jha R. K. (2019). FD-based detector for medical image watermarking. *IET Image Processing*, 13(10): 1773–1782.
10. Wang K., Gao T., You D., et al. (2022). A secure dual-color image watermarking scheme based 2D DWT, SVD and Chaotic map. *Multimedia Tools and Applications*, 81(5): 6159–6190.
11. Fares K., Amine K., Salah E. (2020). A robust blind color image watermarking based on Fourier transform domain. *Optik*, 208: 164562.
12. Sivananthamaitrey P., Kumar P. R. (2022). Optimal dual watermarking of color images with SWT and SVD through genetic algorithm. *Circuits, Systems, and Signal Processing*, 41: 224–248.
13. Liu Z., Li J., Ai Y., et al. (2022). A robust encryption watermarking algorithm for medical images based on ridgelet-DCT and THM double chaos. *Journal of Cloud Computing*, 11(1): 1–20.
14. Liu P., Wu H., Luo L., et al. (2022). DT CWT and Schur decomposition based robust watermarking algorithm to geometric attacks. *Multimedia Tools and Applications*, 1–43.

15. Bhatti U. A., Yu Z., Yuan L., et al. (2022). A Robust Remote Sensing Image Watermarking Algorithm Based on Region-Specific SURF. Proceedings of International Conference on Information Technology and Applications: ICITA 2021. Singapore: Springer Nature Singapore, 75–85.

16. Ahmad R. M., Yao X. M., Nawaz S. A., et al. (2020). Robust image watermarking method in wavelet domain based on SIFT features. Proceedings of the 2020 3rd International Conference on Artificial Intelligence and Pattern Recognition, 180–185.

17. Dwivedi R., Srivastava V. K. (2022). Geometrically Robust Digital Image Watermarking Based on Zernike Moments and FAST Technique. Advances in VLSI, Communication, and Signal Processing: Select Proceedings of VCAS 2021. Singapore: Springer Nature Singapore, 671–680.

18. Pal P., Jana B., Bhaumik J. (2019). Watermarking scheme using local binary pattern for image authentication and tamper detection through dual image. *Security and Privacy*, 2(2): e59.

19. Yuan X. C., Li M. (2018). Local multi-watermarking method based on robust and adaptive feature extraction. *Signal Processing*, 149: 103–117.

20. Tola E., Lepetit V., Fua P. (2009). Daisy: An efficient dense descriptor applied to wide-baseline stereo. *IEEE Transactions on Pattern Analysis and Machine Intelligence*, 32(5): 815–830.

21. Diwakar M., Tripathi A., Joshi K., et al. (2021). A comparative review: Medical image fusion using SWT and DWT. *Materials Today: Proceedings*, 37: 3411–3416.

22. Kanwal S., Inam S., Othman M. T. B., et al. (2022). An effective color image encryption based on Henon map, tent chaotic map, and orthogonal matrices. *Sensors*, 22(12): 4359.

23. Fang Y., Liu J., Li J., et al. (2022). Robust zero-watermarking algorithm for medical images based on SIFT and Bandelet-DCT. *Multimedia Tools and Applications*, 81(12): 16863–16879.

24. Evsutin O., Dzhanashia K. (2022). Watermarking schemes for digital images: Robustness overview. *Signal Processing: Image Communication*, 100: 116523.

25. Kahlessenane F., Khaldi A., Kafi R., et al. (2021). A DWT based watermarking approach for medical image protection. *Journal of Ambient Intelligence and Humanized Computing*, 12(2): 2931–2938.

26. Bhatti, U. A., Huang, M., Neira-Molina, H., Marjan, S., Baryalai, M., Tang, H., Wu, G., & Bazai, S. U. (2023). MFFCG – Multi feature fusion for hyperspectral image classification using graph attention network. *Expert Systems with Applications*, 229: 120496. 10.1016/j.eswa.2023.120496.

27. Bhatti, U. A., Yu, Z., Chanussot, J., Zeeshan, Z., Yuan, L., Luo, W., Nawaz, S. A., Bhatti, M. A., Ain, Q. U., & Mehmood, A. (2022). Local similarity-based spatial–spectral fusion hyperspectral image classification with deep CNN and Gabor filtering. *IEEE Transactions on Geoscience and Remote Sensing*, 60: 1–15. 10.1109/tgrs.2021.3090410.

28. Zhang, P., Li, J., Aslam Bhatti, U., Liu, J., Chen, Y.-W., Li, D., & Cao, L. (2023). Robust watermarking algorithm for medical volume data based on polar cosine transform and 3D-DCT. *Computers, Materials & Continua*, 75: 5853–5870. 10.32604/cmc.2023.036462.

29. Yang, M., Li, J., Aslam Bhatti, U., Shao, C., & Chen, Y.-W. (2023). Robust watermarking algorithm for medical images based on non-subsampled Shearlet transform and Schur decomposition. *Computers, Materials & Continua*, 75: 5539–5554. 10.32604/cmc.2023.036904.

30. Huang, S., Huang, M., Zhang, Y., Chen, J., & Bhatti, U. (2020). Medical image segmentation using deep learning with feature enhancement. *IET Image Processing*, 14: 3324–3332. 10.1049/iet-ipr.2019.0772.

31. Bhatti, U. A., Yu, Z., Li, J., Nawaz, S. A., Mehmood, A., Zhang, K., & Yuan, L. (2020). Hybrid watermarking algorithm using Clifford algebra with Arnold scrambling and chaotic encryption. *IEEE Access*, 8: 76386–76398. 10.1109/access.2020.2988298.

CHAPTER 6

Robust Multi-watermarking Algorithm Based on DarkNet53

Dekai Li[1], Jingbing Li[1,2], and Uzair Aslam Bhatti[1,2]

[1]School of Information and Communication Engineering, Hainan University, Haikou, China
[2]State Key Laboratory of Marine Resource Utilization in the South China Sea,
 Hainan University, Haikou, China

6.1 INTRODUCTION

The increasing prevalence of digital media and the widespread use of the Internet has brought significant attention to the issue of digital copyright, as well as the security and privacy concerns associated with digital media [1]. Digital watermarking technology plays a crucial role in safeguarding the security of digital media. By incorporating specific information into digital media through embedding techniques, functions such as copyright protection, content tracking, and anti-counterfeiting traceability of digital media can be realized [2,3]. In practical applications, digital watermarking technology needs to have the characteristics of robustness, concealment, and fault tolerance, so as to effectively resist various attacks without affecting the quality of the media.

Digital watermarks can be divided into two main types: visible and invisible digital watermarks [4,5]. Visible digital watermarking generally refers to directly embedding visual information in digital media, such as logos and images. Invisible digital watermarking refers to the process of embedding information into non-visible or imperceptible areas of digital media, such as embedding into low-frequency areas of audio or video, high-frequency areas of images, and so on.

Invisible digital watermarking is a more covert digital watermarking technology, and its robustness is also stronger, but because the embedded information is invisible, it is also more difficult to detect and identify. To enhance the robustness and reliability of invisible digital watermarking, numerous researchers have turned to the application of deep learning technology [6]. The convolutional neural network (CNN) is a powerful deep learning model renowned for its exceptional feature extraction and classification

DOI: 10.1201/9781003427674-6

abilities, making it highly effective in the domain of image processing [7]. Therefore, the integration of CNNs into digital-watermarking technology offers several benefits. It not only enhances the robustness of digital watermarking but also helps to reduce false detection rates and false positive rates, thereby improving the overall reliability and practicality of digital-watermarking technology [8].

With the growing trend towards digitization in healthcare, medical information including patient data from images is increasingly being transmitted and stored online [9]. This puts patient information at risk of disclosure and misappropriation. The protection of patients' privacy information is a pressing issue that requires the attention and efforts of researchers to resolve. As an information hiding technology, invisible digital-watermarking technology can effectively hide patients' private information in medical images by using its invisibility, which plays a very good role in protecting patient information.

Embedding watermarks in medical images requires special consideration due to the sensitivity of medical data. The alteration of the original pixel information during the watermark embedding process can potentially impact the accuracy of medical diagnosis conducted by doctors. This has great restrictions on the embedding method of the watermark. Traditional digital-watermarking techniques may leave visible or hidden marks in medical images, which may interfere with doctors' judgments, thereby affecting patients' diagnosis and treatment [10]. Zero watermark technology provides a new option for medical images, which can hide watermark information in medical images without affecting image quality and doctor's judgment [11].

Currently, there are two main methods in digital watermarking technology: spatial domain embedding and transform domain embedding. In the spatial domain, the watermark is directly embedded into the pixel values of the original image. One representative algorithm in this domain is the least significant bit (LSB) substitution method. LSB embedding involves replacing the least significant bit of the original image with the watermark information. While this method is easy to implement, it is also relatively susceptible to detection by attackers. For example, Bamatraf et al. [12] proposed a digital-watermarking algorithm that utilizes LSB embedding by using the third and fourth LSBs for watermarking, resulting in a simple and robust algorithm. Their approach demonstrates improved robustness compared to traditional LSB techniques in concealing data within images.

Transform domain watermark embedding involves embedding watermark information into the coefficients of image transform domains, which can enhance the strength and robustness of the watermark. This makes it more challenging for potential attackers to detect or remove the watermark. Popular transform domain embedding methods include discrete cosine transform (DCT), wavelet transform, and discrete Fourier transform (DFT). Among them, the wavelet transform method has gained popularity due to its favorable properties in both the frequency and time domains.

For example, Barni et al. [13] proposed a watermarking method in the frequency domain. The method uses a sequence of pseudo-random numbers to embed the watermark in selected DCT coefficients. Similarly, Ganic and Eskicioglu [14] proposed a method for embedding watermarked data using DWT-SVD. The method decomposes the cover image into four bands and then implements a watermarking scheme by modifying the singular values of each band. Li et al. [15] proposed a watermarking scheme using the AKAZE-DCT algorithm. The algorithm uses AKAZE-DCT to extract the features of the image and combines perceptual hashing techniques to implement the watermarking scheme. This approach demonstrates good robustness and invisibility. These transform domain watermarking techniques leverage the strengths of specific transforms and provide improved robustness and invisibility for watermarking in various applications, including medical images.

In recent years, the advancements in deep learning technology have offered new avenues and methods for the research and application of digital watermarking. The CNN, as a vital component of deep learning, has made significant breakthroughs in various areas such as image recognition, target detection, and image segmentation [16–18]. In a study by Huang et al. [19], they proposed a robust zero-watermarking scheme that leverages a deep hyper-parameterized VGG network. The method uses a DO-VGG model to feature information from medical images and incorporates watermarking techniques to complete the watermarking scheme. This scheme exhibits good robustness and invisibility. These advancements highlight how deep learning techniques, particularly CNNs, have been successfully integrated into digital-watermarking research. They offer improved performance, robustness, and invisibility in various applications, including medical image watermarking.

DarkNet53 is indeed a deep convolutional neural network model that has gained recognition for its computational efficiency and accuracy. It finds applications in diverse fields such as target detection and face recognition [20]. Building upon the capabilities of DarkNet53, this chapter introduces a robust multi-watermarking algorithm. The proposed algorithm aims to safeguard the copyright and security of digital media while preserving the quality of the media [21–34]. The algorithm can improve the robustness and reliability of watermark by embedding multiple watermarks. Traditional digital-watermarking technology usually only embeds a watermark; once the watermark is attacked or cracked, the copyright protection function of the entire digital media will be invalid. The robust multi-watermark algorithm can embed multiple watermarks, even if one of the watermarks is attacked or cracked, other watermarks can still play a protective role.

This chapter will introduce the principle and implementation method of the robust zero-watermark algorithm in detail, including the specific steps of watermark embedding and extraction, and conduct experimental verification and performance analysis of the algorithm. At the same time, this chapter also summarizes and looks forward to the algorithm in the field of digital watermarking, hoping to provide new ideas and methods for digital media security and copyright protection.

6.2 BASIC THEORY

6.2.1 DarkNet53

DarkNet53 is a lightweight convolutional neural network and is the backbone network in YOLOv3 [22]. It consists of 53 convolutional layers and 5 maximum pooling layers,

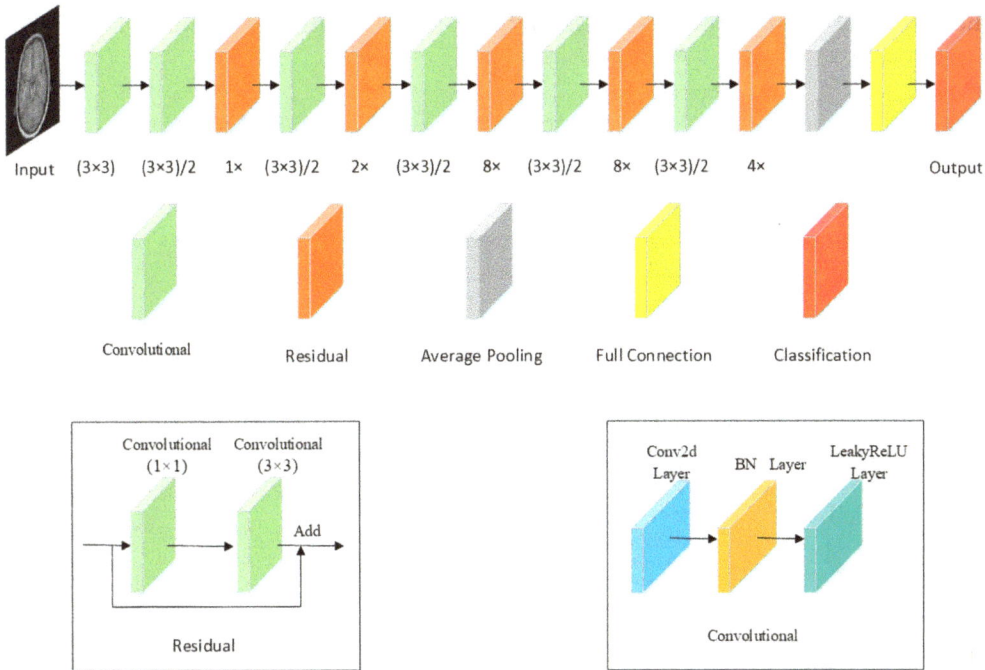

FIGURE 6.1 DarkNet53 network model structure diagram.

which can quickly extract features from images and is suitable for tasks such as target detection and classification.

The main concept behind DarkNet53 is to employ residual structures and skip connections in order to construct deep networks that can mitigate issues such as network degradation and gradient disappearance. To achieve this, DarkNet53 utilizes convolutional kernels of different sizes (1×1, 3×3, and 5×5) to capture features at various scales and combines them through residual connections. Additionally, techniques like batch normalization and linear activation functions are incorporated to enhance the network's performance. The network model structure of DarkNet53 is depicted in Figure 6.1.

6.2.2 Discrete Cosine Transform

The DCT is a mathematical technique commonly employed in signal processing and data compression [23]. It transforms a signal from the time domain into a set of coefficients in the frequency domain, enabling the representation of information contained within the original signal. Similar to the Fourier transform, the DCT also decomposes the signal into a set of sine or cosine waves, but it uses only real operations and is therefore more computationally efficient. DCT is widely used in image and video compression, because it can filter out redundant information while retaining important information of images or videos, thereby achieving compression. In addition, DCT can also be used in audio compression, speech recognition, data encryption, and other fields.

The commonly used formula for one-dimensional DCT transformation is as follows:

$$Y(u) = C(u)\sqrt{\frac{2}{N}} \sum_{m=0}^{N-1} X(m)\cos\left[\frac{(2m+1)u\pi}{2N}\right], \quad u = 0, 1, ..., N-1$$

$$C(u) = \begin{cases} \dfrac{1}{\sqrt{2}} & u = 0 \\ 1 & u \neq 0 \end{cases}$$

(6.1)

The commonly used formula for two-dimensional DCT transformation is as follows:

$$Y(u, v) = \frac{2}{\sqrt{MN}} C(u)C(v) \sum_{m=0}^{M-1} \sum_{n=0}^{N-1} X(m, n)\cos\left[\frac{(2m+1)u\pi}{2M}\right]\cos\left[\frac{(2n+1)v\pi}{2N}\right]$$

$$u = 0, 1, ..., N-1; \quad v = 0, 1, ..., M-1$$

$$C(u) = \begin{cases} \dfrac{1}{\sqrt{2}} & u = 0 \\ 1 & u \neq 0 \end{cases}, \quad C(v) = \begin{cases} \dfrac{1}{\sqrt{2}} & v = 0 \\ 1 & v \neq 0 \end{cases}$$

(6.2)

6.2.3 Logistic Map

Logistic Map is a simple nonlinear dynamic system that can be used to simulate many complex phenomena in nature. The model was first proposed by biologist Robert May in 1976 to study the dynamic process of species change [24]. Since then, this model has been applied in many other fields, including chaos theory, fractal geometry, financial market analysis, etc.

The sub-illustration of the Logistic Map is shown in Figure 6.2, and its mathematical form is

$$x_{n+1} = rx_n(1 - x_n)$$

(6.3)

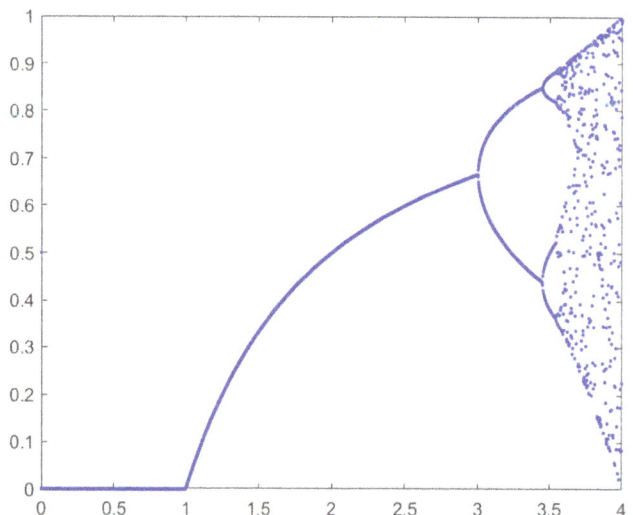

FIGURE 6.2 Logistic map bifurcation diagram.

Among them, x_n represents the population size of the n-th iteration, and r is a parameter controlling the growth rate of the population. In each iteration, the population size will change according to the above formula, and the new population size will be used as the input for the next iteration. When r is less than 1, the population size will eventually tend to be a stable value; when r is between 1 and 3, the population size will converge to a periodic value after several iterations; and when r is greater than 3, the population quantity will be chaotic, which means that its changes will become unpredictable.

The chaotic behavior of Logistic Map makes it widely used in scientific research and engineering applications. For example, it can be used to generate random number sequences, simulate weather changes, and analyze financial market fluctuations. It is also used as a basic algorithm in the fields of digital signal processing, image compression, cryptography, etc.

6.3 PROPOSED ALGORITHM

This chapter presents a robust multi-watermarking algorithm for medical images. The algorithm incorporates improvements in pre-training networks, migration learning, encryption of watermark information, and embedding and extraction of watermarks.

In the network improvement phase, the DarkNet53 CNN is employed to extract image features. This network architecture is chosen for its ability to capture relevant features from medical images effectively.

For transfer learning, the algorithm utilizes pre-trained network parameters obtained from a self-built medical image data set. By leveraging transfer learning, the robustness of the algorithm is enhanced, leading to improved performance in watermark extraction.

The algorithm uses a logistic chaotic graph for encryption of the watermarked information. This encryption method ensures that the watermarked information is secure.

Through extensive experimental verification, the algorithm demonstrates superior watermark extraction performance even under various attack scenarios. This highlights the algorithm's effectiveness in protecting medical image watermarks and maintaining their integrity.

6.3.1 Improvement of DarkNet53 Network Model

6.3.1.1 Improvement of Network Structure

The CNN has proven to be highly effective in extracting meaningful features from images, making it invaluable for the development of medical image watermarking algorithms. For the purpose of this chapter, DarkNet53 was chosen because it has been trained on ImageNet, which ensures its powerful feature extraction capabilities.

To improve the robustness of the algorithm, a simple modification was made to the network structure. Specifically, the Softmax layer was replaced with an fc layer and the classification layer was replaced with a regression layer for the regression task.

These modifications allow the network to adapt its classification-oriented structure into a regression-based framework, thus enabling it to accurately extract and utilize the essential features from medical images for watermarking purposes.

6.3.1.2 Data Set Creation

The data set used in this chapter is sourced from the Medical Imaging Park and America's Research Hospital Clinical Center. It consists of 500 medical images covering different regions of the brain, pelvis, skeletal muscle, colon, and chest, as depicted in Figure 6.3. These images were downloaded and organized into a data set. To ensure proper training and evaluation, the data set was divided randomly.

To enhance the robustness of the neural network, data augmentation techniques were applied to both the training set and validation set. Table 6.1 outlines the specific augmentation methods used in the chapter. In total, 22,100 images were used for training, all resized to a dimension of 256 × 256 × 3 to match the input requirements of the DarkNet53 CNN.

Regarding the labeling process, the chapter employed the DCT on the images from the training and validation sets. The low-frequency part of the DCT coefficients, specifically the 32-bit feature vector, was selected as the label for the images.

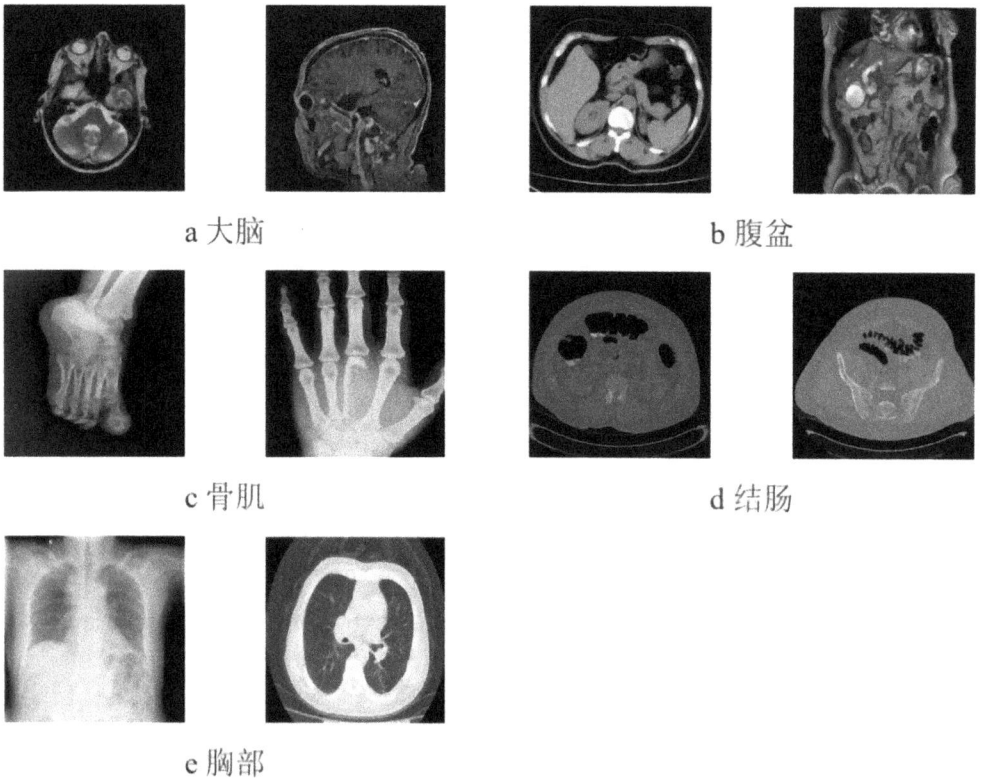

a 大脑   b 腹盆

c 骨肌   d 结肠

e 胸部

FIGURE 6.3   Take two original images for each type of image: (a) image of the brain, (b) image of the pelvis, (c) image of the skeletal muscle, (d) image of the colon, and (e) image of the chest.

TABLE 6.1   The Specific Implementation Operations of Data Enhancement

Enhancement methods	Intensity	New images
Gaussian (%)	5、10、15、20	5
JPEG (%)	2、4、6、8	5
Median Filter (10 times)	3 × 3、5 × 5、7 × 7	3
Clockwise Rotation (°)	5、10、15、20、25, etc.	8
Scaling	0.3、0.6、0.9、1.2	6
Down (%)	5、10、15、20、25、30	6
Y shear (%)	5、10、15、20、25、30	6
X shear (%)	5、10、15、20、25、30	6
Left (%)	5、10、15、20、25、30	6
Right (%)	5、10、15、20、25、30	6

6.3.1.3 Training Network

In the experiment, a computer with the following specifications was used:

- NVIDIA GeForce GTX 1050Ti 4 GB graphics card

- Intel® CoreTM i5-8300H CPU @ 2.30 GHz4

The experiment utilized the neural network toolbox provided by MATLAB 2022a. Specifically, the pre-trained DarkNet53 network was chosen for training. During the training process, the weights of the first 87 layers of the network were frozen by setting their learning rate to 0. This was done to improve the efficiency of training.

The initial learning rate of the training was 0.001 and the MiniBatchSize was 30. Epochs were 4.

6.3.2 Encryption of Watermark

This chapter designs an encryption algorithm for watermarking based on Logistic Maps. The algorithm can effectively improve the security of watermarked information. The process involves several steps, which are outlined below:

1. Coefficient and initial state selection: The algorithm requires selecting suitable coefficients and initial state values for the Logistic Map. These values play a crucial role in generating a chaotic sequence.

2. Chaotic sequence generation: Using the chosen coefficients and initial state values, the chaotic system generates a chaotic sequence. The chaotic sequence exhibits complex and unpredictable behavior, which enhances the security of the watermark encryption process.

3. Watermark encryption: The binary watermark image is subjected to a bit-by-bit XOR (exclusive OR) operation with the generated chaotic sequence. This operation

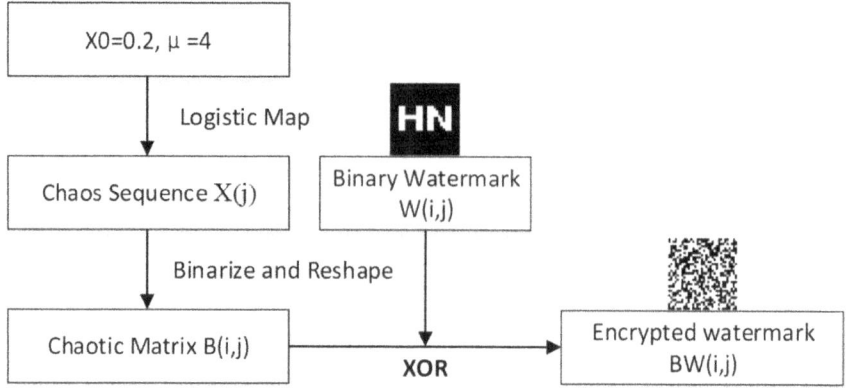

FIGURE 6.4   Watermark encryption process.

combines the randomness and unpredictability of the chaotic sequence with the watermark image, resulting in an encrypted watermark image.

The specific flow of the watermarking chaotic encryption algorithm is illustrated in Figure 6.4.

6.3.3 Watermark Embedding

The watermark embedding process described in this chapter consists of the following steps:

1. Medical images are processed with DarkNet53 to extract features.

2. DCT the extracted features and select the low-frequency portion to represent the essential features of the medical image.

3. Combine with a hashing algorithm to convert the selected low-frequency matrix into a binary feature sequence.

4. Perform XOR operation on the feature sequence and the encrypted watermark.

By adopting this approach, zero watermark embedding is achieved in this chapter. Figure 6.5 illustrates the specific steps in the watermark embedding process.

6.3.4 Extraction of a Watermark

To extract the watermark, the same process is used as in watermark embedding. Figure 6.6 provides a detailed illustration of the watermark extraction process.

6.3.5 Decryption of a Watermark

The following steps are in the watermark decryption process:

1. Chaotic sequence generation: The same chaotic system (Logistic Map) used in the watermark encryption method is utilized to generate the chaotic sequence $X(j)$.

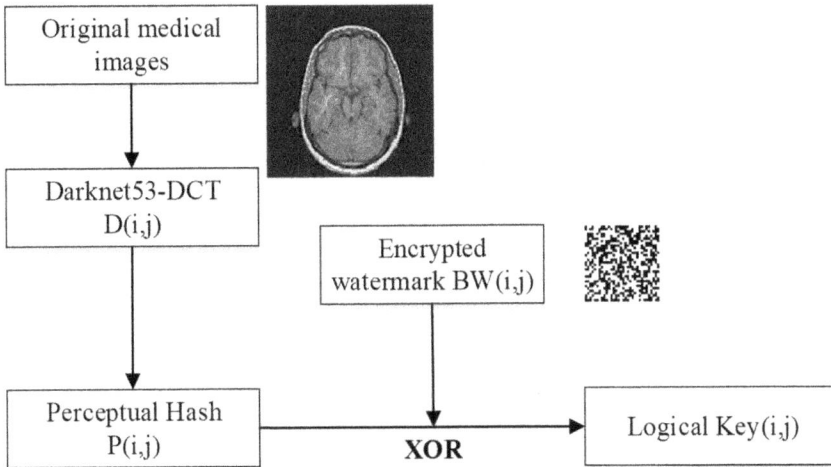

FIGURE 6.5   Flowchart of watermark embedding.

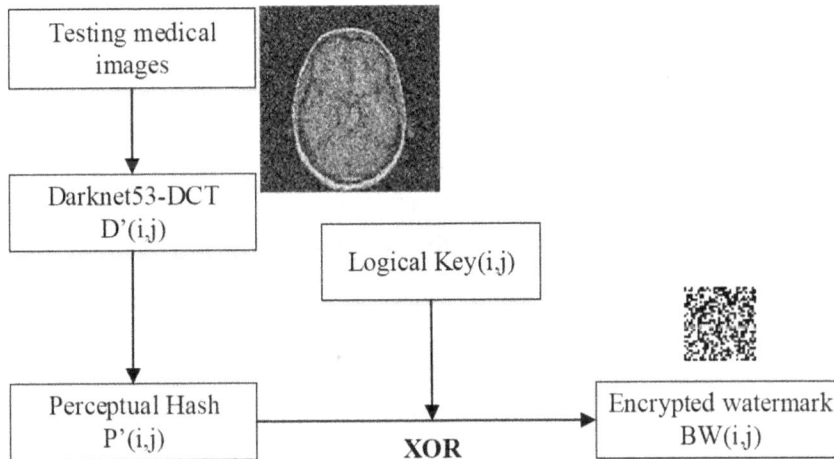

FIGURE 6.6   Flowchart of watermark extraction.

2. XOR operation: The generated chaotic sequence $X(j)$ is XORed with the encrypted watermark, $BW'(i, j)$. This operation combines the chaotic sequence with the encrypted watermark, resulting in the decrypted watermark.

By performing XOR between the chaotic sequence and the encrypted watermark, this chapter achieves decryption of the watermark. Figure 6.7 provides a visual representation of the specific watermark decryption process, illustrating the sequence of operations involved.

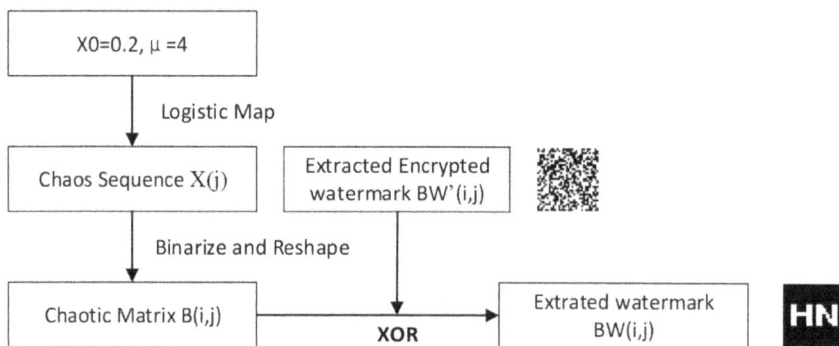

FIGURE 6.7 Decryption process of watermark.

6.4 EXPERIMENTAL RESULTS AND ANALYSIS

This chapter conducts simulation experiments on the MATLAB 2022a platform. The test medical image is a random brain map in the test set, and the watermark is three 32×32 watermark images carrying information. In order to improve security, we use Logistic chaotic encryption to encrypt the watermark image, specifically shown in Figure 6.8.

6.4.1 Performance

In this chapter, the performance of the algorithm is assessed using two evaluation metrics: the normalized correlation coefficient (NC) and peak signal-to-noise ratio (PSNR). NC measures the similarity between the original watermark and the extracted watermark. A higher NC value, closer to 1, indicates a greater degree of

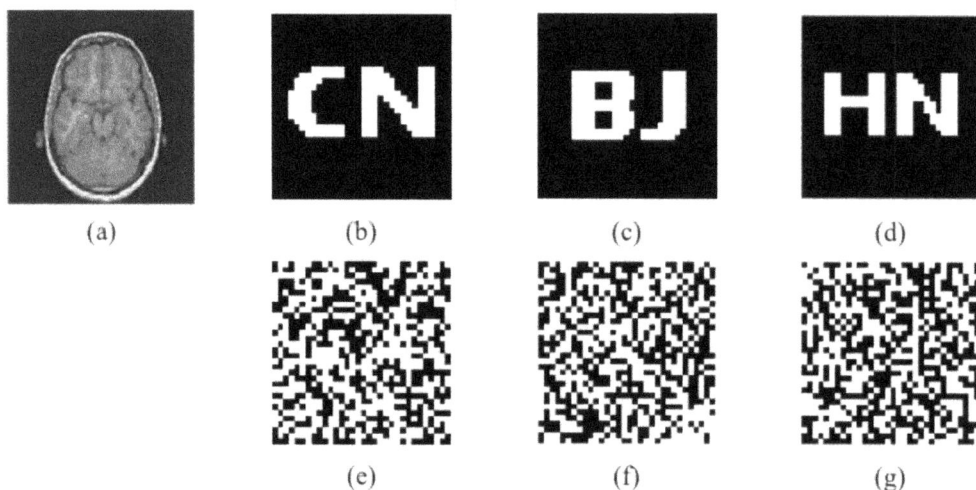

FIGURE 6.8 Medical images and watermarked images: (a) original medical image; (b), (c), (d) original watermark image; (e), (f), (g) encrypted watermark image.

similarity and higher robustness of the algorithm. Equation (6.4) is the formula used to calculate NC. On the other hand, PSNR quantifies the level of distortion in the attacked image, with lower PSNR values indicating more severe distortion. Equation (6.5) represents the formula for calculating PSNR. By employing NC and PSNR, this chapter establishes an objective evaluation criterion for assessing the proposed watermarking algorithm.

$$NC = \frac{\sum_{i=1}^{n}(x_i - \bar{x})(y_i - \bar{y})}{\sqrt{\sum_{i=1}^{n}(x_i - \bar{x})^2}\sqrt{\sum_{i=1}^{n}(y_i - \bar{y})^2}} \tag{6.4}$$

$$PSNR = 10\lg\left[\frac{MN\max_{i,j}(I(i,j))^2}{\sum_i\sum_j(I(i,j) - I'(i,j))^2}\right] \tag{6.5}$$

6.4.2 Reliability Analysis

In this chapter, a total of eight medical images (Figure 6.9) were chosen. The reliability of the proposed model algorithm was evaluated by calculating the normalized cross-correlation (NC) value between each pair of images. The test results are presented in Table 6.2, which clearly demonstrate the model algorithm's capability to effectively differentiate between different medical images. These findings highlight the research significance of the proposed model algorithm in this study.

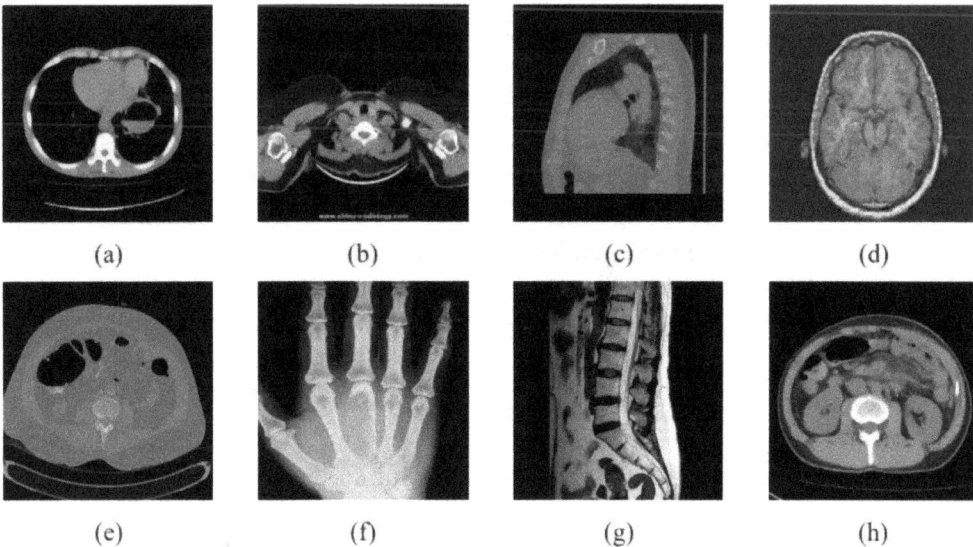

FIGURE 6.9   (a)–(h) Different medical images for testing.

TABLE 6.2 NC Between Different Encrypted Images

Image	(a)	(b)	(c)	(d)	(e)	(f)	(g)	(h)
(a)	1.00							
(b)	0.08	1.00						
(c)	0.43	0.06	1.00					
(d)	0.24	0.17	0.04	1.00				
(e)	0.18	0.00	0.51	0.18	1.00			
(f)	0.04	0.06	0.06	0.30	0.00	1.00		
(g)	0.06	0.12	0.12	0.06	0.25	0.12	1.00	
(h)	0.12	0.21	0.46	0.25	0.31	0.21	0.06	1.00

6.4.3 Traditional Attack

This chapter tests the robustness of the model algorithm against traditional attacks such as Gaussian noise, JPEG compression, and median filtering. Through extensive experimental calculations, it can be concluded that the model algorithm is still able to recover the watermark information well against traditional attacks. In fact, in some cases, the recovery is quite good.

For instance, when the Gaussian attack reaches 13%, the algorithm achieves a normalized correlation (NC) value of 0.8 for NC1, 0.78 for NC2, and 0.82 for NC3. When the JPEG compression quality is reduced to 5%, the NC values are 0.88 for NC1, 0.9 for NC2, and 0.92 for NC3. Similarly, when applying median filtering [5 × 5] for 20 times, the NC values are 0.88 for NC1, 0.91 for NC2, and 0.91 for NC3.

These results indicate that the algorithm can effectively withstand these attacks and recover the watermark information with high fidelity. Figure 6.10 provides a visual representation of some of the experimental results, and Table 6.3 presents a subset of the experimental data, demonstrating the algorithm's robustness.

6.4.4 Geometric Attack

To evaluate the robustness of the algorithm against geometric attacks, this chapter conducts tests on several common geometric transformations, including rotation, scaling, translation, and shearing. The experiments aim to measure the algorithm's ability to recover the watermark information under these attacks.

The results indicate that the algorithm demonstrates effective robustness against these geometric attacks. For instance, when subjected to a 40° clockwise rotation, the algorithm achieves a normalized correlation (NC) value of 0.86 for NC1, 0.72 for NC2, and 0.82 for NC3. At a scaling factor of 0.1, the NC values are 0.84 for NC1, 0.87 for NC2, and 0.87 for NC3. When shifted 40% to the right, the NC values are 0.64 for NC1, 0.58 for NC2, and 0.67 for NC3. Similarly, at a 40% shear along the y-axis, the NC values are 0.54 for NC1, 0.59 for NC2, and 0.59 for NC3. For a 40% shear along the x-axis, the NC values are 0.73 for NC1, 0.70 for NC2, and 0.69 for NC3.

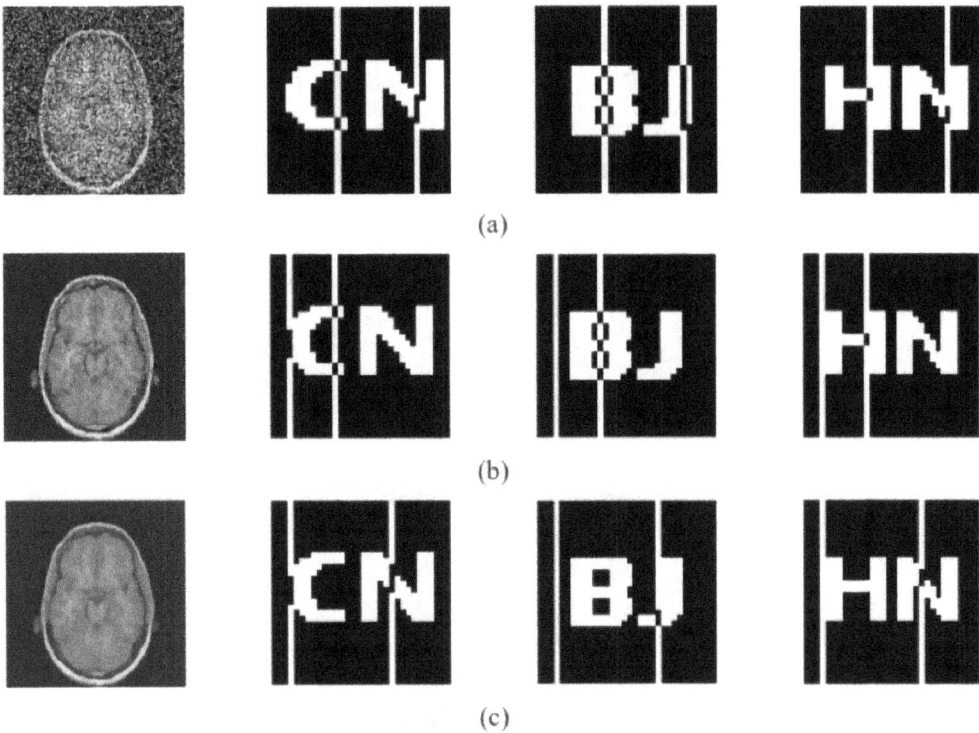

(a)

(b)

(c)

FIGURE 6.10   The image after the attack and the extracted watermark: (a) Gaussian noise 13% and extracted watermark, (b) JPEG compression 5% and extracted watermark, and (c) median filter [5 × 5] 20 times and extracted watermarks.

TABLE 6.3   Experimental Data Under Conventional Attack

Attacks	Intensity	PSNR (dB)	NC1	NC2	NC3
Gaussian (%)	3	16.23	0.96	0.96	0.96
	7	13.07	0.84	0.82	0.86
	11	11.59	0.84	0.80	0.86
	13	11.04	0.80	0.78	0.82
JPEG (%)	5	28.71	0.88	0.90	0.92
	15	33.81	0.94	0.93	0.95
	30	36.62	1.00	1.00	1.00
Median Filter (20 times)	[3 × 3]	35.20	1.00	1.00	1.00
	[5 × 5]	27.76	0.88	0.91	0.91
	[7 × 7]	25.64	0.82	0.86	0.86

These results demonstrate that the algorithm effectively recovers the watermark information, even under these geometric attacks. Figure 6.11 visually presents some of the experimental results, while Table 6.4 provides a subset of the experimental data, confirming the algorithm's robustness against geometric attacks.

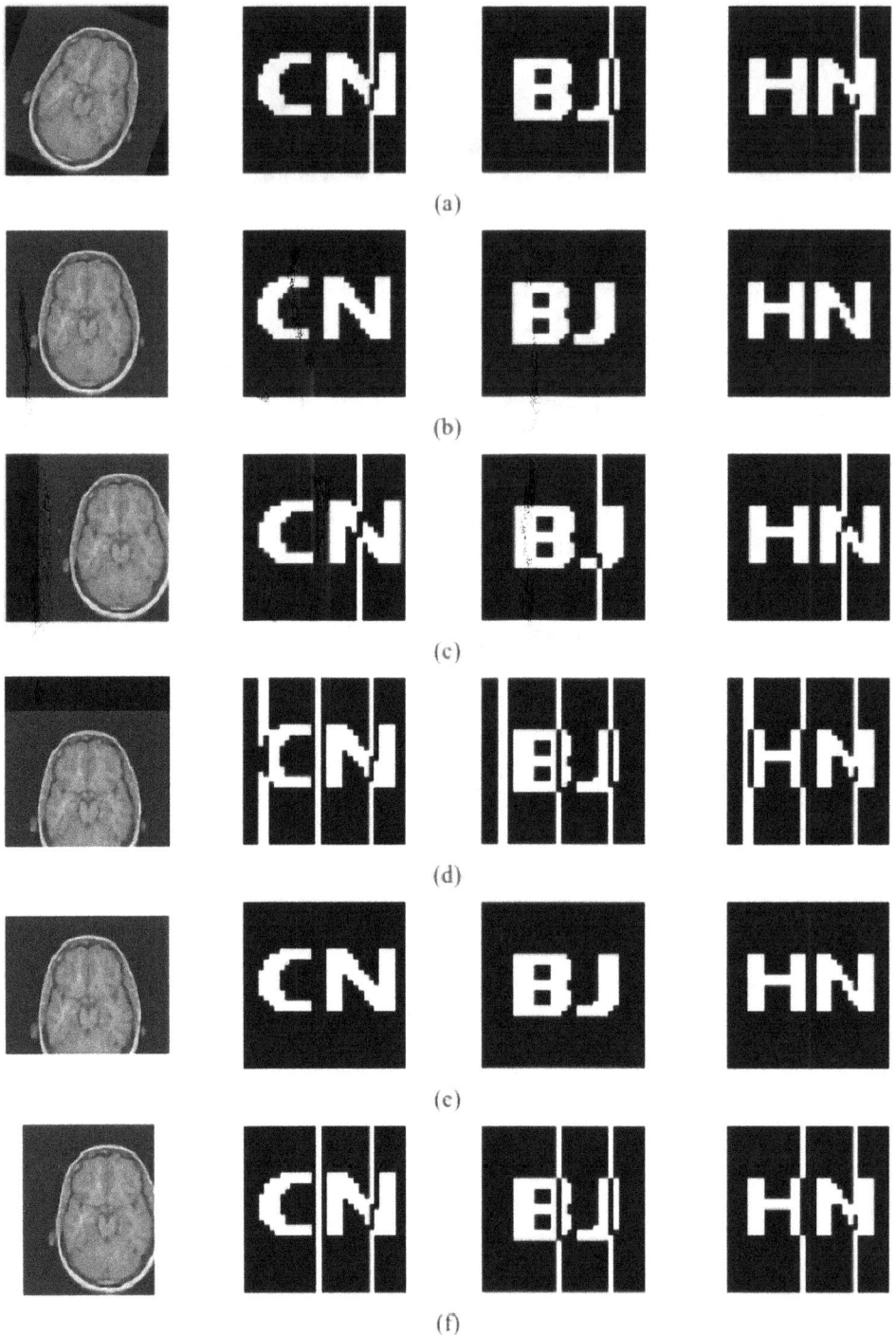

FIGURE 6.11 Part of the image after the geometric attack and the extracted watermark: (a) image rotated 20° clockwise and the extracted watermark, (b) reduced image and extracted watermark by 0.5 times, (c) image after moving 20% to the right and the extracted watermark, (d) image after shifting down by 20% and the extracted watermark, (e) image after 20% cropping along the y-axis and the extracted watermark, and (f) image after cutting 20% along the x-axis and the extracted watermark.

TABLE 6.4   Experimental Data Under Geometric Attack

Attacks	Intensity	PSNR (dB)	NC1	NC2	NC3
Counterclockwise Rotation (°)	10	15.61	0.94	0.91	0.95
	20	14.61	0.90	0.87	0.91
	40	14.06	0.86	0.72	0.82
Clockwise Rotation (°)	10	15.62	0.96	0.96	0.96
	20	14.61	0.75	0.61	0.68
	40	14.06	0.69	0.57	0.64
Scaling	0.1	-	0.84	0.87	0.87
	0.5	-	1.00	1.00	1.00
	2	-	1.00	1.00	1.00
Left (%)	10	13.84	0.94	0.96	0.96
	20	11.96	0.81	0.73	0.78
Right (%)	5	14.53	1.00	1.00	1.00
	20	12.11	0.86	0.87	0.87
	40	10.21	0.64	0.58	0.67
Down (%)	5	15.14	0.93	0.87	0.87
	20	13.80	0.73	0.70	0.69
X Crop (%)	5	-	0.96	0.90	0.91
	20	-	0.86	0.77	0.82
	40	-	0.73	0.70	0.69
Y Crop (%)	10	-	0.96	0.96	0.96
	20	-	0.96	0.96	0.96
	40	-	0.54	0.59	0.59

6.5 CONCLUSION

In this chapter, a zero watermarking algorithm is designed for experimental testing using the DarkNet53 network as a model for feature extraction, together with migration learning, DCT, Logistic Map, and hash transform. Through extensive experimentation and testing against both conventional and geometric attacks, a good robustness is achieved, making it suitable for medical image applications.

However, the algorithm still has potential for improvement. Further optimizations could be explored, such as enhancing the neural network architecture, extracting more representative features, and augmenting the algorithm's overall robustness. As technology continues to advance, this algorithm is expected to play a more significant role in medical image processing, providing reliable and effective solutions for various medical imaging challenges.

REFERENCES

1. Fridrich, J. *Steganography in digital media: principles, algorithms, and applications.* Cambridge University Press, 2009.
2. Cox, I., Miller, M., Bloom, J., et al. Digital watermarking. *Journal of Electronic Imaging*, 2002, 11(3): 414.

3. Xuehua, J. Digital watermarking and its application in image copyright protection. 2010 International Conference on Intelligent Computation Technology and Automation. *IEEE*, 2010, 2: 114–117.
4. Chandra, M., Pandey, S.A. DWT domain visible watermarking techniques for digital images. 2010 International Conference on Electronics and Information Engineering. *IEEE*, 2010, 2: V2-421–V2-427.
5. Savakar, D.G., Ghuli, A. Robust invisible digital image watermarking using hybrid scheme. *Arabian Journal for Science and Engineering*, 2019, 44(4): 3995–4008.
6. LeCun, Y., Bengio, Y., Hinton, G. Deep learning. *Nature*, 2015, 521(7553): 436–444.
7. Anwar, S.M., Majid, M., Qayyum, A., et al. Medical image analysis using convolutional neural networks: A review. *Journal of Medical Systems*, 2018, 42: 1–13.
8. Kandi, H., Mishra, D., Gorthi, S.R.K.S. Exploring the learning capabilities of convolutional neural networks for robust image watermarking. *Computers & Security*, 2017, 65: 247–268.
9. Semmlow, J.L., Griffel, B. *Biosignal and medical image processing*. CRC Press, 2021.
10. Khare, P., Srivastava, V.K. A secured and robust medical image watermarking approach for protecting integrity of medical images. *Transactions on Emerging Telecommunications Technologies*, 2021, 32(2): e3918.
11. Xia, Z., Wang, X., Wang, C., et al. A robust zero-watermarking algorithm for lossless copyright protection of medical images. *Applied Intelligence*, 2022, 52(1): 607–621.
12. Bamatraf, A., Ibrahim, R., Salleh, M.N.B.M. Digital watermarking algorithm using LSB. 2010 International Conference on Computer Applications and Industrial Electronics. *IEEE*, 2010: 155–159.
13. Barni, M., Bartolini, F., Cappellini, V., et al. A DCT-domain system for robust image watermarking. *Signal Processing*, 1998, 66(3): 357–372.
14. Ganic, E., Eskicioglu, A.M. Robust DWT-SVD domain image watermarking: embedding data in all frequencies. Proceedings of the 2004 Workshop on Multimedia and Security. 2004: 166–174.
15. Li, D., Chen, Y., Li, J., et al. Robust watermarking algorithm for medical images based on accelerated – KAZE discrete cosine transform. *IET Biometrics*, 2022, 11(6): 534–546.
16. Henaff, O. Data-efficient image recognition with contrastive predictive coding. International conference on machine learning. *PMLR*, 2020: 4182–4192.
17. Bai, D., Sun, Y., Tao, B., et al. Improved single shot multibox detector target detection method based on deep feature fusion. *Concurrency and Computation: Practice and Experience*, 2022, 34(4): e6614.
18. Minaee, S., Boykov, Y., Porikli, F., et al. Image segmentation using deep learning: A survey. *IEEE Transactions on Pattern Analysis and Machine Intelligence*, 2021, 44(7): 3523–3542.
19. Huang, T., Xu, J., Tu, S., et al. Robust zero-watermarking scheme based on a depthwise overparameterized VGG network in healthcare information security. *Biomedical Signal Processing and Control*, 2023, 81: 104478.
20. Pathak, D., Raju, U.S.N. Content-based image retrieval using feature-fusion of GroupNormalized-Inception-Darknet-53 features and handcraft features. *Optik*, 2021, 246: 167754.
21. Liu, J., Li, J., Ma, J., et al. A robust multi-watermarking algorithm for medical images based on DTCWT-DCT and Henon map. *Applied Sciences*, 2019, 9(4): 700.
22. Mujahid, A., Awan, M.J., Yasin, A., et al. Real-time hand gesture recognition based on deep learning YOLOv3 model. *Applied Sciences*, 2021, 11(9): 4164.
23. Ochoa-Dominguez, H., Rao, K.R. *Discrete cosine transform*. CRC Press, 2019.
24. Weisstein, E.W. Logistic map. https://mathworld.wolfram.com/, 2001.

25. Li, D., Li, J., Bhatti, U.A., Nawaz, S.A., Liu, J., Chen, Y.W., & Cao, L. Hybrid encrypted watermarking algorithm for medical images based on DCT and improved DarkNet53. *Electronics*, 2023, 12(7): 1554.
26. Bhatti, U.A., Tang, H., Wu, G., Marjan, S., & Hussain, A. Deep learning with graph convolutional networks: An overview and latest applications in computational intelligence. *International Journal of Intelligent Systems*, 2023, 2023: 1–28.
27. Sheng, M., Li, J., Bhatti, U.A., Liu, J., Huang, M., & Chen, Y.W. Zero watermarking algorithm for medical image based on Resnet50-DCT. *CMC-Computers Materials & Continua*, 2023, 75(1): 293–309.
28. Liu, J., Li, J., Ma, J., Sadiq, N., Bhatti, U.A., & Ai, Y. A robust multi-watermarking algorithm for medical images based on DTCWT-DCT and Henon map. *Applied Sciences*, 2019, 9(4): 700.
29. Fan, Y., Li, J., Bhatti, U.A., Shao, C., Gong, C., Cheng, J., & Chen, Y. A Multi-watermarking algorithm for medical images using inception V3 and DCT. *CMC-Computers Materials & Continua*, 2023, 74(1): 1279–1302.
30. Li, T., Li, J., Liu, J., Huang, M., Chen, Y.W., & Bhatti, U.A. Robust watermarking algorithm for medical images based on log-polar transform. *EURASIP Journal on Wireless Communications and Networking*, 2022, 2022(1): 1–11.
31. Bhatti, U.A., Huang, M., Wu, D., Zhang, Y., Mehmood, A., & Han, H. Recommendation system using feature extraction and pattern recognition in clinical care systems. *Enterprise Information Systems*, 2019, 13(3): 329–351.
32. Bhatti, U.A., Yu, Z., Chanussot, J., Zeeshan, Z., Yuan, L., Luo, W., … & Mehmood, A. Local similarity-based spatial-spectral fusion hyperspectral image classification with deep CNN and Gabor filtering. *IEEE Transactions on Geoscience and Remote Sensing*, 2021, 60: 1–15.
33. Bhatti, U.A., Yu, Z., Li, J., Nawaz, S.A., Mehmood, A., Zhang, K., & Yuan, L. Hybrid watermarking algorithm using Clifford algebra with Arnold scrambling and chaotic encryption. *IEEE Access*, 2020, 8: 76386–76398.
34. Liu, J., Li, J., Zhang, K., Bhatti, U.A., & Ai, Y. Zero-watermarking algorithm for medical images based on dual-tree complex wavelet transform and discrete cosine transform. *Journal of Medical Imaging and Health Informatics*, 2019, 9(1): 188–194.

CHAPTER 7

Robust Multi-watermark Algorithm for Medical Images Based on SqueezeNet Transfer Learning

Pengju Zhang[1], Jingbing Li[1,2], and Uzair Aslam Bhatti[1,2]

[1]School of Information and Communication Engineering, Hainan University, Haikou, China
[2]State Key Laboratory of Marine Resource Utilization in the South China Sea, Hainan University, Haikou, China

7.1 INTRODUCTION

The traditional healthcare industry is transforming toward an intelligent direction, and more and more new diagnosis and treatment models are emerging, such as remote healthcare and intelligent screening utilizing big data and cloud technology [1]. Meanwhile, various intelligent diagnosis and treatment terminals enable users to conveniently monitor their health status in real time and access diagnosis and treatment information. This intelligent trend further promotes the generation and transmission of medical data. However, as an important part of medical data, medical images are not only an important basis for diagnosis, but also contain personal privacy information of patients. Intelligent medical platforms [2], medical image clouds [3], remote image centers [4], and other technologies facilitate communication between doctors and patients, but also face issues such as privacy protection. Protecting patients' personal information in medical images and electronic medical records (EMRs) has become a crucial issue in the healthcare industry with the development of machine learning, cloud computing, and big data. Medical image digital watermarking technology (MIW) is an effective solution to prevent the leakage of personal information in CT, MRI, and other medical images [5,6].

Digital image watermarking algorithms can be divided into robust watermarks and fragile watermarks based on their characteristics. According to their detection characteristics, they can be classified as blind watermarks, non-blind watermarks, and zero watermarks. Depending on the hiding location, they can be categorized into frequency-domain-based

DOI: 10.1201/9781003427674-7

watermarks and spatial-domain-based watermarks [7]. Blind watermarking involves quantizing the watermark information and embedding it into the host image as a whole, and no original watermark information is required for watermark extraction. In one study [8], a technique was proposed that utilizes SVD to embed the watermark into the low-frequency sub-band of the host image after DWT decomposition. The maximum singular value of each non-overlapping sub-block is quantized for watermark embedding, and SIFT transformation is used to correct the attack image. Non-blind watermarking, which embeds the feature information of the watermark image into the host image, is more robust than blind watermarking as it requires another part of the watermark information for watermark extraction. For example, reference [9] uses two-level LWT-SVD and the Y channel diagonal matrix of the host image and watermark image's YCbCr color model to embed feature information. Both blind and non-blind watermarking belong to embedded watermarking, and there is a trade-off between robustness and invisibility. Wen et al. [10] proposed the idea of zero watermarking, which extracts the host image features and combines them with the watermark information to create a zero-watermark that is stored in a third party. Zero watermarking is often used in medical image digital watermarking technology due to the unique nature of medical image information. This includes medical image frequency-domain zero watermarking, medical image spatial-domain zero watermarking, medical image neural network zero watermarking, etc. For example, in reference [11–15], PHTs combined with zero-watermarking technology are used for two-dimensional medical images. The method reduces redundant information using 2D DCT and extracts image feature information using PHTs-DCT, and then combines perceptual hash technology to obtain a zero watermark. Deep learning has been widely used in various fields, including medical imaging, but research on medical image watermarking is relatively limited [16]. In reference [17], the VGG19 deep convolutional neural network is used to extract the two-dimensional features of medical images and obtain a zero watermark. This method utilizes the feature extraction and feature fusion performance of the VGG19 pre-training network and applies DFT to extract the real part of the feature vector, combined with mean-value perceptual hash technology to generate a zero watermark. However, this method has poor discriminability despite its certain robustness.

Medical images are an important form of digital media with characteristics such as high confidentiality and precision. Compared to general digital media, the watermarking of medical images requires high robustness, meaning that the watermark should still be detectable after the image has been compressed, rotated, cropped, filtered, and so on. Medical images require accurate analysis and diagnosis, and thus digital watermarks should not affect their quality and accuracy. Zero watermarking, which does not affect image quality, is an ideal method for protecting medical images. This chapter proposes a robust multi-watermark algorithm for medical images using SqueezeNet transfer learning. The algorithm obtains the feature map of medical images using SqueezeNet combined with transfer learning technology. The DCT algorithm is used to aggregate the feature information and the zero-threshold perceivable hash technology is used to obtain the binary feature vector. The SPM algorithm is used to scramble and disturb the watermark image during the embedding process. Encrypted watermark is XORed with the binary feature vector to generate a zero watermark, which is stored in a third party. The same process is used for watermark extraction. This approach improves the robustness, reliability, and security of digital watermarks, while also increasing the information transmission capacity, by combining multi-watermark and zero-watermark technologies to embed multiple watermarks separately into the carrier image.

7.2 FUNDAMENTAL THEORY

7.2.1 SqueezeNet Neural Network

SqueezeNet is a lightweight and efficient convolutional neural network (CNN) model proposed by Iandola et al. in 2016 [18], which combines the research ideas of small models, including structural optimization and model compression. Small models are not only conducive to training, but also more effective when deployed on specific hardware such as FPGA. Compared with AlexNet, SqueezeNet has only 1/50 of the number of parameters, and its model size is less than 1 MB, making it a very lightweight network structure.

The design philosophy of SqueezeNet mainly utilizes 1×1 convolutional kernels to compress the input feature maps, thereby reducing the number of parameters in the network. In addition, SqueezeNet proposes a novel structure called the "fire module," which consists of a squeeze layer and an expand layer. The squeeze layer is used for channel compression, while the expand layer is responsible for increasing the number of channels. The design of the fire module can improve the network's expressive power without increasing the number of parameters in the network.

SqueezeNet performed well in the 2014 ImageNet Large Scale Visual Recognition Challenge (ILSVRC2014) competition, with a Top-1 error rate of 37.5% and a Top-5 error rate of 17.8%. Compared to other deep learning models, SqueezeNet has a smaller model size and faster inference speed, making it an efficient convolutional neural network architecture.

7.2.2 Transfer Learning

Transfer learning in neural networks is a technique of applying a neural network that has been trained on one task to another related task. It improves performance on the new task by utilizing the knowledge learned from the previous task, thus reducing the amount of training data and computation time needed for the new task. The original SqueezeNet was trained on the ImageNet data set, and for its application in watermarking medical images, it needs to be retrained on a medical data set. First, a small custom medical data set is collected to train the network. Second, data augmentation is performed on this small data set to obtain a larger augmented medical data set. Third, the network structure is fine-tuned by adding a fully connected layer. Finally, hyperparameters are adjusted to train the network.

7.2.3 SPM Composite Chaotic Mapping

SPM is an efficient one-dimensional composite chaotic mapping. This method combines the Sine mapping and the piecewise linear chaotic mapping (PWLCM) to expand the range of chaotic mappings, ensure the traversal of results, and greatly improve the efficiency of generating chaotic sequences without reducing security [19–30]. When $\eta \in (0,1)$ and $\mu \in (0,1)$, the system is in a chaotic state, and $x\,(1)$ is generated through a random seed, with its values ranging from 0 to 1. See equation (7.1) for the calculation method.

$$x(t+1) = \begin{cases} \mathrm{mod}\left(\dfrac{x(t)}{\eta} + \mu \sin(\pi x(t)) + r,\ 1 \right), & 0 \le x(t) < \eta \\[2ex] \mathrm{mod}\left(\dfrac{x(t)/\eta}{0.5 - \eta} + \mu \sin(\pi x(t)) + r,\ 1 \right), & \eta \le x(t) < 0.5 \\[2ex] \mathrm{mod}\left(\dfrac{(1 - x(t))/\eta}{0.5 - \eta} + \mu \sin(\pi(1 - x(t))) + r,\ 1 \right), & 0.5 \le x(t) < 1 - \eta \\[2ex] \mathrm{mod}\left(\dfrac{1 - x(t)}{\eta} + \mu \sin(\pi(1 - x(t))) + r,\ 1 \right), & 1 - \eta \le x(t) < 1 \end{cases} \tag{7.1}$$

7.3 PROPOSED ALGORITHM

We propose a robust medical image watermarking algorithm based on SqueezeNet transfer learning. By fine-tuning the SqueezeNet network structure on a self-built medical data set, the network is retrained to extract features, which are combined with perceptual hashing to obtain the watermark. The algorithm consists of four main parts: network retraining, watermark encryption, zero-watermark generation and extraction, and watermark decryption.

7.3.1 Retraining the Network

Firstly, the data set is constructed. The medical data set used in this chapter comes from the National Institutes of Health Clinical Center. We selected five categories of medical images to form a small self-built data set of size 80, and performed data augmentation to obtain an augmented data set of size 4,400. To expand the size and diversity of the data set, prevent model overfitting, and improve the robustness of the model, we used data augmentation techniques during the training process, and the various parameters of data augmentation are shown in Table 7.1. At the same time, we divided the data set into a training set and a

TABLE 7.1 Data Augmentation Parameters Setting

Type of enhancement	Strength of enhancement			Quantity
	Start	Step	End	
Gaussian noise (%)	5	5	20	4
JPEG compression (%)	2	2	8	4
Median filter (10 times)	3	2	7	3
Clockwise rotation (°)	5	5	45	9
Scaling	0.2	0.2	0.8	4
Down shift (%)	5	5	30	6
Left shift (%)	5	5	30	6
Right shift (%)	5	5	30	6
Y-axis shear (%)	5	5	30	6
X-axis shear (%)	5	5	30	6

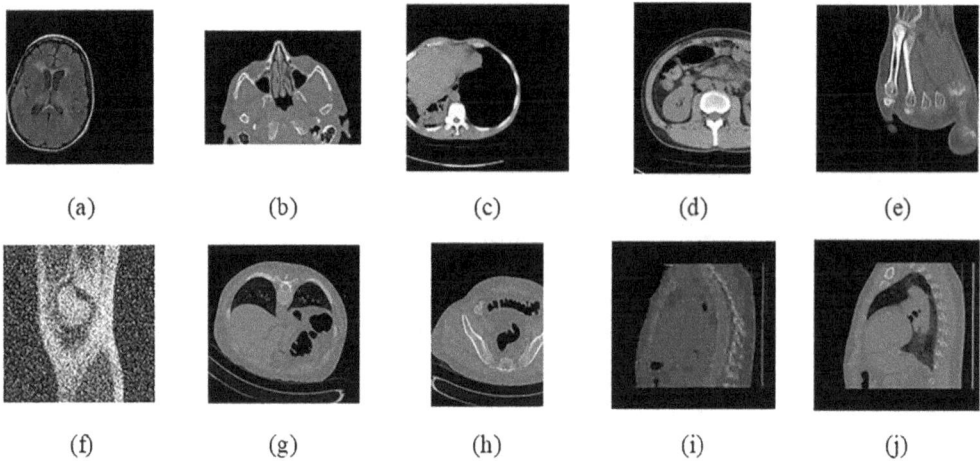

FIGURE 7.1   Examples of augmented data set images.

validation set in a ratio of 8:2. The training set is used for model training, and the validation set is used for adjusting model hyperparameters and evaluating model performance.

To demonstrate the effectiveness of data augmentation, we randomly selected two images from each of the five categories of medical images, including brain, abdomen, skeletal muscle, colon, and chest, in the augmented data set for display, as shown in Figure 7.1.

Next, we adjusted the network structure to better suit medical image sets. As the fully connected layer can map the learned distributed feature representation to the sample label space, we added a fully connected layer between the pool10 pooling layer and the softmax layer and set the output parameter of the new fully connected layer to 5. The network structure before and after fine-tuning is shown in Figure 7.2.

Next, the network was trained. Compared with traditional gradient descent algorithms, the Adam algorithm has a faster convergence speed, and usually only requires a few iterations to achieve good results. Additionally, the Adam optimizer occupies less memory compared to other optimizers. Therefore, we used the Adam optimizer with an initial learning rate of 0.001. The data set was shuffled in each training round, and a piecewise learning rate strategy was adopted, with the learning rate decreasing by a factor of 0.1 every 10 rounds. Other training parameters are shown in Table 7.2.

Using the training set of the enhanced data set, the network structure was adjusted and trained with the set training parameters. After 1,650 iterations with a minimum batch size of 64, the trained network achieved a validation accuracy of 99.20% on the validation set after a training duration of 72 minutes and 43 seconds. This result indicates that the trained network has excellent classification ability for the five types of medical images in the data set, and also demonstrates its good feature extraction capability. Other training parameters are shown in Table 7.2.

7.3.2 Watermark Encryption

Using the SPM chaos method, the key coefficients r, η, μ are selected to generate a pseudo-random sequence. In this experiment, r = rng(7), η = 0.4, μ = 0.3 (where rng(7) is

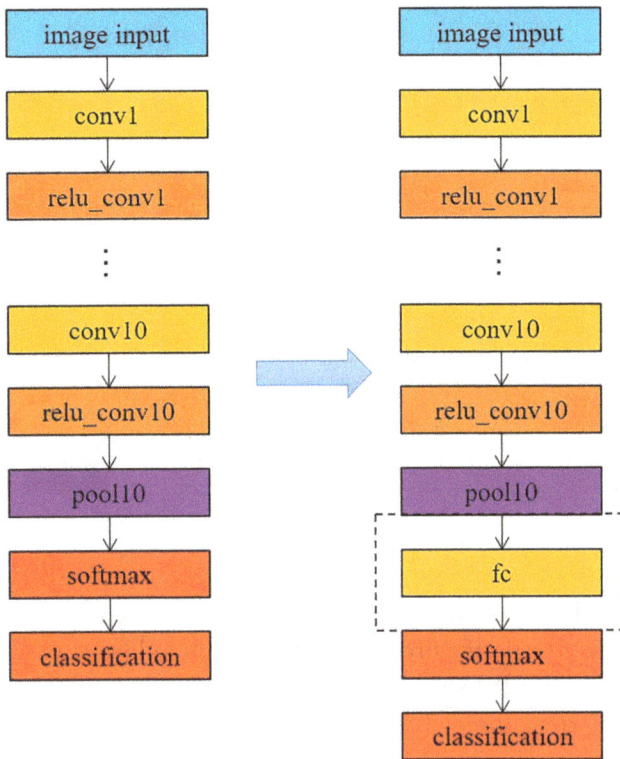

FIGURE 7.2 Fine-tuning network architecture.

TABLE 7.2 Setting of Network Training Parameters

Solver	Adam
Initial learn rate	0.001
Validation frequency	50
Max epochs	30
Mini batch size	64
Shuffle	Every epoch
Learn rate schedule	Piecewise
Learn rate drop factor	0.1
Learn rate drop period	10

the random number seed) were selected, and the iteration number was set to 4,095. The generated pseudo-random sequence spm(i) was sorted in ascending order of value, and then the corresponding pixels were scrambled using the spatial changes in position before and after sorting to obtain the encrypted watermark Scrambled (i,j). The specific encryption process is shown in Figure 7.3.

7.3.3 Generation and Extraction of Zero Watermark

To input medical images into the trained Squeezenet network for feature extraction, pre-processing of these images is required. First, the original single-channel images with a

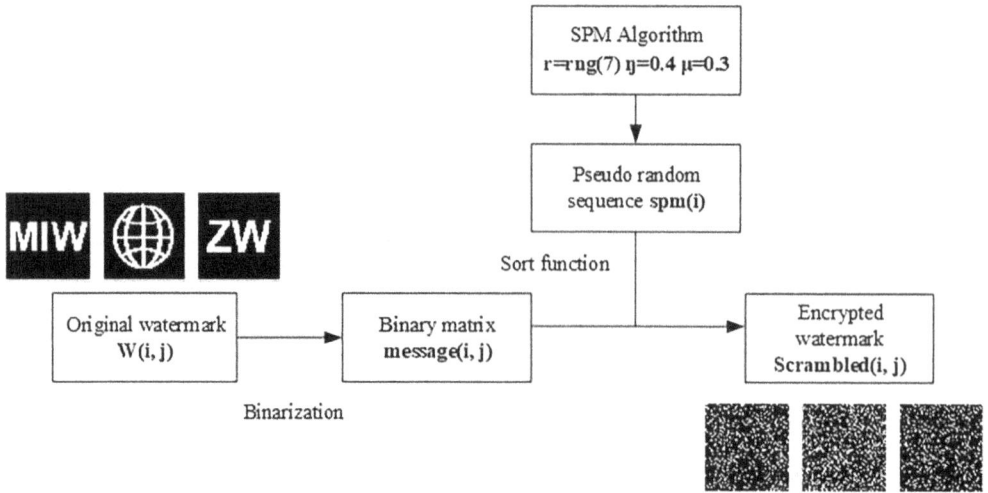

FIGURE 7.3   Watermark encryption flowchart.

size of 512 × 512 needs to be normalized to form a three-channel image with a specified size of 227 × 227 × 3. The specific processing steps are as follows: first, the original single-channel image is resized to the specified size of 227 × 227, and then the resized image is duplicated twice to form a three-channel image.

The convolutional layer contains rich feature information when mapping raw data to hidden feature space. Additionally, the deeper the convolutional layer, the more advanced feature representations it contains. In the Squeezenet network, there is a convolutional layer named "conv10" at the end. This layer not only learns the parameters of the target medical dataset in transfer learning, but also retains the parameters learned in the source domain ImageNet dataset, resulting in stronger feature representation ability. Therefore, after normalizing medical images, the output of the conv10 convolutional layer is selected as the feature extractor, resulting in a feature map of size 14 × 14 × 1,000.

The obtained 14 × 14 × 1,000 feature map is flattened into a one-dimensional vector and undergoes a one-dimensional DCT transformation. Since DCT has the function of energy concentration, only the coefficients with higher energy need to be preserved to obtain an effective feature representation. Therefore, after the one-dimensional DCT transformation, the first 64 coefficients are selected and combined with perceptual hashing technology to obtain a feature vector that represents the original image. This approach not only reduces the feature dimension but also improves the robustness and generalization performance of the features.

The 64 coefficients obtained are subjected to thresholded perceptual hashing with a threshold set at 0. The coefficients are converted to binary numbers and quantized to 0 or 1. If the coefficient value is greater than 0, it is quantized to 1; otherwise, it is quantized to 0. This results in a 64-bit binary feature vector. Through thresholded perceptual hashing, the original 64 coefficients are transformed into a binary feature vector with a smaller storage space, which facilitates the generation and extraction of zero watermark.

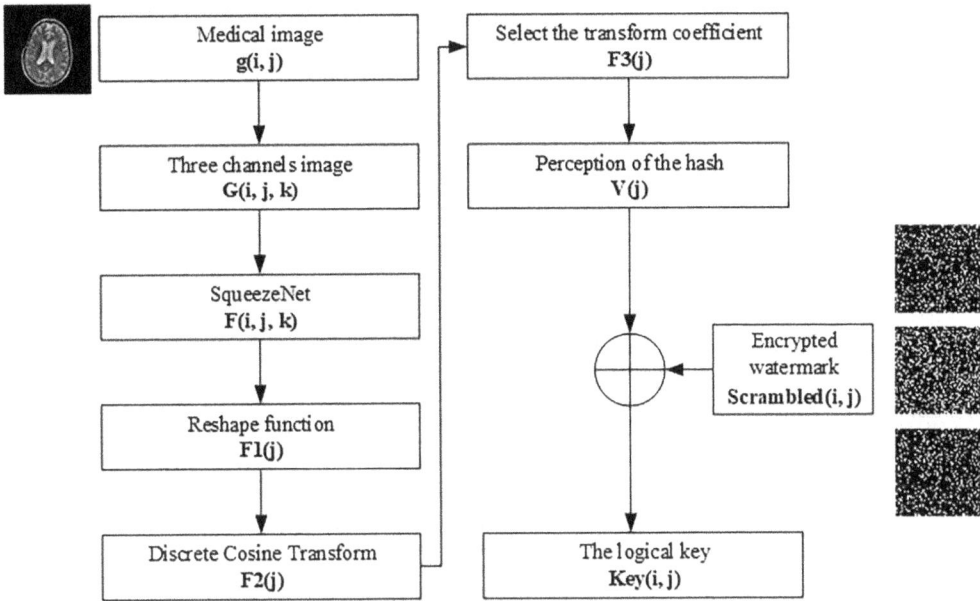

FIGURE 7.4   Watermarking generation flowchart.

The following is an explanation of the zero-watermark generation process using the example of brain image g(i,j). See Figure 7.4 for the detailed generation flowchart.

1. Resize the brain image g(i,j) of size 512 × 512 and expand its channels to obtain a three-channel image G(i,j,k) of size 227 × 227 × 3.

2. Input the three-channel image G(i,j,k) into the trained SqueezeNet network to extract the output of the conv10 convolutional layer, denoted as F(i,j,k).

3. Unroll the output of the convolutional layer, F(i,j,k), to obtain a one-dimensional coefficient vector, denoted as F1(j).

4. Perform DCT transformation on the coefficient vector F1(j) to obtain the transformed coefficient vector F2(j).

5. Select the first 64 bits of the transformed coefficient vector F2(j) to form a selected coefficient vector, denoted as F3(j).

6. Perform threshold-based perceptual hashing on the selected coefficient vector F3(j) using a threshold of 0 to obtain a binary hash vector V(j) representing the image feature.

7. This chapter adopts a multi-watermark algorithm to sequentially XOR the encrypted watermark Scrambled(i,j) with the hash vector, V(j), to obtain the logical key, Key(i,j), i.e., the zero-watermark, which is then stored in a third-party repository.

The extraction process of zero-watermark is similar to the generation process. The tested medical image is subjected to the above steps 1–6 to obtain the hash vector, V'(j), which is then XORed with the logical key stored in the third party to extract the watermark.

7.3.4 Decryption of Watermark

Using the same method as watermark encryption, the same binary pseudo-random sequence, spm(i), is obtained. First, the scrambled position space is restored using the spm(i) sequence generated by the SPM chaotic scrambling algorithm and the Sort function, and the pixels corresponding to the restored position space are obtained, resulting in the recovered watermark W'(i,j). Then, by calculating the correlation coefficient NC between W(i,j) and W'(i,j), the ownership of the medical image and the embedded watermark information can be determined.

7.4 EXPERIMENTAL RESULTS

7.4.1 Evaluation Metrics

PSNR is used to measure the degree of image attack, with a lower PSNR value indicating a greater difference between the original and attacked images. The calculation method can be expressed as equation (7.2), where M and N represent the dimensions of the medical image, and g(i,j) and g'(i,j) represent the pixel values at the corresponding positions of the original medical image and the tested medical image, respectively.

$$PSNR = 10 \lg \left[\frac{MN \max_{i,j}(g\,(i,\,j))^2}{\sum_i \sum_j (g\,(i,\,j) - g'(i,\,j))^2} \right] \tag{7.2}$$

The NC method uses the Pearson correlation coefficient to accurately measure the similarity between the original and extracted watermark images. This is achieved by calculating the correlation coefficient using equation (7.3), which takes into account the mean and standard deviation of the original and extracted images. The Pearson correlation coefficient ranges from −1 to 1, where a value closer to 1 indicates a stronger correlation. A high NC value obtained through the correlation coefficient indicates better algorithm robustness when the original medical image is under attack.

$$NC = \frac{1}{MN - 1} \sum_{i=1}^{M} \sum_{j=1}^{N} \left(\frac{\overline{W\,(i,\,j) - \mu_W}}{\sigma_W} \right) \left(\frac{W'(i,\,j) - \mu_{W'}}{\sigma_{W'}} \right) \tag{7.3}$$

To evaluate the algorithm's generalization ability, i.e., its ability to distinguish between different medical images, we calculate the correlation between the feature vectors of different medical images. The formula for calculating NC changes to equation (7.4). The range of NC obtained in this case is also [−1,1], where a value of −1 represents complete negative correlation, a value of +1 represents complete positive correlation, and a value of 0 represents no correlation. A smaller NC value indicates that the algorithm has a lower correlation between the extracted feature vectors of different images, which means that the algorithm has a better ability to distinguish between images, higher discrimination ability, and lower false alarm rate of watermark [20].

$$NC = \frac{\sum_{j=1}^{K}(V(j) - \mu_v)(V'(j) - \mu_{v'})}{\{\sum_{j=1}^{K}(V(j) - \mu_v)^2 \sum_{j=1}^{K}(V'(j) - \mu_{v'})^2\}^{1/2}} \tag{7.4}$$

7.4.2 Discrimination Testing

Five medical images were randomly selected using the proposed algorithm in this chapter to extract 64-bit perceptual hash feature vectors for each image. Then, using equation (7.4) to calculate the correlation coefficient between feature vectors, the results are presented in Figure 7.5. The figure shows that the correlation coefficient between the same image is 1. There are seven groups with weak correlation, ranging from 0.2 to 0.4, accounting for 70% of the test data; two groups with extremely weak correlation, ranging from 0 to 0.2, accounting for 20% of the test data; and one group with negative correlation less than 0, accounting for 10% of the test data. It can be seen that the NC values of the perceptual hash feature vectors extracted by the proposed algorithm for the tested medical images are all less than 0.5, indicating a high degree of discrimination.

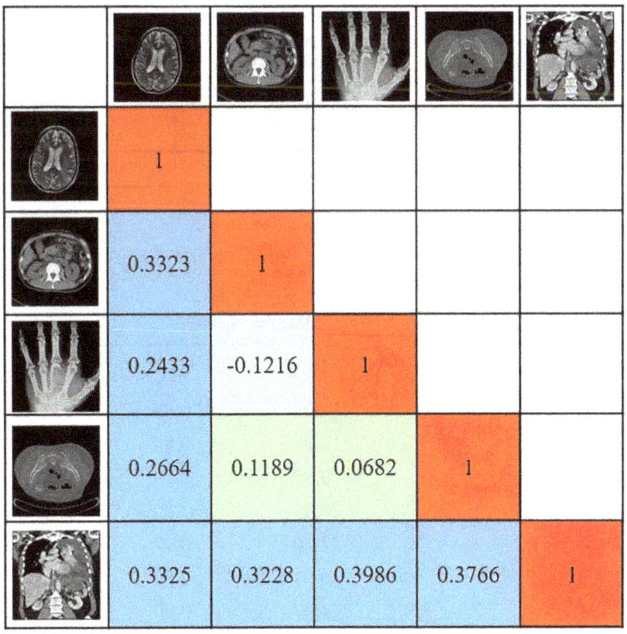

FIGURE 7.5   Results of discriminability.

7.4.3 Robustness Testing

Image attacks are mainly divided into conventional attacks and geometric attacks. Here, we will test the robustness of the algorithm against these two types of attacks. We selected three medical images for algorithm robustness testing, with the size of each test image being 512×512. Multiple watermarks were embedded in each medical image, with three different watermarks embedded in each image to increase the watermark's information capacity. The size of each watermark is 64*64. Watermark 1 contains the information "MIW", which is the abbreviation for Medical Image Watermark, watermark 2 contains simplified earth image information, and watermark 3 contains the abbreviation for zero watermark, "ZW". The test images and watermark images used in the experiment are shown in Figure 7.6.

The robustness test results of the algorithm against conventional attacks are shown in Table 7.3. It should be noted that the PSNR values in the table are the cumulative average of the PSNR after intensity attacks on the three test images. The NC1 value is also the cumulative average of the extracted watermark NC value after the three test images are attacked, and the calculation method for NC2 and NC3 is the same. The NC value in the table is the average of NC1, NC2, and NC3.

When testing Gaussian noise attacks, we used the imnoise function in Matlab for noise addition. Although the intensity of noise addition is the same each time, the masking

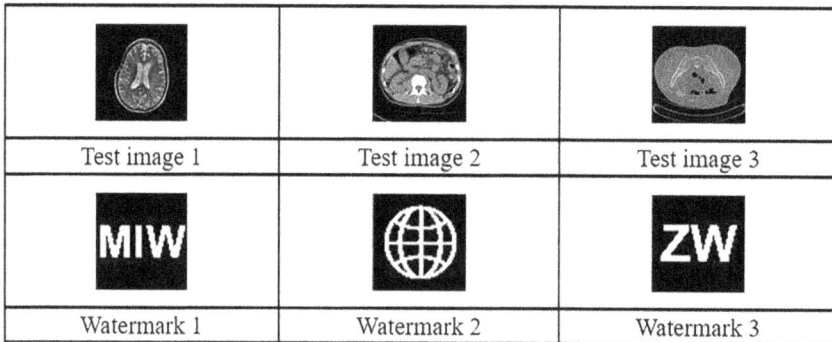

FIGURE 7.6   The test images and watermark images.

TABLE 7.3   Experimental Results of Conventional Attacks

Attacks	Intensity	PSNR (dB)	NC1	NC2	NC3	NC
Gaussian noise (%)	8	12.96	0.88	0.91	0.86	0.88
	14	10.99	0.84	0.88	0.83	0.85
	20	9.87	0.79	0.84	0.79	0.81
JPEG compression (%)	1	26.49	1.00	1.00	1.00	1.00
	3	27.20	0.97	0.97	0.96	0.97
	5	28.61	0.92	0.93	0.92	0.92
Median filter (10 times)	7	24.87	0.90	0.91	0.89	0.90
	9	22.89	0.88	0.90	0.87	0.88
	11	21.90	0.80	0.83	0.79	0.81

effect on image features is different because the noise points are randomly placed in the image during noise addition. To reduce the masking effect of random noise positions, we repeated the same intensity of Gaussian noise attacks on each test image three times and obtained the average NC value.

We carried out three conventional attacks on the test images, including Gaussian noise, JPEG compression, and median filtering. From the table, we can see that even when the intensity of the Gaussian noise attack is 20%, the NC value is still as high as 0.81, far higher than 0.50. When the quality of JPEG compression is lower, the block effect of the image will be more obvious, and the image quality will also be worse. When the compression quality is 1%, the NC value is 1, and when the compression quality is 5%, the NC value is as high as 0.92. As the compression quality decreases, the image quality gradually deteriorates, and this change can be measured by PSNR. At the same time, theoretically, the NC value should also decrease as the compression quality decreases. However, in the experiment, we found that the NC value when the compression quality is 1% is slightly higher than that when the compression quality is 5%. This may be because some key feature points' information is compressed and lost at a compression quality of 5%, leading to poorer results. We also carried out median filtering attacks, with the filtering times set to 10, and conducted experiments by changing the filtering window size. The results showed that even when the window size is 11 × 11, the NC value is still 0.81.

In summary, the algorithm proposed in this chapter has good robustness against conventional attacks.

Geometric attacks such as rotation, scaling, translation, and shearing were performed on medical test images. Anti-geometric attack is a challenge in watermarking algorithms, and as a type of geometric attack, rotation is generally difficult to resist with existing algorithms. The experimental results in Table 7.4 show that when rotated counter-clockwise from 15° to 55°, the extracted watermark NC slowly decreases. When rotated 15°, the NC is 0.97, close to 1, indicating that the proposed algorithm has excellent resistance to low-intensity rotation. When rotated 35° and 55°, the NC values are 0.87 and 0.75, respectively, indicating good resistance to medium- to high-intensity rotation.

For scaling attacks, the NC values of the test intensities are all above 0.90. When right and down scaling attacks of 12% to 36% are performed, the NC values are all above 0.80, and the NC value is as high as 0.86 when scaled down by 36%. When shearing 30% to 50% along the y- and x-axes, the NC values are all above 0.75. It is worth noting that when the shearing intensity along the y- and x-axes reaches 50%, half of the image information has been lost. However, the proposed algorithm still achieves an NC value of 0.78 and 0.77, far higher than 0.50.

The attacked image and the extracted watermark are shown in Figure 7.7.

7.4.4 Comparison

To further test the performance of the proposed algorithm, comparative experiments were conducted with two existing algorithms. The first comparative algorithm (algorithm 1) is the VGG-DFT algorithm proposed by Han et al. [17], which extracts image features

TABLE 7.4   Experimental Results of Geometric Attacks

Attacks	Intensity	PSNR (dB)	NC1	NC2	NC3	NC
Counterclockwise rotation (°)	15	16.33	0.97	0.97	0.96	0.97
	35	14.85	0.87	0.89	0.86	0.87
	55	13.99	0.74	0.78	0.72	0.75
Scaling factor	0.25	-	0.92	0.93	0.92	0.92
	0.5	-	1.00	1.00	1.00	1.00
	0.75	-	1.00	1.00	1.00	1.00
Translation right (%)	12	13.92	0.90	0.92	0.89	0.90
	24	12.18	0.87	0.89	0.86	0.87
	36	11.28	0.80	0.83	0.78	0.80
Translation down (%)	12	13.34	0.90	0.92	0.90	0.91
	24	11.83	0.87	0.89	0.86	0.87
	36	11.06	0.86	0.88	0.84	0.86
Y-axis crop (%)	30	-	0.88	0.91	0.87	0.89
	40	-	0.84	0.87	0.83	0.85
	50	-	0.77	0.81	0.75	0.78
X-axis crop (%)	30	-	0.92	0.93	0.91	0.92
	40	-	0.83	0.86	0.82	0.84
	50	-	0.76	0.80	0.75	0.77

based on a pre-trained VGG19 network, and then performs DFT transformation on the feature coefficients. The real part of the transformation result is used to generate a binary feature sequence, which is then used to generate the zero watermark. The second comparative algorithm (algorithm 2) is the PHTs-DCT algorithm proposed by Yi et al. [11], which uses PHT moment coefficients of images combined with DCT transformation to generate the zero watermark. In the comparative experiments, the length of the binary feature sequence for both algorithms was set to 64 bits, and the original watermark image used was watermark 1.

The performance comparison between the proposed algorithm and the compared algorithms is shown in Figure 7.8 and Table 7.5. The experimental results show that under the tested attack intensities, the proposed algorithm outperforms the comparative algorithm 1. The poor performance of algorithm 1 may be due to its poor generalization ability, as well as the fact that the VGG19 network used to extract image features was not retrained on medical images. Therefore, its performance was poor on the medical image set used in this experiment.

The original paper of comparative algorithm 2 showed good resistance to conventional attacks, but in this case, the proposed algorithm still performed better under corresponding attack intensities. For example, under the Gaussian noise testing intensity, the proposed algorithm's NC value was higher by 0.16, and under the JPEG compression testing intensity, NC value was higher by 0.18. Under the median filtering testing intensity, NC value was higher by 0.14. In terms of anti-geometric attacks, the proposed algorithm has a significant advantage over comparative algorithm 2, with NC values higher by 0.21 and 0.22 under right-shift and y-axis shear attack testing intensities, respectively.

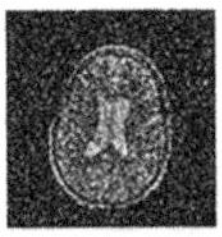			
Medical image1 (Gaussian noise 8%)	Extracted watermark1 (NC=0.95)	Extracted watermark2 (NC=0.96)	Extracted watermark3 (NC=0.94)
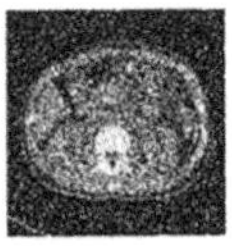			
Medical image2 (Gaussian noise 8%)	Extracted watermark1 (NC=0.90)	Extracted watermark2 (NC=0.88)	Extracted watermark3 (NC=0.89)
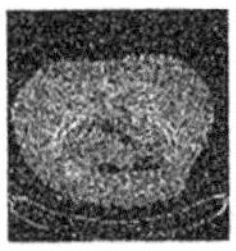			
Medical image3 (Gaussian noise 8%)	Extracted watermark1 (NC=0.80)	Extracted watermark2 (NC=0.88)	Extracted watermark3 (NC=0.75)
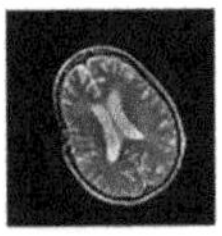			
Medical image1 (Counterclockwise rotation 35°)	Extracted watermark1 (NC=0.90)	Extracted watermark2 (NC=0.92)	Extracted watermark3 (NC=0.89)
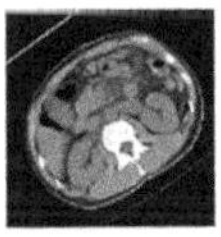			
Medical image2 (Counterclockwise rotation 35°)	Extracted watermark1 (NC=0.90)	Extracted watermark2 (NC=0.92)	Extracted watermark3 (NC=0.89)
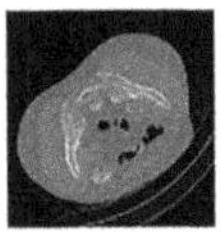			
Medical image3 (Counterclockwise rotation 35°)	Extracted watermark1 (NC=0.81)	Extracted watermark2 (NC=0.84)	Extracted watermark3 (NC=0.80)

FIGURE 7.7 The attacked image and the extracted watermark.

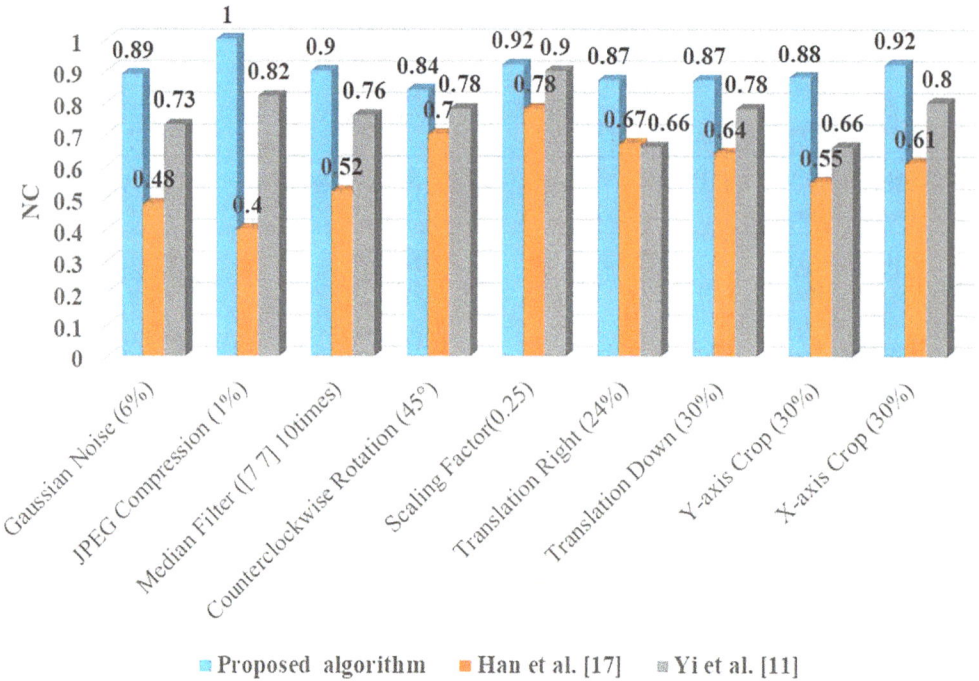

FIGURE 7.8 The performance comparison between the proposed algorithm and the compared algorithms.

TABLE 7.5 Comparison Results

Attacks	Intensity	Proposed algorithm	Han et al. [17]	Yi et al. [11]
Gaussian noise (%)	6	0.89	0.48	0.73
JPEG compression (%)	1	1.00	0.40	0.82
Median filter (10 times)	7	0.90	0.52	0.76
Counterclockwise rotation (°)	45	0.84	0.70	0.78
Scaling factor	0.25	0.92	0.78	0.90
Translation right (%)	24	0.87	0.67	0.66
Translation down (%)	30	0.87	0.64	0.78
Y-axis crop (%)	30	0.88	0.55	0.66
X-axis crop (%)	30	0.92	0.61	0.80

7.5 CONCLUSION

This chapter proposes a robust multi-watermark algorithm for medical images based on transfer learning using SqueezeNet. The algorithm uses a fine-tuning network structure and retraining strategy to obtain the feature extraction network SqueezeNet and adopts multi-watermark technology to increase the information capacity. Experimental results show that the proposed algorithm has a certain robustness against conventional attacks and good robustness against geometric attacks. Additionally, the feature vectors extracted from different medical images are low in correlation, indicating that the algorithm has strong generalization ability and low false alarm rate. Moreover, using multi-watermark technology, the NC values for watermark extraction are highest for watermark 2, followed by watermark 1, and then watermark 3 under the same conditions. All three watermarks are binary watermarks composed of 0 and 1, with differences in the average value distribution. Watermark 2 has the most even distribution, which visually reflects the proportion of black and white parts in the watermark being the closest. Therefore, under the experimental conditions of this article, a regularity can be summarized that the more evenly distributed the binary watermark values are, the slightly higher the NC values for watermark extraction. In comparison experiments, the proposed algorithm also performs well. Compared with the zero-watermark algorithm using the pre-trained VGG19 network, it has a significant performance advantage, and it also has a certain advantage over the zero-watermark algorithm using PHTs image moments. In summary, the proposed algorithm has good robustness and a certain generalization ability, and has a certain application prospect in the field of medical data information protection.

REFERENCES

1. Liu, J., Ma, J., Li, J., Huang, M., Sadiq, N., & Ai, Y. (2020). Robust watermarking algorithm for medical volume data in internet of medical things. *IEEE Access*, 8, 93939–93961.
2. Jiang, F., Jiang, Y., Zhi, H., Dong, Y., Li, H., Ma, S., ... & Wang, Y. (2017). Artificial intelligence in healthcare: past, present and future. *Stroke and Vascular Neurology*, 2(4).
3. Marsland, M. J., Tomic, D., Brian, P. L., & Lazarus, M. D. (2018). Abdominal anatomy tutorial using a medical imaging platform. *MedEdPORTAL*, 14, 10748.
4. Wang, L., Wang, X. L., & Yuan, K. H. (2013, October). Design and implementation of remote medical image reading and diagnosis system based on cloud services. In *2013 IEEE International Conference on Medical Imaging Physics and Engineering* (pp. 341–347). IEEE.
5. Qiao, Z., Zhang, F., Lu, H., Xu, Y., & Zhang, G. (2023). Research on the medical knowledge deduction based on the semantic relevance of electronic medical record. *International Journal of Computational Intelligence Systems*, 16(1), 38.
6. Hisham, S. I., Liew, S., & Zain, J. (2013). A quick glance at digital watermarking in medical images. *Biomedical Engineering Research*, 2(2), 79–87.
7. Rey, C., & Dugelay, J. L. (2002). A survey of watermarking algorithms for image authentication. *EURASIP Journal on Advances in Signal Processing*, 2002(6), 1–9.
8. Ye, X., Chen, X., Deng, M., & Wang, Y. (2014, October). A SIFT-based DWT-SVD blind watermark method against geometrical attacks. In *2014 7th International Congress on Image and Signal Processing* (pp. 323–329). IEEE.
9. Patsariya, S., & Dixit, M. (2022). A new block based non-blind hybrid color image watermarking approach using lifting scheme and chaotic encryption based on Arnold Cat Map. *Traitement du Signal*, 39(4).
10. Wen, Q., Sun, T., & Wang, S. (2003). Concept and application of zero-watermark. *Acta Electronica Sinica*, 31(2), 214–216 [in Chinese].
11. Yi, D., Liu, J., Li, J., Zhou, J., Bhatti, U. A., Fang, Y., & Nawaz, S. A. (2021). A Robust digital watermarking for medical images based on PHTs-DCT. In *Cyberspace Safety and Security:*

12th International Symposium, CSS 2020, Haikou, China, December 1–3, 2020, Proceedings 12 (pp. 95–108). Springer International Publishing.

12. Ren, S., He, K., Girshick, R., & Sun, J. (2015). Faster r-cnn: Towards real-time object detection with region proposal networks. *Advances in Neural Information Processing Systems*, 28.

13. Devlin, J., Chang, M. W., Lee, K., & Toutanova, K. (2018). Bert: Pre-training of deep bidirectional transformers for language understanding. *arXiv preprint arXiv:1810.04805*.

14. Amodei, D., Ananthanarayanan, S., Anubhai, R., Bai, J., Battenberg, E., Case, C., ... & Zhu, Z. (2016, June). Deep speech 2: End-to-end speech recognition in English and Mandarin. In *International Conference on Machine Learning* (pp. 173–182). PMLR.

15. Shen, D., Wu, G., & Suk, H. I. (2017). Deep learning in medical image analysis. *Annual Review of Biomedical Engineering*, 19, 221–248.

16. Gong, C., Liu, J., Gong, M., Li, J., Bhatti, U. A., & Ma, J. (2022). Robust medical zero-watermarking algorithm based on Residual – DenseNet. *IET Biometrics*, 11(6), 547–556.

17. Han, B., Du, J., Jia, Y., & Zhu, H. (2021). Zero-watermarking algorithm for medical image based on VGG19 deep convolution neural network. *Journal of Healthcare Engineering*, 2021.

18. Iandola, F. N., Han, S., Moskewicz, M. W., Ashraf, K., Dally, W. J., & Keutzer, K. (2016). SqueezeNet: AlexNet-level accuracy with 50x fewer parameters and < 0.5 MB model size. *arXiv preprint arXiv:1602.07360*.

19. Duo-han, B., Xin, L. & Xin-yuan, W. (2020). Efficient Image Encryption Algorithm Based on 1D Chaotic Map. *Computer Science*, 47(4), 278–284 [in Chinese].

20. Wang, W., Liu, F., Gong, D., & Liu, S. (2020). Analysis of false alarm problem of watermarking method based on singular value decomposition. *Computer Engineering*, 46(11), 273–278 [in Chinese].

21. Li, D., Li, J., Bhatti, U. A., Nawaz, S. A., Liu, J., Chen, Y. W., & Cao, L. (2023). Hybrid encrypted watermarking algorithm for medical images based on DCT and Improved DarkNet53. *Electronics*, 12(7), 1554.

22. Bhatti, U. A., Tang, H., Wu, G., Marjan, S., & Hussain, A. (2023). Deep learning with graph convolutional networks: An overview and latest applications in computational intelligence. *International Journal of Intelligent Systems*, 2023, 1–28.

23. Sheng, M., Li, J., Bhatti, U. A., Liu, J., Huang, M., & Chen, Y. W. (2023). Zero watermarking algorithm for medical image based on Resnet50-DCT. *CMC-Computers Materials & Continua*, 75(1), 293–309.

24. Liu, J., Li, J., Ma, J., Sadiq, N., Bhatti, U. A., & Ai, Y. (2019). A robust multi-watermarking algorithm for medical images based on DTCWT-DCT and Henon map. *Applied Sciences*, 9(4), 700.

25. Fan, Y., Li, J., Bhatti, U. A., Shao, C., Gong, C., Cheng, J., & Chen, Y. (2023). A Multi-watermarking algorithm for medical images using inception V3 and DCT. *CMC-Computers Materials & Continua*, 74(1), 1279–1302.

26. Li, T., Li, J., Liu, J., Huang, M., Chen, Y. W., & Bhatti, U. A. (2022). Robust watermarking algorithm for medical images based on log-polar transform. *EURASIP Journal on Wireless Communications and Networking*, 2022(1), 1–11.

27. Bhatti, U. A., Huang, M., Wu, D., Zhang, Y., Mehmood, A., & Han, H. (2019). Recommendation system using feature extraction and pattern recognition in clinical care systems. *Enterprise Information Systems*, 13(3), 329–351.

28. Bhatti, U. A., Yu, Z., Chanussot, J., Zeeshan, Z., Yuan, L., Luo, W., … & Mehmood, A. (2021). Local similarity-based spatial–spectral fusion hyperspectral image classification with deep CNN and Gabor filtering. *IEEE Transactions on Geoscience and Remote Sensing*, 60, 1–15.
29. Bhatti, U. A., Yu, Z., Li, J., Nawaz, S. A., Mehmood, A., Zhang, K., & Yuan, L. (2020). Hybrid watermarking algorithm using Clifford algebra with Arnold scrambling and chaotic encryption. *IEEE Access*, 8, 76386–76398.
30. Liu, J., Li, J., Zhang, K., Bhatti, U. A., & Ai, Y. (2019). Zero-watermarking algorithm for medical images based on dual-tree complex wavelet transform and discrete cosine transform. *Journal of Medical Imaging and Health Informatics*, 9(1), 188–194.

CHAPTER 8

Deep Learning Applications in Digital Image Security

Latest Methods and Techniques

Saqib Ali Nawaz[1], Jingbing Li[1,2], Uzair Aslam Bhatti[1,2], Muhammad Usman Shoukat[3], and Raza Muhammad Ahmad[4]

[1]*School of Information and Communication Engineering, Hainan University, Haikou, China*
[2]*State Key Laboratory of Marine Resource Utilization in the South China Sea, Hainan University, Haikou, China*
[3]*School of Automotive Engineering, Wuhan University of Technology, Wuhan, China*
[4]*School of Cyberspace Security, Hainan University, Haikou, China*

8.1 INTRODUCTION

A digital watermark is a distinguishable digital signal or pattern embedded in other information (host data), and should not affect the availability of the host data [1–5]. The application of watermarking technology mainly includes copyright protection, data monitoring, and data tracking [6–9]. With the development of computer technology and the change and increase of objects to be protected, digital watermarking knowledge has experienced the progress process of multimedia watermarking, software watermarking, machine learning algorithm, and model watermarking.

Digital images have become a ubiquitous form of communication in our society, with a wide range of applications in various fields such as healthcare, education, entertainment, and social media. However, with the rise of digital images, the issue of image security has become increasingly important. Digital images are helpless to numerous sorts of attacks, such as unauthorized access, modification, and distribution, which can result in privacy violations, financial losses, and reputational damage. To address these security concerns, researchers and practitioners have turned to deep learning as a powerful tool for developing robust security mechanisms for digital images. Deep learning techniques such as convolutional neural networks (CNNs), recurrent neural networks (RNNs), and generative adversarial networks (GANs) have exposed great possibilities in various image security applications such as encryption, watermarking, steganography, and authentication [10].

DOI: 10.1201/9781003427674-8

For instance, in recent studies, deep learning–based tactics have been employed for image steganography, where hidden information is embedded in images, without any significant distortion. Moreover, the combination of deep learning and other traditional cryptographic methods has also shown promising results. The combination of deep neural networks and chaotic maps has been used for image encryption, providing high security and efficiency [11]. Therefore, adding watermarks can protect its intellectual property rights during illegal copying, redistribution, and misappropriation [12]. Multimedia watermarking is a watermarking technology mainly for images, videos, audios, text documents, and other media. Considering their robustness and security, it is often necessary to randomize and encrypt the watermark [13–15]. Two key technologies commonly used in multimedia watermarking include spread spectrum watermarking and quantization watermarking [16]. Spread spectrum watermarking is through spread spectrum communication technology, where the carrier signal is regarded as a wideband signal and the watermark signal is regarded as a narrowband signal, and the energy spectrum of a watermark is extended to a wide frequency band so as to be allocated to each frequency component. The watermark signal energy is small and difficult to detect [17,18]. Quantized watermarking divides the original carrier data into different quantization intervals based on the different watermark information. During detection, the watermark information is identified by the quantization interval to which the data belongs [19,20].

With the popularization of software applications, the problem of software code reuse becomes more and more prominent, and software watermarking is a significant means to explain the problematic of software copyright. Software watermark mainly refers to implanting a special identifier (watermark) in the code, which can carry information such as software author, copyright, etc., and then using a special extractor to identify or extract it from the defendant's software as evidence to achieve the purpose of detection [21–23]. The code watermark is hidden in the instruction part of the program, and the information watermark is hidden in the facts including header files, strings, and debugging information. According to the way the watermark is loaded, software watermarks can be divided into static watermarks and dynamic watermarks. Static watermarking mainly implants the watermark into the code or data of the executable program, and the extraction process does not need to run the program [24]. The goal of dynamic watermarking is to insert the watermark into the program's execution process or running state; that is, to encode information based on the program's running state at any given time, primarily using thread-based, graph-based, path-based, and other watermarking technologies [25].

Because of its widespread applicability in the realm of digital imaging, digital image watermarking has developed into a highly convergent study subject in recent years. Each application has unique criteria for the watermarking systems that must be used. As a result, the development of watermarking procedures that are suitable for a wide variety of applications may be challenging and may differ from more typical versions. Use control (transactional, replication), signature authentication and tamper detection, copyright protection, and medical uses (telemedicine data transfer, integrity of medical imaging), to name a few, all make use of comparable technologies [26,27].

With the efforts of many scholars, digital-watermarking technology has come out as a new type of information-hiding technology. This technology can provide further protection for the decrypted data, and it can be embedded in a large volume of watermark data in the novel carrier. The main use of digital-watermark technology is mainly used. In the copyright safety and complete certification of digital images, as a main technology that protects digital images, copyright information is embedded in the host image in a hidden form through the watermark algorithm. The extraction algorithm extracts copyright information as the main evidence of the attribution of digital images [28–30]. At present, the geometric invariant is an effective anti-geometric attack method. The main purpose of this type of algorithm is to extract the geometric invariant of the original image and embed the watermark information through the feature extraction algorithm. Since the geometric invariant feature has geometric invariance, the geometric invariant will not change when the watermarked image suffers numerous attacks during the propagation process, so the validity of the embedded watermark information can be guaranteed [31,32].

Digital watermarking, a copyright protection technique [33–36], effectively addresses copyright protection and verification issues. Its elementary clue is to insert the copyright logo into the carrier data over the embedding algorithm. The reverse operation extracts the copyright information in it to approve the copyright attribution of the image. Nevertheless, due to the constant updating of image processing tools, it has brought great tasks to the copyright safety of digital images. Traditional encryption methods, such as the info encryption algorithm (DES) [37], RSA [38], and 'Hash [39], can assurance the security of image material to a firm extent, but this encryption technique can only avoid the image data from animation used and cannot security the copyright title of the material. Digital watermarking know-how can well block the deficiencies of out-of-date encryption procedures. Consequently, it is very essential to training the watermarking method for digital images.

Text- or video-based digital-watermarking algorithms usually choose to place the copyright logo on the surface of the text to indicate the attribution of copyright [40–44]. Although this copyright protection method can clearly display the copyright mark, it cannot guarantee the security of the information. Attackers can use text or video processing software to remove the apparent copyright and add their own copyright information to it. This non-transparent watermarking algorithm makes it difficult to determine the copyright ownership of information when a copyright dispute occurs. Digital image information is different from other text information. In the process of embedding copyright data, it is essential to ensure the transparent effect of copyright, and at the same time, it cannot affect the normal use of the image. For this reason, the existing watermarking algorithms often choose to embed a small-capacity binary copyright image. Although the carrier image can achieve a good visual effect, with increasing user demand, it will become more and more difficult to choose a colored logo as copyright information. Therefore, increasing the capacity of embedded information under the premise of ensuring the transparency of the image is a serious challenge in the pitch of digital image watermarking. We present a few suggestions that are pertinent to the ongoing work in the field and ought to be taken into consideration. Table 8.1 provides an illustration of a summary of some digital image watermarking algorithms.

TABLE 8.1 Summary of Some Digital Image Watermarking Algorithms

References	Purposes	Methods	Size	Notes
[45]	Anonymous image watermarking	DCT, HVS, and SVM	$(256 \times 256)/(32 \times 32)$	The improved integrity and resistance to multiple attack types provided by this algorithm makes it a promising choice.
[46]	Secure watermarking	MWT and SVM	$(512 \times 512)/(32 \times 32)$	This method improves visual quality and is resistant to several attacks.
[47]	With a view to ensuring the design for medical use	DCT, SVM, and Spread Spectrum	$(256 \times 256)/(256 \times 256)$	Accuracy has been demonstrated experimentally for the suggested technique.
[48]	Protect your data using watermarking that is both undetectable and incredibly strong.	PDTDFB, LSSVM	$(512 \times 512)/(32 \times 32)$	This strategy is more effective in resiliency in the face of a skewing of geometry.
[49]	Copyright protection via reliable and secure watermarking	LWT, SVM, and PCA	$(512 \times 512)/(32 \times 16)$	Here, LWT strengthens the watermark's reliability.
[50]	Intuitive and trustworthy watermarking of images	DWT, SVD, and SVM	$(256 \times 256)/(128 \times 128)$	The primary goal of this method is to hold firm in the face of hostile forces.
[51]	Method of encrypted watermarking for use in situations requiring more safety	DRT and SVM	(512×512)	This method has improved robustness and raised the bar for security.
[52]	Image watermarking for privacy and security	DWT, SVD, SVM, and ML	$256 \times 256)/(128 \times 128)$	This plan guarantees that it gives you access to a protected medium for transmission using LTE network.

At present, there are many new investigate algorithms in the copyright safety know-how of digital images, but maximum of them are based on one feature of practical research, lacking an efficient kind of the content of digital image watermarking. This chapter sorts out and analyzes the research on new copyright protection technologies and algorithms in the field of digital images in recent years. Compared with other literature, the main contributions of this chapter are as follows:

1. The most recent developments in image watermarking techniques are discussed here.

2. The most common evaluation criteria used in the field of digital image water-marking are summed up, and the most effective attacks on images are categorized and summed up. The pros and cons of the digital image watermarking algorithms that are already in use are looked at in more detail.

3. The application of deep learning in digital image security has summarized to open up new possibilities for developing more robust and efficient security mechanisms. The latest methods and techniques in this field are continuously evolving and improving, leading to better protection for digital images.

4. Finally, some possible future approaches in image watermarking are presented along with the challenges and ongoing research. According to the existing problems in existing digital image watermarking, the research direction of this field has more development potential in the future.

The remaining parts of this article are structured as follows: Section 8.1 of this chapter introduces the basic model, basic characteristics, and evaluation indicators of digital image watermarking; Section 8.2 organizes the research framework of image water-marking algorithms; Section 8.3 introduces the classification of digital watermarking, mainly in three aspects: characteristics, detection methods, and hidden positions; Section 8.4 introduces performance evaluation and algorithms; Section 8.5 summarizes attacks, including robust attack, no attack, and non-geometric attacks; Section 8.6 discusses learning-based watermarking; Section 8.7 discusses the application of digital image watermarking in other fields, such as medical field, remote sensing field, map copyright, copyright protection, content authentication, infringement tracking, radio monitoring, and copy control; and finally, section 8.8 organizes the conclusion and describes some future directions and challenges of this article.

8.2 BACKGROUND

As an effective way of hiding information through technology, digital watermarking technology refers to the procedure of embedding watermark data into multimedia original data by means of digital watermarking technology [53–56]. Watermark data is a set of data with identification and rights information. The watermark data's validity can be used to check the rights information of the carrier data and identify sabotage. In general, the watermark info embedded in the original data is invisible unless there are special requirements. Digital watermarking technology already has a mature application system. In order to improve security, today's digital watermarking know-how is often used in combination with encryption technology.

8.2.1 Basic Model

The general model of a digital watermarking system includes two parts: watermark embedding and watermark extraction. The main goal pursued by researchers during watermark embedding is how to choose the optimal relationship among imperceptibility and robustness. Figure 8.1 presents a simple watermark embedding process.

Conferring to the digital watermark embedding flowchart, the watermark embedding process is:

$$I_W = E(I, W, K) \tag{8.1}$$

In formula (8.1), E represents the digital-watermark embedding algorithm, W signifies the watermark information, K denotes the key, I is the creative carrier information, and I_W is the watermark carrier information.

Figure 8.2 introduces the basic flow of digital-watermark extraction and detection. During watermark extraction, it can be separated into two categories: blind extraction

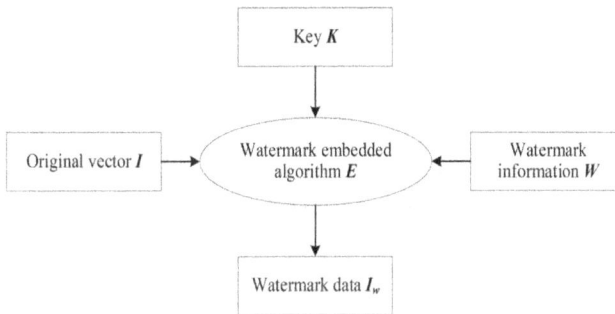

FIGURE 8.1   Flowchart of digital watermark embedding.

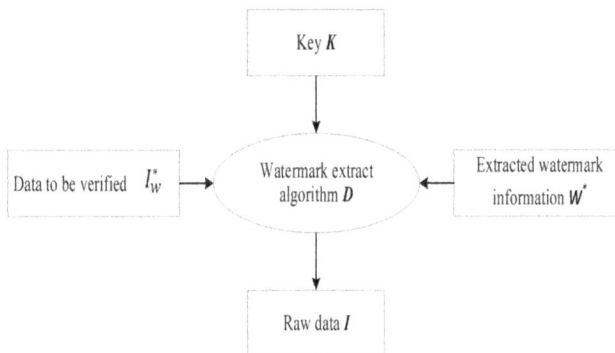

FIGURE 8.2   Digital watermark extraction.

and non-blind extraction, according to whether the original carrier information is needed as input data. Because in daily life, it is very difficult to obtain the original carrier information, to some extent, the digital-watermarking algorithm founded on blind extraction has more important research value. By comparison, the copied watermark info to the original watermark data, the watermark extraction algorithm figures out who owns the copyright to the original carrier data.

The non-blind extraction algorithm that needs to rely on the original carrier data I is defined as follows:

$$W^* = D(I_W^*, I, K) \tag{8.2}$$

The extraction method without original carrier data I can be defined as:

$$W^* = D(I_W^*, K) \tag{8.3}$$

In the formula, D is the watermark extraction algorithm, K is the same key as the watermark embedding method, I_W^* represents the attacked watermark carrier data, and W^* represents the extracted watermark data.

8.2.2 Learning-based Model

The use of deep learning in image authentication has been a growing area of research. With the help of GANs, it is possible to generate realistic images that can be used to verify the authenticity of an image by comparing it with the original image [57]. GAN network includes two subnetworks: generator network and discriminator network, which can be constructed by convolutional neural network (CNN), recursive neural network (RvNN) or self-coder. The idea behind the GAN network is a two-person zero-sum game. The sum of the interests of mutually sides of the game is a constant. When one side has the upper hand, the other side is bound to have the lower hand. Generally, the GAN network involves of two parts: the generator network G and the discriminator network D. The basic structure of the GAN network is shown in Figure 8.3.

The input of the network is random noise, which is recorded as f. First, it is processed by the generator network G, and the output signal f_G is obtained. Then, f_G and the real data f_R are input into the discriminator network D to obtain p_G and p_R, respectively. The target of discriminator network D is relatively simple. Its task is to identify whether the input data is real data or generated data output by G. Generally, the output of subnetwork D is a number between 0–1, which indicates the probability that the input value

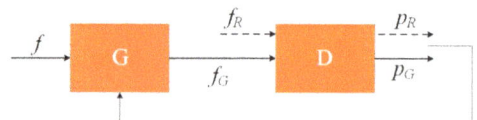

FIGURE 8.3   Basic structure of GAN network.

is real data. The sigmoid function is often used to achieve this function. When training the discriminator, if the input is f_R, the corresponding label is set to 1. If the input is f_G, the corresponding label is set to 0. Therefore, in theory, the optimal state of subnetwork D is $p_R = 1$, $p_G = 0$, which means that it can perfectly distinguish the real data from the generated data.

8.3 CLASSIFICATION OF DIGITAL WATERMARKING

8.3.1 Divided by Characteristics

8.3.1.1 Robust Watermarking

When discussing robust watermarking, it is important to note that general image processing operations (noise reduction, filtering, sharpening, etc.) can extract all copyright information from the carrier carrying the copyright information [58–60]. Increasing the algorithm's stability is a primary focus in resilient watermarking. Wavelet and discrete wavelet transform, as well as singular value decomposition and orthogonal triangular decomposition, are common examples of frequency domain transforms used by modern, secure watermarking methods (orthogonal-triangular decomposition, QR). In order for image data to withstand conventional image processing processes, matrix decomposition tools such as are used to remove redundant and sensitive information. Therefore, the watermarking algorithm based on frequency domain transformation is also included in the category of robust watermarking algorithms. For example, the literature [61,62] uses singular value decomposition to find the maximum singular value of each sub-block of an image.

$$I = U\Sigma VT^T = U \begin{bmatrix} \lambda_{1,1} & \cdots & \lambda_{1,n} \\ \vdots & \ddots & \vdots \\ \lambda_{n,1} & \cdots & \lambda_{n,n} \end{bmatrix} V^T \tag{8.4}$$

Among them, U represents the left singular matrix, V represents the right singular matrix, A represents the singular value, and I represents the size of the matrix. Since the largest singular value changes less after conventional image processing, the robustness of the algorithm can be improved. Begum et al. [63] modified the original singular value decomposition process and proposed a robust watermarking algorithm of double singular value decay. In addition to maintaining the traditional singular rate decomposition characteristics, the double singular value decomposition also adds more key matrices to recover the security of the algorithm. This local representation ability and rotation invariance can effectively reduce noise, filter, rotation, and other attacks on the carrier image, thereby ensuring the integrity and security of the watermark information.

8.3.1.2 Fragile Watermark

Fragile watermarks, which do not reveal any data about the original digital content [64–66], are used to determine whether digital content has been interfered with and to differentiate tampered zones from non-tampered regions. Kumar et al. [67] proposed a chaotic, fragile watermark in the discrete cosine transform area. According to the sensitivity of the high-frequency information in the DCT area, the algorithm will map the chaotic watermarks. The watermark info is embedded in the high-frequency information, which can obtain a good tampering location ability and at the same time realize blind detection in the copyright authentication stage, but this method cannot locate the image tampering position after a filtering attack and a JPEG compression attack. After two chaotic mappings, the watermark information is embedded in the least significant bits of the carrier image. Since the chaotic mapping is sensitive to the initial value, when using the wrong When extracting the copyright information with the secret key, the tampering position can be accurately located. Azeroual and Afdel [68] used the Faber-Schauder discrete wavelet transform (FSDWT) to transform the least significant bit (LSB) to make the watermark algorithm more fragile. The image after FSDWT is divided into blocks, and the maximum coefficient of the sub-block is selected, and finally the coefficient and the copyright watermark are XORed. The algorithm embeds the watermark generated by the FSDWT coefficients into the LSB of the original image to accurately locate the tampered area. Shen et al. [69] planned a fragile watermarking algorithm using the relationship between the left and right singular value matrices after singular value decomposition. The algorithm directly performs singular value analysis on each sub-block of the carrier image. It is decomposed, and a binary feature matrix is constructed according to the product of the first column of the left and right singular value matrices, and finally the feature matrix is embedded in the least significant bits of the carrier image. Wang and Men [70] used the vector map fragile watermarking algorithm of coordinate grid division to divide the map by using the geographic coordinates of the map. This way of dividing the map ensures the integrity of the map coordinate information. In order to prevent data authentication failure caused by overflow of data points, the algorithm ensures that data points remain in data sub-blocks by modifying the shortest distance between data points and sub-block margins. Compared with traditional authentication methods, this algorithm has good stability and positioning accuracy.

8.3.2 Divided by Detection Method
8.3.2.1 Blind Watermark
To complete the watermark withdrawal procedure without revealing any information about the original watermark, blind watermarking involves embedding the entirety of the copyright watermark's information into the carrier image and then quantifying it. The blind watermark embedding model is as follows:
when $W_{i,j} = 1$, there is

$$\lambda = \begin{cases} \lambda - T_{i,j} - \delta/4 & T_{i,j} \leqslant \alpha/4 \\ \lambda - T_{i,j} + 3 \times \alpha/4 & \text{"other"} \end{cases} \tag{8.5}$$

When $W_{i,j} = 0$, there is

$$\lambda = \begin{cases} \lambda - T_{i,j} + 5 \times \alpha/4 & T_{i,j} \geqslant 3 \times \alpha/4 \\ \lambda - T_{i,j} + \alpha/4 & \text{other} \end{cases} \tag{8.6}$$

In the formula, $T_{i,j} = \mod(\lambda_{i,j}, \alpha)$, λ represents the maximum singular value, α is the maximum quantization parameter, and $W_{i,j}$ is the pixel value in the binary watermark.

Because blind watermarking is convenient and practical, it is widely used in image copyright protection. Yadav et al. [71] uses DWT and SVD to obtain the maximum singular value of the carrier image, and then cyclically moves the last seven bits of the singular value matrix to embed the watermark information into the luminance factor of the carrier image. This method has good resistance to small-scale clipping attacks, but the extracted watermark images contain more noise points. Thanki et al. [72] used the Curvelet transform to obtain the high-frequency Curvelet coefficients of the carrier image and then used the redundant discrete wavelet transform (RDWT) to embed the watermark information into each sub-block, respectively, in the RDWT. Because the Curvelet transform has a strong expressiveness for the curve characteristics of the image and RDWT uses a non-subsampling mechanism to transform the image, which can efficiently ensure the translation invariance of the image, it overcomes the problem that the wavelet coefficients change quickly after the wavelet transform is down-sampled.

One major drawback of blind watermarking is that the shearing attack will only be able to partly extract the watermark. For example, the blind watermarking algorithm proposed in the [73–76] has a local missing watermark at the image shearing position. This partial missing phenomenon has little impact on common watermark copyright identification, but it may cause the copyright information to be unreadable for watermark algorithms that use two-dimensional codes as the copyright image. Reference [77,78] future a blind watermarking algorithm based on QR codes, which cannot quotation the complete watermark info under a shear attack. In order to overcome the shortcomings of blind watermarking in geometric attacks, [79] proposes a blind watermarking algorithm against geometric attacks. He et al. [80] uses the oblique anisotropy of the Directionlet transform to construct the synchronization information of the watermark, and directly uses the direction of the image edge as the reference direction to embed the watermark. Since the selected edge slope is not easy to change, it is resistant to attacks and strong geometric performance. Ye et al. [81] corrects the attacked image using the scale-invariant feature transformation (SIFT). When the image is rotated by 45°, the obtained image appears with values are as high as 0.9980. It shows that this strategy can greatly improve the anti-geometric attack performance of the algorithm. In summary, blind watermarking mainly has the following advantages.

1. The blind watermark is selected by the quantizer to be the closest to the original carrier data. The close data replaces the original carrier data, so it has better fidelity.

2. The procedure of removing the watermark from the blind watermark goes to the blind extraction method. So, the watermark extraction procedure can be finished without giving any copyright information during copyright authentication. This makes copyright authentication easy and useful.

8.3.2.2 Non-blind Watermarking

Non-blind watermarking mentions the procedure of removing watermark information that requires the user to provide another part of the watermark data when extracting copyright information. Only a subset of the watermark information will be compromised if the carrier image is targeted. Therefore, compared with blind watermarking, its robustness will be better, as in formulas (8.7) to (8.9):

$$[U \quad S \quad V] = \text{svd}(I) \tag{8.7}$$

$$[u \quad s \quad v] = \text{svd}(w) \tag{8.8}$$

$$S' = S + \alpha \times s \tag{8.9}$$

where SVD is the singular value decomposition operation; I is the carrier image; w is the watermark image; U, S, and V denote the left singular matrix, singular rate matrix, and right singular matrix of the carrier image and copyright watermark, respectively; S' is the singular value matrix after embedding the watermark information; and α is the embedding strength. It can be seen from the non-sturgeon watermark model that the embedding process only embeds the singular values of the watermark into the singular values of the carrier image, so when extracting the copyright watermark, it is necessary to provide the left singular value matrix, U, and the right singular value of the original watermark. Matrix V, the authentication process of this type of watermarking algorithm is cumbersome, and its practical performance is very low.

Reference [76,82–84] utilizes the stability of low-frequency information after wavelet transformation and the robustness of singular value decomposition and embeds copyright information in the low-frequency domain of carrier images. The wavelet transform has strong robustness in common non-geometric attacks because it has a good filtering effect on high-frequency signals (such as noise), but in geometric attacks, the pixel position changes significantly and the wavelet coefficients change. The impact is large, and the resistance performance is poor. At the same time, although the extreme singular value of the block can represent the energy info of a sub-block, when the pixels in the local position of the carrier image are clipped, the block singular value changes from the maximum value to 0, and the FS information of the image sub-block cannot be expressed at this time.

Because singular value decomposition has great volatility under high-intensity geometric attacks, the [85,86] uses SIFT to perform geometric improvement to recover the performance of the algorithm in geometric attacks. In the [87], the carrier image was first transformed by non-subsampled contourlet transform (NSCT) to obtain low-frequency information, and then SIFT was used to extract the feature points of the carrier image, and finally the watermark data was embedded in the low-frequency information. It has strong resistance and robustness to common geometric attacks and combined attacks.

8.3.2.3 Zero Watermark

Although both robust and blind watermarking offer significant improvements in resilience, most embedded watermarking techniques suffer from a conflict between robustness and transparency. Copyright watermarks, when added to a carrier image, not only leave a visible trace on the image's transparency but also corrupt the image's underlying data. Robust watermarking and blind watermarking are challenging ways to satisfy the demand for information with high content integrity.

Based on the lack of content integrity and transparency of robust watermarking and blind watermarking, Wen et al. [88] projected the idea of a zero watermark. The concept is to extract features from both the carrier image and the copyright watermark, and use those features to create a one-of-a-kind zero watermark. Ultimately, as depicted in Figure 8.4, the generated zero-watermark data is filed with the Copyright Protection Center. Figure 8.5 shows that most of the characteristics that the carrier image uses to build its representation are binary feature matrices. The purpose is to facilitate logical operations with binary copyright watermarks. At the same time, the generated zero watermarks are usually disorganized binary info helpful to refining the security of data. It is sufficient to conduct logical operations on the generated zero watermark and the attacked image features in the

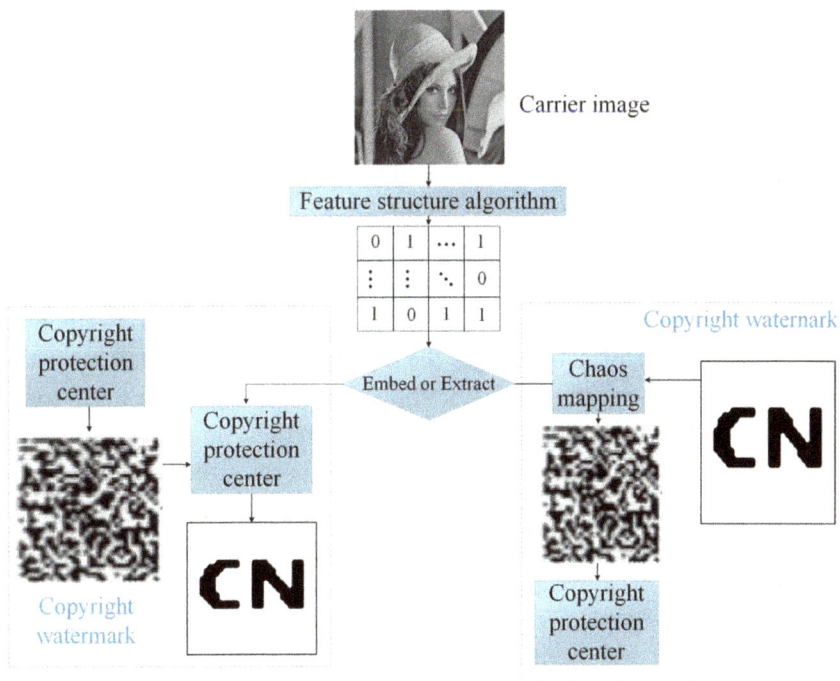

FIGURE 8.4 Zero-watermark model.

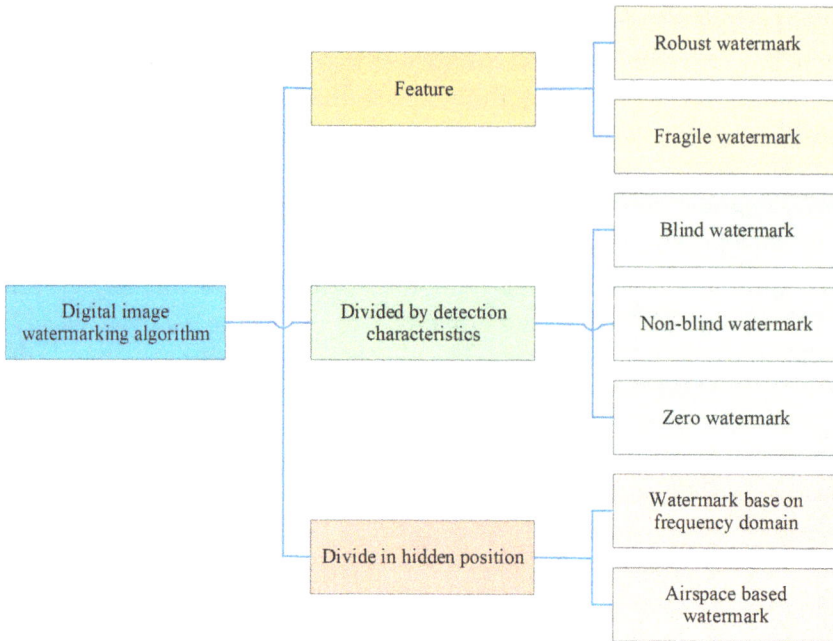

FIGURE 8.5   Algorithm-based classification of digital image watermarking.

case of copyright disputes. This technique of authentication is more practical than the alternatives, such as blind watermarking and robust watermarking.

Because zero watermark solves the contradiction between robustness and transparency of embedded watermarks, it is widely used in the copyright protection of digital images [36,89,90]. The zero-watermark algorithm can be separated into two types based on the feature matrix created by zero watermark: zero watermark based on frequency domain transformation and zero watermark based on spatial domain transformation. The rate domain-based zero-watermarking transforms the image from the longitudinal domain to the frequency domain information by using DCT, DWT, Curvelet transformation, principal component analysis (PCA) [91], and other transformation tools to obtain the carrier image's low-frequency information, and then using singular value decomposition to construct the carrier image's feature matrix. For example, in 2014, Prathap et al. [92] used the Contourlet transform to extract the low-frequency information of the carrier image and constructed the feature information through PCA, which has strong robustness against non-geometric attacks. In the [93], the copyright image is processed by visual cryptography, the copyright image is divided into two parts, and only one key map share and feature matrix are used to generate zero-watermark information, while the other key map share is used for the copyright authentication process. This algorithm is characterized by high security of copyright information.

The zero-watermarking algorithm based on rate domain transformation requires frequent time-frequency transformation, which increases the complexity of the algorithm. For this reason, Xiyao et al. [94] directly constructed a zero watermark based on geometric correction in the airspace of the image and used SIFT to correct the rotated image, which achieved good robustness in noise attack and rotation attack but poor performance against JPEG compression. Zheng et al. [95] proposed a robust zero-watermarking algorithm in the spatial domain, which directly uses the relationship between the overall mean of the image and the mean of each sub-block to construct a feature matrix in the spatial domain. The mean of the matrix is proportional, so it shows strong robustness to non-geometric attacks. In summary, zero watermark has the following advantages:

1. Good transparency. Since zero watermark doesn't need to embed copyright information into the carrier image, this kind of algorithm has unique advantages in terms of content integrity and transparency.

2. High security. The zero watermark generated by the zero-watermark algorithm contains unique and disorganized image information. Even if it is intercepted, the real copyright information cannot be identified. That is, the zero-watermark generation process is equivalent to the secondary encryption operation of the copyright information, so the security is higher.

8.3.3 Divided by Hidden Domain

8.3.3.1 Watermarking Algorithm Based on Frequency Domain

To incorporate a watermark in an image, a frequency-domain watermarking algorithm must first transform the image from its spatial domain to its frequency domain, then extract the low-frequency sub bands in accordance with the watermark's requirements. Most algorithms choose to embed the watermark in the low-frequency domain of the carrier image, encoding the watermark information into the low-frequency coefficients of each sub-block and then finishing the embedding with the inverse transformation of DCT due to the superior concealment act of low-frequency information. Since DCT is a linear transformation tool, it has strong energy compression characteristics, but its transformation process is to transform the overall spatial domain of the carrier image into the frequency domain, which belongs to a global transformation operation and does not have local transformation characteristics and poor ability to handle local mutation signals.

In view of the shortcomings of DCT, Lu et al. [96] planned a blind watermarking algorithm based on wavelet and cross-wavelet trees, using discrete multi-wavelets to perform three-level wavelet decomposition on the carrier image to get the LL3 low-frequency sub-band, and then embed the watermark into the carrier according to the energy coefficient in the image. Since DWT can divide the image signal into different frequency band info features, the local characteristic of DWT can well overcome the shortcoming that DCT cannot perform local division and can well solve the problem of signal mutation. Sadreazami and Amini [97] proposed a Ridgelet transform watermarking algorithm with high approximation accuracy and sparse expression performance for straight line features. SVD is achieved on each sub-block to construct the feature matrix of the carrier image. Different from DWT, the barren wave transform needs to perform a Radon transform on the image and then use a one-dimensional DWT transform to extract low-frequency coefficients. The superposition of the degree values, therefore, increases the choice of direction in the frequency domain transformation and retains the rich image feature information in the constructed features.

Although the Ridgelet transform has more advantages than DWT in the direction of straight-line approximation, the Ridgelet transform is difficult to approximate the singular characteristics of curves in images. Therefore, in order to make the constructed feature matrix contain more image curve features, [98] carried out the contourlet change on the carrier image and then selected the important coefficients in the sub-bands in different directions as the position of the watermark embedding. Due to the full use of the multi-scale, localized, and directional characteristics of the Contourlet, it has good robustness against geometric attacks such as rotation, flipping, and scaling. Bazargani et al. [99] compared the robust performances based on DWT, Curvelet, and Contourlet transforms. The experimental results show that the robustness of the watermarking algorithm based on the Contourlet domain is better than that of the DWT-based watermarking algorithm in resisting noise, filtering, rotation, and scaling attacks, but the resistance to the shearing attack is poor. The main reason is that the local characteristics of DWT only consider the signal in the approximate direction of the image, while the approximate components obtained by Curvelet and Courselet contain information with the largest energy value in other directions. Therefore, during local clipping, Curvelet and Contourlet lose more information than DWT. In order to overcome the defect that the last layer of low-frequency sub-bands of the Contourlet transform is not divided, in 2015, Sadreazami et al. [100] planned the concept of a virtual tree structure in the Contourlet domain, and the watermark embedding position was selected by calculating the mean square error of each virtual tree. Linked with similar watermarking algorithms, this method has both transparency and robustness.

8.3.3.2 Watermarking Algorithm Based on Spatial Domain

The design of watermarking algorithms can be complex, despite the fact that algorithms based on frequency domain revolution have excellent benefits in transparency and resistance to noise, filtering, and other attacks. For example, the least significant bit (LSB) of the carrier image can be used to covertly insert the watermark information [101]. Because a change in LSB has no effect on the visual effect of the image, the information capacity of the watermark that LSB allows to embed is small, and LSB's robustness when the carrier image is disturbed by noise is poor. Therefore, Nana [102] proposed a spatial watermarking algorithm based on SVD decomposition. The algorithm embeds the preprocessed watermark information into the singular value of the carrier image, and the WC value obtained by cutting the upper left corner is still 1.0000. In 2018, Su and Chen [103] proposed a spatial color watermarking algorithm. This method has two advantages. One is that it does not require frequency domain transformation and can resist robust attacks with greater intensity; the other is to use color. The copyright mark constructs a colored zero watermark, which is more secure and richer in information expression than the traditional binary zero watermark. In 2015, Chan et al. [104] proposed a high-transparency spatial watermarking algorithm. This method embeds the watermark in the blue component of the color image. Since the color image has 24-bit capacity information, when one of the color components changes, the influence on the color carrier image is small, so the watermark information embedded in it has better concealment. According to the analysis of the algorithms in the above literature, the technologies used by various algorithms and their advantages and disadvantages are shown in Table 8.2. According to the characteristics of the watermarking algorithm given in Table 8.2, the overall context of digital image watermarking about the technology used and copyright information is summarized as follows:

1. From the perspective of technical characteristics, most of the algorithms choose DWT and its improved technology as the frequency domain transformation tool, mainly because the low-frequency information decomposed by the wavelet transform can effectively recover the robustness of the algorithm. The algorithm has good performance in resisting geometric attacks.

2. Judging from the type of watermark images used, most of the existing watermarking algorithms use binary images with a size of 64 × 64 as copyright information, mainly because embedding smaller copyright information will not destroy the visual outcome of the image.

3. From the perspective of algorithm types, non-blind watermarking has more watermark information than blind watermarking algorithms, robust watermarking algorithms, fragile watermarking algorithms, and zero watermarking algorithms, mainly because non-blind watermarking only embeds part of the watermark. So, to get the information out of the watermark, you need more watermark information, which is also a weakness of non-blind watermarking.

4. From the perspective of the compensations and difficulties of the algorithm, different types of watermarking algorithms have their own characteristics, which mainly depend on the use of the algorithm. For example, compared with the blind watermark, the fragile watermark mainly focuses on the accurate positioning of the tampering position. Blind watermarking focuses on the process of blind detection of information after being attacked. For different watermarking algorithms of the same type of watermarking technology, robustness and transparency are mainly considered. For example, although some algorithms can resist large-angle rotation attacks, their transparency is poor.

8.3.4 Other Classifications

There are many classifications of digital watermarking technology, which are classified in this chapter as follows:

1. According to whether the watermark can be supposed by the social eye, it can be separated into two classes: visible watermark and invisible watermark. The watermark information of the visible watermark does not need to be extracted and can be

TABLE 8.2 Algorithm Analysis of Watermarking Technology

Watermarking technology	References	Technical features	Advantage	Shortcoming	Watermark type (size)
Robust watermark	[61]	DWT, SVD	Safety is good at robustness, especially for anti-rotation attacks.	Poor shear resistance, cumbersome copyright certification process.	Binary watermark (64 × 64)
	[105]	Double singular value decomposition, wavelet transform	Anti-noise, filtering, rotation attack	----	----
	[106]	DCT transform, Logistic chaotic map	Simple algorithm, accurate tampering positioning	Weak anti-robust attack ability	Binary watermark (4 × 4)
Fragile watermark	[107]	Arnold scrambling, Logistic encryption, LSB	Good security and robustness	Unable to locate the tampered position after filtering attack and JPEG compression attack	Binary watermark (256 × 256)
	[68]	Faber-Schauder, DCT, LSB	Tampering positioning is accurate and real time	---	Binary watermark (64 × 64)
	[108]	DWT, SVD	Good resistance to small-scale non-geometric attacks	Can't resist rotating attacks	Binary watermark (32 × 32)
blind watermark	[72]	DCuT, RDWT	Fast calculation and good transparency	Anti-large angle rotation attack poor with the filter attack	Binary watermark (64 × 64)
	[80]	Directionlet transform	Good anti-geometric attack performance	---	Binary watermark (64×64)
non-blind watermark	[109]	Wavelet transform, SIFT correction	Strong anti-geometric attack ability	---	Binary watermark (64 × 64)
	[92]	Contourlet transform, PCA	Strong robustness to both geometric and non-geometric attacks	Unable to extract clear copyright information for zoom attack	-----
Zero watermark	[110]	Curvelet transform, DSVD, visual cryptography	Robbery and security are all better	The algorithm attack is more complicated	Binary watermark (64 × 64)
	[111]	Relationship between sub-block mean and overall mean	Strong robustness against non-geometric attacks	Unable to resist shear attack rotate attack	Binary watermark (64 × 64)

directly recognized in the human visual system, which is also the inertial under-standing of "watermark" by most of us. The watermark data of the invisible wa-termark is embedded in the multimedia data, and the multimedia data provides it with a certain camouflage. It is generally tough for the human visual scheme to directly obtain watermark information. Therefore, when the copyright information is threatened, it can be directly extracted. watermark information to demonstrate copyright attribution.

2. According to the features of a watermark, it can be divided into three sorts: fragile watermark, semi-fragile watermark, and robust watermark. The fragile watermark can be used as a mark when the original carrier information is tampered with or damaged and can be used to accurately locate the specific location of the tampered data. It is primarily used for content authentication as well as the integrity and authenticity identification of digital works, and it is extremely vulnerable to various conventional image processing and geometric attacks. Semi-fragile watermarks are in the middle of the two, with certain imperceptibility and resistance to attacks. A robust watermark means that after the watermark data is attacked, the watermark information can still be correctly extracted and used to identify the copyright attribution.

3. According to different watermark embedding regions, it is divided into three-dimensional watermarks and transform domain watermarks. Spatial watermarking hides the watermark information in the carrier by modifying the spatial pixel value of the host carrier. The robustness has been greatly improved.

4. According to different host types, it can be divided into text watermarks, image watermarks, audio watermarks, and video watermarks. When the watermark infor-mation is put into any type of carrier, the copyright of the item is protected.

5. According to whether the host information is used when extracting the watermark, it can be divided into non-blind watermarks and blind watermarks. The non-blind watermark is dependent on the host information in the extraction process. The watermark extraction operation can only be realized when the host information is obtained. However, in the actual transmission process, it is difficult to obtain the original carrier data, which limits the non-blind watermark. the application value of blind watermarking. The blind watermarking algorithm is different because it can extract the watermark even without the original carrier information. This gives it a wider range of practical uses.

8.4 PERFORMANCE EVALUATION AND ALGORITHMS

8.4.1 Performance Evaluation

Generally speaking, performance evaluation indicators in any field are basically divided into two categories: subjective and objective. However, because subjective thinking is used as a performance evaluation indicator, there are too many uncertain factors and subjective wills. The image quality must be maintained during the digital watermarking process. Therefore, two sets of metrics are needed: one to test the quality of the images and another to estimate the accuracy of the extracted watermark, in order to evaluate both the watermarked image and the watermark. Table 8.3 also displays a performance comparison of the various methods that have been explored in this study in relation to these indices.

Therefore, in this section, we only introduce two evaluation features to amount the act of a digital watermarking algorithm: watermark imperceptibility and robustness.

1. Imperceptibility means that human eyes can't see any changes in the carrier data or the host data after the watermark information has been hidden.

    Peak signal-to-noise ratio (PSNR) can measure the visual quality of watermarked images, and it is recognized as one of the indicators for perceptual objective evaluation. It can be calculated as follows:

$$\text{PSNR} = 10 \times \log_{10} \frac{255^2}{\text{MSE}} \tag{8.10}$$

$$\text{MSE} = \frac{1}{N^2} \sum_{x=1}^{N} \sum_{y=1}^{N} [f(x, y) - f^*(x, y)]^2 \tag{8.11}$$

    In the formula, $f(x, y)$ and $f'(x, y)$, respectively, represent the original carrier data and watermark data of size N × N. The greater the obtained PSNR value, the smaller the visual quality change of the carrier image, which directs that the projected digital watermarking algorithm has better imperceptibility. Mean squared error (MSE) can assess the degree of image change.

    Structural similarity index (SSIM) evaluates target images through comprehensive indicators of structure, contrast and brightness. It is also a perceptual objective evaluation index, which is defined as follows:

$$\text{SSIM}(f, f^*) = \frac{(2\mu_x \mu_y + C_1) \times (2\sigma_{xy} + C_2)}{\left[\mu_x^2 + \mu_y^2 + C_1\right] \times (\sigma_x^2 + \sigma_y^2 + C_2)} \tag{8.12}$$

    where μ_x and σ_x represent the mean and average deviation of f; individually, μ_y and σ_y are the mean and standard deviation of f^*; σ_{xy} represent the covariance of f and f^*; and C_1 and C_2 are two constants. This chapter sets $C_1 = 0.01$ and $C_2 = 0.03$. The value of SSIM falls in the interval [0,1].

2. Robustness is a significant evaluation standard for detecting the performance of watermarking algorithms. The better the robustness of the watermarking structure,

TABLE 8.3 Different Methods in the Literature Compared to One Another in Terms of Performance

Ref.	Types and numbers	Capacity (C)/length (L)	Findings
[112]	4 color images, and 6 grayscale images	C up to 1 bpp, and grayscale: C up to 0.5 bpp	Images in color should have a signal-to-noise ratio (PSNR) of 33 to 67 dB, and black-and-white images should have a PSNR of 32 to 65 dB.
[113]	2167 DICOM images (CT, MRI, CR)	C up to 0.1 bpp	The optimal PSNR for CT is 66 dB, the optimal MRI PSNR is 65.5 dB, and the optimal CR PSNR and SSIM are 70 dB and Inf, respectively.
[114]	8 grayscale medical images (CT, MRI, MRA, Radio)	L = 256 Byte	All channels have an average PSNR of about 39.26 dB and a BER of 0.
[115]	4 DICOM images (CT, MRI, CR,US)	CT: C = 0.888 bpp MRI: C = 0.2009 bpp CR: C = 0.0808 bpp US: C = 0.3091 bpp	PSNR = 85.50 dB and SSIM = 0.9765 for CT scans; PSNR = 69.71 dB and SSIM = 0.9287 for MRI scans; PSNR = 65.22 dB and SSIM = 0.9997 for CR scans; and PSNR = 36.71 dB/ SSIM = 0.7708 for US scans.
[116]	16 DICOM images (CT, MRI, US, X-ray)	CT: C among 0.515 & 0.525 bpp MRI: C among 0.418 & 0.438 bpp US: C among 0.570 & 0.617 bpp X-ray: C among 0.491 & 0.654 bpp	CT: PSNR among 72.41 and 77.71 dB; MRI: PSNR between 75.6 and 82.34 dB; PSNR between 39.02 and 43.93 dB and SSIM between 0.944 and 0.991 for US and X-ray.
[117]	CT grayscale of 11 affected role (60–100 slices each patient)	L among 125 and 600 Byte	PSNR among 52.5 and 69.5 dB
[118]	25 PET images	L about 1,000 Byte	PSNR > 47.43 dB, SSIM > 0.93
[119]	4 DICOM images (CT, MRI, US, X-ray)	L among 9,274 and 72,690 Byte	PSNR 34.8 dB, MSE 21.5.
[120]	Grayscale US images	C less than 1 bpp	Avg. PSNR about 47.5 dB
[121]	6,000 medical images (dissimilar modalities and formats)	C among 0.1 and 1 bpp	Average PSNR and SSIM approximately 81 dB and 0.999974, respectively.

the stronger its ability to attack numerous occurrences. Overall, the projected watermarking scheme will be judged by the algorithm based on the bit error rate or the standardized link coefficient.

Bit error rate (BER) is a parameter for measuring robustness, which refers to the degree of difference among the watermark data removed from the attacked watermark carrier information and the innovative watermark facts. Similarity among the innovative and de-watermarked versions of a data set is determined using the normalized correlation coefficient (NCC). The value ranges of BER and NCC are both 0 to 1. The smaller the BER value and the larger the NCC value, the better the robustness of the algorithm. We use BER and NCC to measure the robustness of the proposed algorithm. BER is defined as follows:

$$\text{BER} = \frac{B}{P \times Q} \times 100\% \tag{8.13}$$

In the formula, B represents the number of wrong bits of the removed watermark info, and P × Q represents the total amount of bits of the original watermark data. The NCC calculation formula is:

$$\text{NCC} = \frac{\Sigma_{i=1}^{P}\Sigma_{j-1}^{Q}\left[w\left(i, j\right) \times w'\left(i, j\right)\right]}{\Sigma_{i=1}^{P}\Sigma_{j=1}^{Q}\left[w\left(i, j\right)\right]^2} \tag{8.14}$$

where $W = \{w(i, j), 1 \le i \le P, 1 \le j \le Q\}$ and $W' = \{w'(i, j), 1 \le i \le P, 1 \le j \le Q\}$ is the original watermark image and the extracted watermark image.

8.4.2 Algorithms

With the deep research done on watermarking algorithms, many excellent digital image watermarking algorithms have been proposed in recent years [122]. Based on the existing research, this chapter divides the digital image watermarking methods into three categories. Figure 8.5 depicts the specific method of subdividing watermarks into robust watermarks and delicate watermarks based on their characteristics, detection method (blind watermarks, non-blind watermarks, and zero watermarks), and hidden position (frequency domain–based watermarking algorithm, spatial domain–based watermarking algorithm).

8.5 ATTACKS

The three main methods of watermarking attack are robustness attack, representation attack, and interpretation attack. Robustness attack is the most common attack method and the most effective verification method. indicates that the attack can make the detector fail to notice the presence of the watermark. Interpretation attack is the opposite of robust attack and uses the inverse method of watermarking technology to forge watermarks to achieve the purpose of forging the ownership of watermarks. The purpose of any attack is to find a way to circumvent the security that a watermark offers to digital

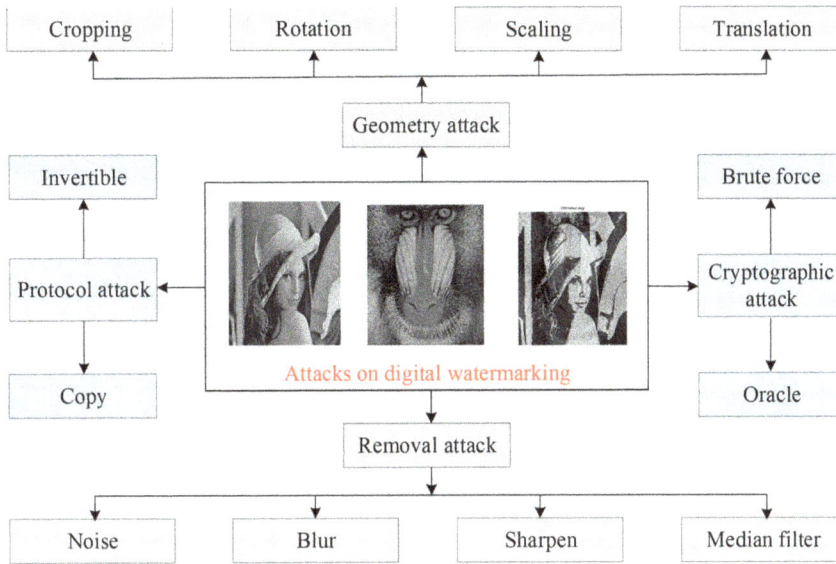

FIGURE 8.6   Different types of digital-watermarking attacks.

content. Attacks on watermarking can be broken down into several categories, including the following: geometry attack, protocol assault, cryptographic attack, removal attack, etc. [123–125]. Figure 8.6 shows the various attacks that can be used to escape digital watermarking from original and watermark images.

The robustness of a watermarking technique is typically evaluated based on its resistance to 15 different attacks. Table 8.4 gives a brief overview of these 15 assaults, whereas it provides a visual depiction of these assaults in action.

8.5.1 Robust Attack

Robustness attack is one of the effective methods often used to verify the algorithm. The image with embedded watermark information may be tampered with or attacked via a network, which will destroy the synchronization of extracting the watermarking algorithm. Therefore, after the design of the watermarking algorithm is completed, it is necessary to carry out corresponding attacks. experiment. According to the nature of the attack, image attacks can be divided into three classes, namely non-geometric attacks, geometric attacks, and combined attacks.

8.5.1.1 Non-geometric Attacks

A non-geometric attack is to perform small calculation changes or add interference value to image pixels. This kind of attack is mainly to weaken the signal strength. Common non-geometric attacks mainly include noise attacks, filtering attacks, and JPEG compression attacks. The specific attack types are classified in Figure 8.7.

As shown in Figure 8.7, the same attack can contain two or more attack types. Different types of attacks have diverse effects on the robustness of watermarks. When Gaussian noise attacks, for example, the noise points follow the Gaussian distribution, and the noise points of a certain intensity are concentrated and the intensity is high, so the

TABLE 8.4   Listed the Attacks, Both Geometric and Arithmetic, and Their Descriptions [126]

Types of attacks	Explanation
Cropping	Images with watermarks will have their pixels replaced with zeros.
Cutting	When working with watermarked images, replace all non-zero numbers in a row or column with a zero.
Gamma correction	Adjust the gamma of the watermarked image.
Gaussian filter	Distort watermarked images with Gaussian noise among [0.0001, 1].
Histogram equalization	To modify the contrast of the watermarked images, a histogram equalization filter is applied.
JPEG solidity	Compress watermarked images using JPEG with a ratio among (30%) and (90%).
Median filter	Watermarked images can have pixels swapped out for the median value of neighboring pixels using a 3 × 3 mask.
Rotating	Turn the watermarked images through a predetermined angle.
Salt pepper noise	Add salt and pepper noise with a density between [0.001, 0.04] to the watermarked image.
Scaling	Scale the size of images with watermarks using scaling factors among [0.6, 2.0].
Speckle noise	Using various concentrations of speckle noise, distort watermarked images.
Translation	Adjust the translation coordinates to read watermarked images in a different language.
Wiener filter	Put the watermarked picture through a Wiener filter.

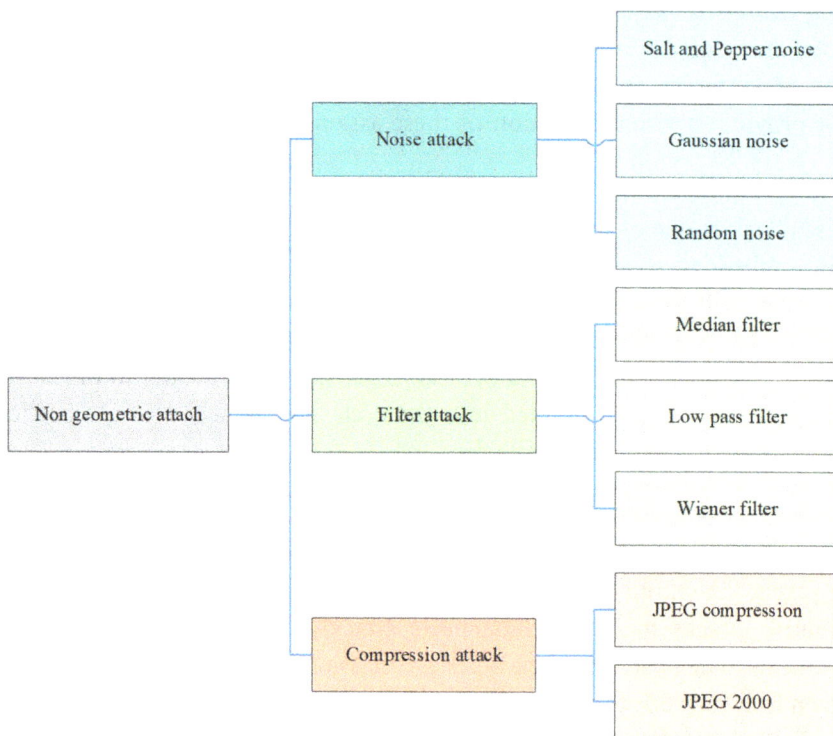

FIGURE 8.7   Non-geometric attack-based classification.

watermark is unaffected. The sharpness has a greater impact; the salt and pepper noise can be expressed as a kind of logical noise, which is characterized by the noise distribution like salt and pepper sprinkled on the surface of the image, which makes the image appear black with white spots and generally has less impact on the algorithm.

8.5.1.2 Geometric Attack

A geometric attack mainly refers to destroying the synchronization of watermark embedding and extraction. This type of attack achieves the effect of destroying the watermark by changing the original pixel position of the image. Geometric attacks have a greater impact on images than non-geometric attacks because they change the local or overall position of image pixels through operations such as occlusion, movement, and rotation. When extracting watermarks, if the original pixel position is still extracted, it will lead to the copyright information is incomplete or cannot be extracted. Common geometric attacks are shown in Figure 8.8.

It can be seen from Figure 8.8.(a) that the characteristic of the shearing attack is to make the pixel of the local site of the mover image to be 0, which belongs to the local attack operation; as shown in Figure 8.8.(b), the characteristic of the rotation attack is The overall pixel position of the image is changed, and the rotation process is accompanied by a shearing operation, so this attack is more destructive to the watermark; the row-column offset attack just moves the overall pixel of the image up and down, and the position is used after being moved. 0 to fill, as shown in Figure 8.8.(c); a scaling attack is to enlarge or reduce the image, as shown in Figure 8.8.(d).

8.5.1.3 Combination Attack

The transmission of digital data in complex networks is often subject to several categories of attacks, such as non geometric attacks + non-geometric attacks, geometric attacks + non-geometric attacks, and geometric attacks + geometric attacks. This attack type not only affects the pixel value of the carrier image, but also changes the image geometry, which is very destructive to the information.

8.5.1.4 Removal Attack

The digital image will have the added data removed from it as a result of this attack. If it is unable to, despite this, they attempt to delete the information that was embedded [127].

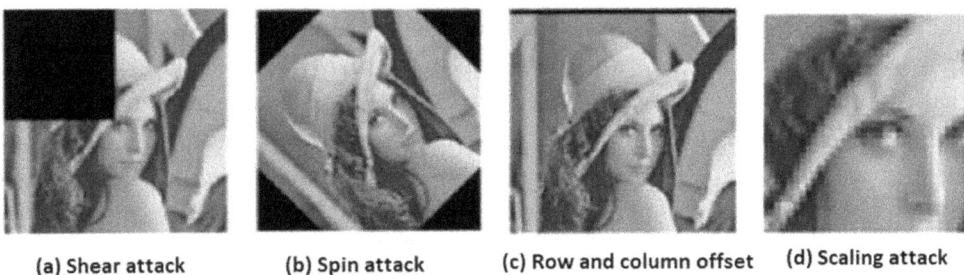

(a) Shear attack   (b) Spin attack   (c) Row and column offset   (d) Scaling attack

FIGURE 8.8 Classification of geometric attacks.

8.5.1.5 Protocol Attack

The attacks that are classified as falling into this sort do not do any damage to the embedded data. There are two distinct varieties of procedure attacks: invertible and copy attacks. Therefore, a watermark should not be able to be replicated and should not be invertible. When an attacker deletes his or her own watermark from the host facts, the watermark is considered to be invertible. After then, the attacker will act as though they are the owner of the information. This demonstrates that non-invertible watermarks are preferable for the purpose of copyright protection [128].

8.5.1.6 Copy Attack

It is also a type of assault known as a protocol attack. The watermark is preserved in this instance as well. The attacker will instead derive an estimate of the watermark from the host's data. After that, it is replicated to many different data [129].

8.5.1.7 Cryptographic Attack

Among these several kinds of assaults are those that circumvent the security provided by watermarking systems. Using this, they are able to retrieve the data that was entered as a watermark or create their own deceptive watermark. Both the Brute-force and Oracle attacks are examples that can be found in this category [130].

8.5.2 No Attack

The purpose of the representation attack is to make the infringement detector fail to detect the presence of the watermark. Usually, a smaller watermark image is used to avoid detection by the detector in order to "evade" detection. The representation attack exploits this feature to segment the watermarked image into fine fragments, making it impossible for the detector to detect the watermark, making it difficult to determine the ownership of the information. The original information can be recovered during copyright authentication by splicing the watermark information of each sub-block.

8.5.3 Explaining the Attack

The interpretation attack is to use the inverse algorithm of the watermark to embed the desired watermark into the original image and ensure the transparency of the original image. The forged watermark information is very similar to the watermark information obtained by the original watermark algorithm, so it is difficult to identify the ownership of the image. But explaining the attack requires knowing the details of the original watermarking algorithm and how it is embedded in order to forge fake watermarks and virtual original images. The way to prevent this attack is to construct an irreversible watermarking algorithm or an irreversible hashing process. At the same time, a trustworthy third-party verification agency for watermarks can be set up to protect the ownership of watermark information and make digital watermark technology safer. However, the comparative analysis is presented in Table 8.5, which provides a summary of the watermarking systems that have been anticipated by various explore groups over the past few years.

TABLE 8.5 An In-depth Overview of Existing Image Watermarking Techniques

Types	Used	Inputs	Visuality	Robustness
Image watermarking [131]	SIRD estimates the best image block region.	Size: 512 × 512 × 3 pixels, and Size: 64 × 64 × 64 pixels	PSNR = 47.6 dB, SSIM = 0.9904	NCC = Range [0.9917 – 1], BER = Range [0.7500-0]
Robust image watermarking [132]	Watermark restoration uses simplified precise kernel instants of the polar multipart exponential change.	Size: 256 × 256 × 3 pixels, and Size: 32 × 32 × 32 pixels	PSNR = Range [40.597–53.64] dB, SSIM = Range [0.933–0.980]	NCC = Range [0.9100 – 1.0], BER = Range [0–0.0156], Rotation at several angles, scaling, translation, and grading + turning.
Semi-blind watermarking [133]	DWT-CT-Schur-SVD	Size: 64 × 64 × 64 pixels	SSIM range = [0.9709–0.9989], PSNR range =[27.63–36.16] dB	Salt and Pepper, Gaussian Noise, and NCC = 1.0
Blind image watermarking [134]	Blocks undergo two-dimensional DCT. Embedded middle-frequency coefficients.	Size: 512 × 512 × 3 pixels, and Size: 32 × 32 × 32 pixels	PSNR = range [36.3189–38.2472], SSIM = range [0.9149–0.9441]	NCC = Range [0.9997–1] JPEG (40), JPEG 2000, Gaussian white noise, salt and pepper noise, Butterworth lowpass, median filtering, and cropping.
Digital image watermarking [135]	Double encryption for embedding uses fractal encoding and DCT.	Size: 1024 × 1024 pixels, and Size: 256 × 256 pixels	PSNR Range = [41–45] dB	White noise, Gaussian filter, and JPEG compression attacks considered.
Robust digital image watermarking [136]	DCT + SVD Hybrid	Size: 512 × 512 pixels, and Size: 50 × 20 pixels	PSNR = Range [51.68–64.54] dB, SSIM = Range [0.9989–1.0000]	Different types of noise, such as Gaussian, Speckle, salt and pepper, and Poisson, as well as rotation and JPEG compression, were taken into account.

Past research and analysis of digital watermarking techniques are listed in Table 8.5. Some of the most well-known methods previously investigated include those in the 3-D domain and the frequency domain. The spatial domain digital watermarking method has also been found to be less secure, which is why it is rarely used. The effectiveness of a watermarked image is measured in terms of its durability, invisibility, security, and storage capacity. The most widely favored criteria were the watermark's invisibility to the naked eye and its durability. In reality, future work has potential by joining techniques and employing them in hybrid form, which not only strengthens the watermarked image but also potentially lessens the negatives of each method taken individually.

8.6 LEARNING-BASED WATERMARKING

In recent years, watermarking has benefited greatly from the application of deep-learning models like CNNs, GAN, and deep neural networks (DNNs). Figure 8.9 illustrates how and where deep learning can be applied. The initial steps of watermark creation, preprocessing, and encoding all make use of deep-learning models. In the long run, it can be put to work extracting watermarks and simulating attacks. This section provides a comprehensive overview of current literature centered on GAN deep-learning model. The learning nature of the GAN network is unsupervised. It does not need to label data or simply generate labels, which makes the application scenarios of the GAN network more extensive.

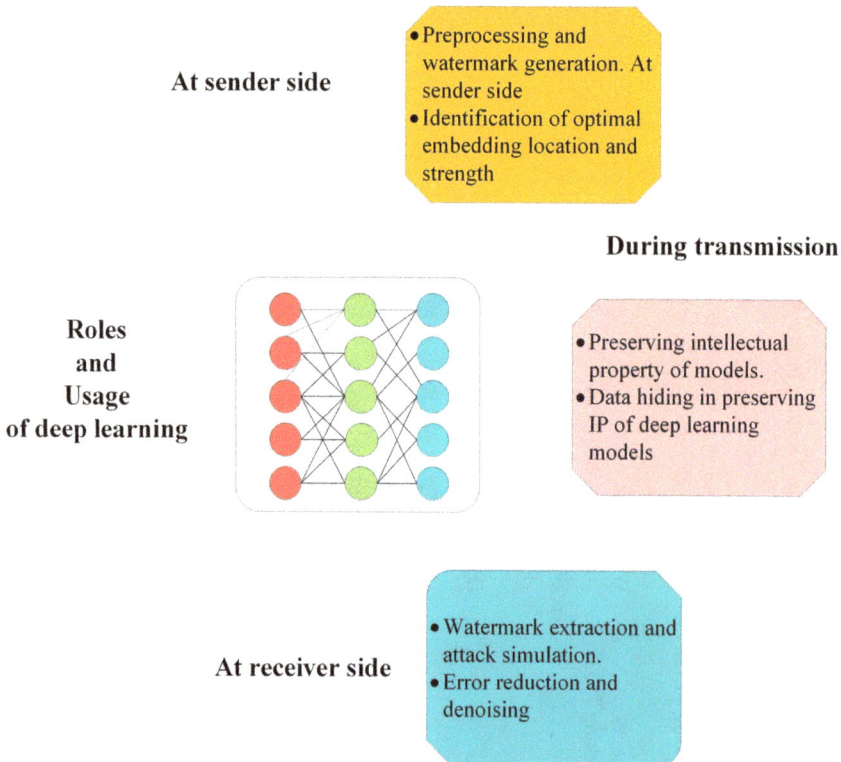

At sender side
- Preprocessing and watermark generation. At sender side
- Identification of optimal embedding location and strength

During transmission

Roles and Usage of deep learning

- Preserving intellectual property of models.
- Data hiding in preserving IP of deep learning models

At receiver side
- Watermark extraction and attack simulation.
- Error reduction and denoising

FIGURE 8.9 Deep learning's various roles and applications throughout the watermarking process are discussed.

The GAN network adopts the alternate training strategy, and the parameters of D will not be updated when training G. First, obtain random noise data, and then enter the training stage of generator network G: input the noise data into the network to get the noise data. Then input the noise data into the D to get the output data, and set the corresponding labels as $label_G = 1$ and $label_R = 0$ and then calculate the loss. Next, enter the training phase of discriminator network D: input the noise data into the network to get the noise data, cut off the gradient of the variable, and then input the noise data into D to get the output data, and set the corresponding labels as $label_G = 0$ and $label_R = 1$ and calculate the loss, and update the parameters of discriminator network D through back propagation. Because of the previous gradient cut-off operation, the back propagation will not update the parameters of G. The loss function is as follows:

$$\min_{G} \max_{D} V(D, G) = E_{x \sim P \, data(x)} [\log(D(X))] + E_{x \sim p(z)} [\log(1 - (D(G(z))))] \quad (8.15)$$

The first item denotes the decision of the discriminator on the real information, and the second item represents the decision of the discriminator on the generated information. The G and D network are alternately trained until the Nash equilibrium point is reached. To check if images have been tampered with by dense watermarks, [137] suggested a deep-learning model. The model includes a generator and discriminator to boost the authenticity and quality of the recovered pictures through improved verification. The generator is an autoencoder that converts a representation vector of a heavily watermarked, corrupted picture into a standard red, green, and blue (RGB) image. The discriminator regulates the generated images' substance and reduces the loss of features. Features and ground truth pictures for the recovered images are extracted using the ResNet 46 model. At a false positive rate (FPR) of 1%, it is able to verify data with a high degree of confidence (96.36%). Literature [138] proposes a GAN-based data hiding scheme for protecting authentic papers. As a first step, the paper is brought into conformity with the necessary form via geometric correction. The adversarial network then generates the document, fortifying it with a watermark containing the confidential information embedded within a pseudo-random number. The method is secure and can spot tampering, but it needs to be explored for other forms of image processing assaults. Some works based on the GAN model are listed below and are summarized in Table 8.6.

8.7 APPLICATIONS OF LEARNING-BASED WATERMARKING

Extensive study on watermarking technology has led to its widespread adoption in other specialized digital image fields such as medical images, remote sensing images, digital maps, and so on. Due to the unique structure of these digital pictures, strict adherence to the original information's transparency and authenticity is essential.

8.7.1 Medical Field

Radiological pictures can be broken down further into subcategories like "B-ultrasound pictures," "X-ray pictures," "CT scan pictures," etc. As a result of the prevalence of disease

TABLE 8.6 Summary of GAN-Based Watermarking

Ref No.	Objective	Goal	Strategies	Role of GAN	Embedding location	Results	Cover/mark size	Noticed weakness	Applications oriented
[137]	Remove dense watermarking	Recover grayscale image	MSE loss	Denoising the image	—	PSNR = 23.37 dB TPR@FPR=1%= 96.36%	—	PSNR is not up to the mark	Grayscale images
[138]	Robust data hiding scheme	High robustness	Image denoising	Improve robustness	Cover image	PSNR = 35.82 dB SSIM = 0.9988	$512 \times 512 \times 512$	Need a detailed analysis of robustness	DSSE dataset
[139]	Robust image watermarking using C-GAN	Copyright and ownership protection	Variational autoencoder	Embedding and extraction of watermark	Cover image	PSNR= 34.97 db SSIM = 0.979	—	Limited capacity	Color image
[140]	Blind watermarking scheme	Robustness and high visual quality	MSE loss	Embedding and extraction	Cover image	PSNR = 41.02 dB in V channel	—	High complexity	General image
[141]	Multiscale robust watermarking	High security	—	Watermarking and image denoising	Cover image	Accuracy = 0.999	—	Tested on limited attacks	DWI images
[142]	Watermarking for distortion-free real images	High watermarked image quality	JND-mask-based loss	Image enhancement	Cover image	PSNR - 36.25 dB Accuracy = 84.9%	—	High complexity	Medical imaging
[143]	Watermarking for media authentication	Identify deep fakes images	Encoder-decoder network	Extraction and classification	Encoded cover image	PSNR = 36.38 dB	$256 \times 256 \times 256$	Limited capacity	General color images

knowledge in digital data, it is increasingly used as a foundation for medical diagnosis. Malicious alteration of patient records can cloud a doctor's decision making. Therefore, ensuring the integrity and security of medical data is a task that the field of medical images has always faced. To prevent malicious modification of medical data, it is often required to have valid authentication. For example, the literature [144,145] proposes a robust non-blind medical image watermarking arrangement that performs an I-level fractional wavelet transform (fractional wave, package, transform, or FR-WPT) on the image and embeds the watermark into the human body. In the modified reference images, it is originated that the test results of this method on mammograms have good robustness.

8.7.2 Remote-sensing Field

Remote-sensing imagery is typically put to use in real-time tracking and statistical surveys. The confidentiality and security of remote-sensing maps must take into account the transparency and integrity of the data they contain. Using the DCT transform and the principle of orthogonal decomposition coefficient invariance, Jiang et al. [146] suggested an encryption-based watermarking technology applied to remote sensing images. There is a high degree of robustness in the algorithm, making it resistant to attacks that are not geometric in nature. Li et al. [147] projected a non-blind watermarking algorithm based on quaternion wavelet transform (QWT) and tensor decomposition. The QWT method can better reserve image features and is useful to color remote-sensing images. Better transparency can be obtained. Tong et al. [148] proposed an improved compressed sensing watermarking algorithm applied to remote sensing images, using a lifting wavelet transform, Hadamard matrix, and ternary watermarking sequence to improve the robustness of the algorithm, and achieved good results.

8.7.3 Map Copyright

As technology and society advance, the range of uses for maps expands to accommodate them. You can think of Baidu Maps' and Google Maps' flat maps as two-dimensional vector maps. Copyright protection of digital pictures like maps primarily takes into account the openness and precision of map information, as these images often contain some information like coordinates, locations, and directions. safety. Two-dimensional vector graph reversible watermarking using reversible contrast mapping was suggested by Fei et al. [149]. To insert an encrypted watermark into an image, it first chooses the vertices' coordinates and the location where the watermark can be embedded based on the data accuracy requirements, and then uses reversible contrast to do so. In the chosen relative coordinate system, the mapping change is implemented. After the watermark is embedded in the map coordinate data, it is still compatible with the original data.

8.7.4 Copyright Protection

In this open network era, people can more and more easily obtain digital works illegally in cyberspace, which seriously damages the legitimate rights and interests of copyright owners. hot issues to be resolved. When there is a conflict of interest, the copyright owner can use the extracted watermark data to prove his identity. This protects the copyright owner's rights and interests [150–152].

8.7.5 Content Authentication

Fragile watermarking is an ingenious method for digital work content authentication. When embedding a watermark into a digital work, even if one bit of the original carrier data is missing, the watermark data in it cannot be completely extracted. Therefore, the use of this feature of fragile watermarks can be used to test whether digital works have been illegally tampered with in the process of dissemination. Only when the watermark information can be completely extracted can the integrity of digital content effectively be verified.

8.7.6 Infringement Tracking

This technology is mostly used to identify the identity of illegal disseminators. Each time a digital work is copied, the identity of the legal licensor is embedded, and the identity will locate the only legal licensor. Find the illegal leaker of the work [153].

8.7.7 Radio Monitoring

Radio and television, like daily necessities, play an important role in our daily life. In order to ensure the accuracy of each broadcast, the digital watermark is hidden in the video clip according to the designed algorithm. Before the program is broadcast, the watermark information is extracted and the video clip is recovered at the same time, and the watermark info is used to verify the accuracy of the video. Avoid interfering with normal program broadcast work [154].

8.7.8 Copy Control

Digital watermarking technology is also often used in the control of copying operations. Before the digital work begins to circulate, it is embedded with a watermark. The watermark information includes the allowable copying and playing times of the digital work. Before the work is played or copied, the information from the watermark is taken out to see if the remaining playback times meet the copy conditions. This is done to control copying [155].

8.7.9 Electronic Field

Internet access is now available in almost every part of the country, from the largest cities to the smallest villages. The conduct of elections is facilitated by electronic voting, which also takes into account concerns regarding security [156].

However, there is not a singular set of properties that can be guaranteed to be satisfied by every watermarking system. There are currently certain methods available that concurrently fulfill a number of the conditions that were previously described and used for certain applications. Table 8.7 summarizes the various methods used for some important daily life applications.

TABLE 8.7 Techniques Used for Some Important Daily Life Applications

Allocations	References	Techniques	Features	Results
Protection of microscopy images	Pizzolante et al. [157]	3D domain method	Robustness and computational complication	Assurances of security and robustness
Authenticity and integrity of copyright protection	Zhang et al. [158]	APBT-based procedure and SVD	Noiselessness and robustness	Better noiselessness and good robustness
Healthcare	Zear et al. [159]	DWT, DCT, and SVD	Robustness, noiselessness, size, and security	Suitable visual excellence for diagnosis
Copyright protection	Zhou et al. [160]	DWT, APDCBT, and SVD	Noiselessness and robustness	Achieves better than current methods
Protection of digital contents	Singh et al. [161]	BPNN	Robustness, security, and capacity	Better robustness and refuge
Image authentication	Takore et al. [162]	Hybrid transform domain and Particle Swarm Optimization (PSO) algorithm	Inaudibility and robustness	Achieves better than standing approaches
DICOM images	Phadikar et al. [163]	Lifting and compounding	Imperceptibility, size, and security	Achieves better than existing reversible watermarking methods

8.8 CONCLUSION

In recent years, the copyright protection of digital data has garnered an increasing amount of attention, particularly in the field of digital images. One of the areas that people focus on is the copyright protection and security of image info, which is one of the fields in which digital images are used. This study primarily focuses on the watermarking algorithms and the current research status of digital images. We look closely at the deep-learning model, especially GAN, that has become so widely adopted in watermarking, and provide a complete overview based on that model. The related technologies of wavelet transform are contrasted and examined, the model formulas of various watermarking technologies are summarized, and the theory of zero watermarking, as well as its merits and limitations, are specifically discussed. This not only serves as an essential position for subsequent explore but also contributes, in some small way, to the advancement of technologies that can safeguard digital images from unauthorized copying. Based on an analysis of the research that has already been done, a summary of the problems and challenges in this field is made.

1. Watermarking algorithm based on color copyright images. Most of the copyright images used by the image watermarking algorithm at this stage are binary copyright marks. Although it is beneficial to the fidelity of the image, with the rapid development of image information and the update of photographing equipment, customers have more and more demand for diverse images and the use of color images, so the use of color logos as copyright watermarks is a serious challenge to image fidelity.

2. Performance of the GAN model arises because of inconsistencies in the network that produces and evaluates data. Because of instability and imbalance in the network parameters, convergence is impossible.

3. More robust neural network modeling for watermarking. Current watermarking methods, especially black-box methods, are also limited by attacks such as fine-tuning of models. In the future, model watermarking should make this part of the model more secure so that it can withstand traditional attacks, especially those that are easier to use.

4. The watermark should reduce the influence on the original model. The current watermark has a certain influence on the original model, such as reducing the accuracy of the original model task. Although some methods can achieve an insignificant reduction in accuracy on a certain test set, it will inevitably affect the original task of the model. Due to the inexplicability of the model, it is difficult to describe the influence of the theory on the model. Compared with the traditional book watermark, it will not reduce the reading effect of the book. Therefore, figuring out how to make the model watermark complete the watermarking task without reducing the original task is one of the important directions in the future.

5. Public watermark. For traditional book watermarks, the watermarks are public and easily judged and identified by users. However, the robustness of the existing neural network model for watermarking depends on the concealment of the watermarking. Once published, it can be easily removed. Future research will give a lot of support for the verifiability of the watermark if it can find a way to make a watermark that can be seen (and not hidden).

6. Theoretical proof of watermarking. The current watermarking methods lack theoretical support. Most of them add tasks and features to the original model, but cannot theoretically prove their various security features. If future research can make progress in theory, it will promote the security guarantee capability of various characteristics, such as the robustness of watermarking technology.

7. Various watermarks. The current watermarking methods are relatively simple, and most of them are added by directly adding features or model backdoors. If we can try out more kinds of watermarks and see how they work together, it will make future neural network watermarking algorithms more secure and better able to protect against attacks.

8. Multi-domain combined watermarking algorithm. One of the important indicators of digital watermarking is to ensure the security of copyright information. However, most of the existing algorithms focus on robustness and transparency, and there are few research studies on the security of watermarking. Therefore, the combination of digital watermarking technology and encryption technology in the future can make copyright protection more easily applied to complex network environments.

9. Efficient watermarking algorithm. High transparency and strong robustness have always been two important indicators of watermarking algorithms. However, as the size of the image continues to increase, it takes more time to embed and extract the watermark.

10. Perform matrix operations. The introduction of multi-core parallel processors can effectively solve the efficiency problem of large matrix operations. Therefore, using the parallel computing characteristics of GPUs in digital image watermarking technology can improve the efficiency of embedding and extracting watermarks.

In short, more powerful attacks are always accompanied by advances in defense methods, and they complement each other and grow together. Therefore, in the future research process, the two aspects of research will be developed simultaneously. The theoretical improvement will promote the progress of the neural network model.

FUNDING

This work was supported in part by Key Research Project of Hainan Province under Grant ZDYF2021SHFZ093, the Natural Science Foundation of China under Grants

62063004 and 62162022, the Hainan Provincial Natural Science Foundation of China under Grants 2019RC018, 521QN206, and 619QN249, the Major Scientific Project of Zhejiang Lab under Grant 2020ND8AD01, and the Scientific Research Foundation for Hainan University under Grant KYQD(ZR)-21013.

REFERENCES

1. Panah, A. S., Van Schyndel, R., Sellis, T., & Bertino, E. (2016). On the properties of non-media digital watermarking: A review of state of the art techniques. *IEEE Access*, 4, 2670–2704.
2. Chang, J., & Liu, D. (2015, August). The Application of Chaos in Digital Water-marking and Information Hiding. In *International Conference on Materials Engineering and Information Technology Applications (MEITA 2015)* (pp. 899–902). Atlantis Press.
3. Belferdi, W., Behloul, A., & Noui, L. (2019). A Bayer pattern-based fragile watermarking scheme for color image tamper detection and restoration. *Multidimensional Systems and Signal Processing*, 30(3), 1093–1112.
4. Nawaz, S. A., Li, J., Bhatti, U. A., Mehmood, A., Shoukat, M. U., & Bhatti, M. A. (2020). Advance hybrid medical watermarking algorithm using speeded up robust features and discrete cosine transform. *PLOS One*, 15(6), e0232902.
5. Bhatti, U. A., Yuan, L., Yu, Z., Li, J., Nawaz, S. A., Mehmood, A., & Zhang, K. (2021). New watermarking algorithm utilizing quaternion Fourier transform with advanced scrambling and secure encryption. *Multimedia Tools and Applications*, 80(9), 13367–13387.
6. Nawaz, S. A., Li, J., Bhatti, U. A., Li, H., Han, B., Mehmood, A., … & Zhou, J. (2020, July). A Robust Color Image Zero-Watermarking Based on SURF and DCT Features. In *International Conference on Artificial Intelligence and Security* (pp. 650–660). Springer, Singapore.
7. Ganapathy, A. (2019). Image association to URLs across CMS websites with unique watermark signatures to identify who owns the camera. *American Journal of Trade and Policy*, 6(3), 101–106.
8. Sheng, D., Xiao, M., Liu, A., Zou, X., An, B., & Zhang, S. (2020, August). CPchain: A copyright-preserving crowdsourcing data trading framework based on blockchain. In *2020 29th International Conference on Computer Communications and Networks (ICCCN)* (pp. 1–9). IEEE.
9. Wang, B., Jiawei, S., Wang, W., & Zhao, P. (2022). Image copyright protection based on blockchain and zero-watermark. *IEEE Transactions on Network Science and Engineering*, 9(4), 2188–2199.
10. Fang, P., Liu, H., & Wu, C. (2020). A novel chaotic block image encryption algorithm based on deep convolutional generative adversarial networks. *IEEE Access*, 9, 18497–18517.
11. Sharma, K., Aggarwal, A., Singhania, T., Gupta, D., & Khanna, A. (2019). Hiding data in images using cryptography and deep neural network. arXiv preprint arXiv:1912.10413.
12. Li, D., Deng, L., Gupta, B. B., Wang, H., & Choi, C. (2019). A novel CNN based security guaranteed image watermarking generation scenario for smart city applications. *Information Sciences*, 479, 432–447.
13. Cox, I. J., Miller, M. L., Linnartz, J. P. M., & Kalker, T. (2018). A review of watermarking principles and Practices 1. *Digital Signal Processing for Multimedia Systems*, 461–485.
14. Alromih, A., Al-Rodhaan, M., & Tian, Y. (2018). A randomized watermarking technique for detecting malicious data injection attacks in heterogeneous wireless sensor networks for internet of things applications. *Sensors*, 18(12), 4346.
15. Darwish, S. M., & Al-Khafaji, L. D. S. (2020). Dual watermarking for color images: a new image copyright protection model based on the fusion of successive and segmented watermarking. *Multimedia Tools and Applications*, 79(9), 6503–6530.

16. Xiang, Y., Natgunanathan, I., Rong, Y., & Guo, S. (2015). Spread spectrum-based high embedding capacity watermarking method for audio signals. *IEEE/ACM transactions on audio, speech, and language processing*, 23(12), 2228–2237.
17. Zhang, P., Xu, S., & Yang, H. (2012, March). Robust and transparent audio watermarking based on improved spread spectrum and psychoacoustic masking. In 2012 IEEE International Conference on Information Science and Technology (pp. 640-643). IEEE.
18. Soderi, S., Mucchi, L., Hämäläinen, M., Piva, A., & Iinatti, J. (2017). Physical layer security based on spread-spectrum watermarking and jamming receiver. *Transactions on Emerging Telecommunications Technologies*, 28(7), e3142.
19. Hosseini, S., & Mahdavi, M. (2021). Image content dependent semi-fragile watermarking with localized tamper detection. arXiv preprint arXiv:2106.14150.
20. Das, T. K. (2018). Anti-forensics of JPEG compression detection schemes using approximation of DCT coefficients. *Multimedia Tools and Applications*, 77(24), 31835–31854.
21. Alrehily, A., & Thayananthan, V. (2017). Software watermarking based on return-oriented programming for computer security. *International Journal of Computer Applications*, 166(8), 21–28.
22. Hou, J. U., Kim, D., Ahn, W. H., & Lee, H. K. (2018). Copyright protections of digital content in the age of 3d printer: Emerging issues and survey. *IEEE Access*, 6, 44082–44093.
23. Ma, H., Lu, K., Ma, X., Zhang, H., Jia, C., & Gao, D. (2015, April). Software watermarking using return-oriented programming. In *Proceedings of the 10th ACM Symposium on Information, Computer and Communications Security* (pp. 369–380). ACM.
24. Kumar, K., & Kaur, P. (2019). A Comparative analysis of static and dynamic Java Bytecode watermarking algorithms. In *Software Engineering* (pp. 319–334). Springer, Singapore.
25. Begum, M., & Uddin, M. S. (2022). Towards the development of an effective image watermarking system. *Security and Privacy*, 5(2), e196.
26. Zeng, C., Liu, J., Li, J., Cheng, J., Zhou, J., Nawaz, S. A., … & Bhatti, U. A. (2022). Multiwatermarking algorithm for medical image based on KAZE-DCT. *Journal of Ambient Intelligence and Humanized Computing*, 1–9.
27. Mustafa, D. T. (2022). *Incorporating a watermarking with an iris code to ensure the copyright protection of the image* (Master's thesis, Altınbaş Üniversitesi/Lisansüstü Eğitim Enstitüsü).
28. Khan, J. S., Kayhan, S. K., Ahmed, S. S., Ahmad, J., Siddiqa, H. A., Ahmed, F., … & Al Dubai, A. (2022). Dynamic S-box and PWLCM-based robust watermarking scheme. *Wireless Personal Communications*, 1–18.
29. Zhou, J., Li, J., Li, H., Liu, J., Liu, J., Dai, Q., & Nawaz, S. A. (2019, December). Multiwatermarking algorithm for medical image based on NSCT-RDWT-DCT. In *International Symposium on Cyberspace Safety and Security* (pp. 501–515). Springer, Cham.
30. Pourhadi, A., & Mahdavi-Nasab, H. (2020). A robust digital image watermarking scheme based on bat algorithm optimization and SURF detector in SWT domain. *Multimedia Tools and Applications*, 79(29), 21653–21677.
31. Roy, R., Ahmed, T., & Changder, S. (2018). Watermarking through image geometry change tracking. *Visual Informatics*, 2(2), 125–135.
32. Nawaz, S. A., Li, J., Bhatti, U. A., Li, H., Han, B., & Mehmood, A. (2020, February). Improved watermarking algorithm based on SURF and SVD with wavelet transformation against geometric attacks. In *Proceedings of the 2020 3rd International Conference on Image and Graphics Processing* (pp. 49–56).
33. Shi, J., Yi, D., & Kuang, J. (2019, December). A blockchain and SIFT based system for image copyright protection. In *Proceedings of the 2019 2nd International Conference on Blockchain Technology and Applications* (pp. 1–6).

34. Darwish, S. M., & Hassan, O. F. (2021). A new colour image copyright protection approach using evolution-based dual watermarking. *Journal of Experimental & Theoretical Artificial Intelligence*, 33(6), 945–967.

35. Chen, G. (2022). Discussion on the application and supervision strategy of digital copyright protection from the perspective of blockchain. *The Frontiers of Society, Science and Technology*, 4(2).

36. Singh, D., & Singh, S. K. (2017). DWT-SVD and DCT based robust and blind watermarking scheme for copyright protection. *Multimedia Tools and Applications*, 76(11), 13001–13024.

37. Yun-Peng, Z., Wei, L., Shui-Ping, C., Zheng-Jun, Z., Xuan, N., & Wei-di, D. (2009, October). Digital image encryption algorithm based on chaos and improved DES. In *2009 IEEE International Conference on Systems, Man and Cybernetics* (pp. 474–479). IEEE.

38. Nadiya, P. V., & Imran, B. M. (2013, February). Image steganography in DWT domain using double-stegging with RSA encryption. In *2013 International Conference on Signal Processing, Image Processing & Pattern Recognition* (pp. 283–287). IEEE.

39. Norouzi, B., Seyedzadeh, S. M., Mirzakuchaki, S., & Mosavi, M. R. (2014). A novel image encryption based on hash function with only two-round diffusion process. *Multimedia Systems*, 20(1), 45–64.

40. Bhattacharya, S., Chattopadhyay, T., & Pal, A. (2006, June). A survey on different video watermarking techniques and comparative analysis with reference to H. 264/AVC. In *2006 IEEE International Symposium on Consumer Electronics* (pp. 1–6). IEEE.

41. Shojanazeri, H., Adnan, W. A. W., Ahmad, S. M. S., & Saripan, M. I. (2011, December). Analysis of watermarking techniques in video. In *2011 11th International Conference on Hybrid Intelligent Systems (HIS)* (pp. 486–492). IEEE.

42. Sabri, A. Q. M., Mansoor, A. M., Obaidellah, U. H., & Faizal, E. R. M. (2017). Metadata hiding for UAV video based on digital watermarking in DWT transform. *Multimedia Tools and Applications*, 76(15), 16239–16261.

43. Mahmood, A. M., Jawad, M. J., & Naser, M. A. (2018). Copyright protection and content integrity for digital video based on the watermarking techniques. *International Journal of Pure & Applied Mathematics*, 119(15), 487–504.

44. Hassan, N. F., Ali, A. E., Aldeen, T. W., & Al-Adhami, A. (2021). Video mosaic watermarking using plasma key. *Indonesian Journal of Electrical Engineering and Computer Science*, 22(2), 11–20.

45. Meng, F., Peng, H., Pei, Z., & Wang, J. (2008, December). A novel blind image watermarking scheme based on support vector machine in DCT domain. In *2008 International Conference on Computational Intelligence and Security* (Vol. 2, pp. 16–20). IEEE.

46. Peng, H., Wang, J., & Wang, W. (2010). Image watermarking method in multiwavelet domain based on support vector machines. *Journal of Systems and Software*, 83(8), 1470–1477.

47. Ramly, S., Aljunid, S. A., & Shaker Hussain, H. (2011, July). SVM-SS watermarking model for medical images. In *International Conference on Digital Enterprise and Information Systems* (pp. 372–386). Springer, Berlin, Heidelberg.

48. Yang, H. Y., Wang, X. Y., Zhang, Y., & Miao, E. N. (2013). Robust digital watermarking in PDTDFB domain based on least squares support vector machine. *Engineering Applications of Artificial Intelligence*, 26(9), 2058–2072.

49. Verma, V. S., Jha, R. K., & Ojha, A. (2015). Digital watermark extraction using support vector machine with principal component analysis based feature reduction. *Journal of Visual Communication and Image Representation*, 31, 75–85.

50. Rai, A., & Singh, H. V. (2017). SVM based robust watermarking for enhanced medical image security. *Multimedia Tools and Applications*, 76(18), 18605–18618.

51. Vairaprakash, S., & Shenbagavalli, A. (2018). A discrete Rajan transform-based robustness improvement encrypted watermark scheme backed by Support Vector Machine. *Computers & Electrical Engineering*, 70, 826–843.

52. Rai, A., & Singh, H. V. (2018). Machine learning-based robust watermarking technique for medical image transmitted over LTE network. *Journal of Intelligent Systems*, 27(1), 105–114.

53. Alarood, A. A. (2022). Improve the efficiency for embedding in LSB method based digital image watermarking. *Journal of Theoretical and Applied Information Technology*, 100(15), 1–15.

54. Wazirali, R., Ahmad, R., Al-Amayreh, A., Al-Madi, M., & Khalifeh, A. (2021). Secure watermarking schemes and their approaches in the IoT technology: An overview. *Electronics*, 10(14), 1744.

55. Shan, C., Zhou, S., Zhang, Z., & Li, M. (2022). Digital watermarking method for image feature point extraction and analysis. *International Journal of Intelligent Systems*, 37(10), 7281–7299.

56. Jhade, A., & Singh, K. N. (2022). An efficient digital image watermarking based on DCT and advanced image data embedding method. *International Journal of Advanced Computer Technology*, 11(1), 11–16.

57. Neff, T., Payer, C., Štern, D., & Urschler, M. (2018, May). Generative adversarial networks to synthetically augment data for deep learning based image segmentation. In *Proceedings of the OAGM Workshop* (pp. 22–29).

58. Usama, M., & Yaman, U. (2022). Embedding information into or onto additively manufactured Parts: A review of QR codes, steganography and watermarking methods. *Materials*, 15(7), 2596.

59. Zade, A. R., Konde, S., Patil, P., Navasare, P., & Dhanani, R. (2017). A survey on various techniques used to add watermark to multimedia data for digital copyrights protection. *International Research Journal of Engineering and Technology*, 4(3), 1635–1640.

60. Garg, P., & Kishore, R. R. (2020). Performance comparison of various watermarking techniques. *Multimedia Tools and Applications*, 79(35), 25921–25967.

61. Nguyen, T. H., Duong, D. M., & Duong, D. A. (2015, January). Robust and high capacity watermarking for image based on DWT-SVD. In *The 2015 IEEE RIVF International Conference on Computing & Communication Technologies-Research, Innovation, and Vision for Future (RIVF)* (pp. 83–88). IEEE.

62. Zheng, W., Mo, S., Jin, X., Qu, Y., Deng, F., Shuai, J., … & Long, S. (2018, May). Robust and high capacity watermarking for image based on DWT-SVD and CNN. In *2018 13th IEEE Conference on Industrial Electronics and Applications (ICIEA)* (pp. 1233–1237). IEEE.

63. Begum, M., Ferdush, J., & Uddin, M. S. (2021). A Hybrid robust watermarking system based on discrete cosine transform, discrete wavelet transform, and singular value decomposition. *Journal of King Saud University-Computer and Information Sciences*, 34(8), 5856–5867.

64. Wei, X., Zhang, W., Yang, B., Wang, J., & Xia, Z. (2021). Fragile watermark in medical image based on prime number distribution theory. *Journal of Digital Imaging*, 34(6), 1447–1462.

65. Kim, C., & Yang, C. N. (2021). Self-embedding fragile watermarking scheme to detect image tampering using AMBTC and OPAP approaches. *Applied Sciences*, 11(3), 1146.

66. Su, G. D., Chang, C. C., & Chen, C. C. (2021). A hybrid-Sudoku based fragile watermarking scheme for image tampering detection. *Multimedia Tools and Applications*, 80(8), 12881–12903.

67. Kumar, S., Panna, B., & Jha, R. K. (2019). Medical image encryption using fractional discrete cosine transform with chaotic function. *Medical & Biological Engineering & Computing*, 57(11), 2517–2533.

68. Azeroual, A., & Afdel, K. (2017). Real-time image tamper localization based on fragile watermarking and Faber-Schauder wavelet. *AEU-International Journal of Electronics and Communications*, 79, 207–218.

69. Shen, J. J., Lee, C. F., Hsu, F. W., & Agrawal, S. (2020). A self-embedding fragile image authentication based on singular value decomposition. *Multimedia Tools and Applications*, 79(35), 25969–25988.

70. Wang, N., & Men, C. (2013). Reversible fragile watermarking for locating tampered blocks in 2D vector maps. *Multimedia Tools and Applications*, 67(3), 709–739.

71. Yadav, B., Kumar, A., & Kumar, Y. (2018). A robust digital image watermarking algorithm using DWT and SVD. In *Soft Computing: Theories and Applications* (pp. 25–36). Springer, Singapore.

72. Thanki, R., Kothari, A., & Trivedi, D. (2019). Hybrid and blind watermarking scheme in DCuT–RDWT domain. *Journal of Information Security and Applications*, 46, 231–249.

73. Rawat, S., & Raman, B. (2012). A blind watermarking algorithm based on fractional Fourier transform and visual cryptography. *Signal Processing*, 92(6), 1480–1491.

74. Feng, L. P., Zheng, L. B., & Cao, P. (2010, July). A DWT-DCT based blind watermarking algorithm for copyright protection. In *2010 3rd International Conference on Computer Science and Information Technology* (Vol. 7, pp. 455–458). IEEE.

75. Zeng, F., Bai, H., & Xiao, K. (2022). Blind watermarking algorithm combining NSCT, DWT, SVD, and HVS. *Security and Privacy*, 5(4), e223.

76. Tian, C., Wen, R. H., Zou, W. P., & Gong, L. H. (2020). Robust and blind watermarking algorithm based on DCT and SVD in the contourlet domain. *Multimedia Tools and Applications*, 79(11), 7515–7541.

77. Gao, M., & Sun, B. (2011). Blind watermark algorithm based on QR barcode. In *Foundations of Intelligent Systems* (pp. 457–462). Springer, Berlin, Heidelberg.

78. Gong, X., & Li, W. (2021). A color image blind digital watermarking algorithm based on QR code. *Computer Science & Information Technology (CS & IT)*, 67–75.

79. Wang, C., Zhang, Y., & Zhou, X. (2018). Robust image watermarking algorithm based on ASIFT against geometric attacks. *Applied Sciences*, 8(3), 410.

80. He, W., Liu, J., Duan, J., & Wang, J. (2012, October). A novel watermarking schem using directionlet. In *2012 5th International Congress on Image and Signal Processing* (pp. 557–561). IEEE.

81. Ye, X., Chen, X., Deng, M., & Wang, Y. (2014, October). A SIFT-based DWT-SVD blind watermark method against geometrical attacks. In *2014 7th International Congress on Image and Signal Processing* (pp. 323–329). IEEE.

82. Zhu, T., Qu, W., & Cao, W. (2022). An optimized image watermarking algorithm based on SVD and IWT. *The Journal of Supercomputing*, 78(1), 222–237.

83. Fan, J., & Wu, Y. (2011, September). Watermarking algorithm based on kernel fuzzy clustering and singular value decomposition in the complex wavelet transform domain. In *2011 International Conference of Information Technology, Computer Engineering and Management Sciences* (Vol. 3, pp. 42–46). IEEE.

84. Wang, Y., & Li, T. (2012). Digital watermarking algorithm based on singular value decomposition and wavelet transformation. IET, 911–914.

85. Sun, J., & Lan, S. (2010, January). Geometrical attack robust spatial digital watermarking based on improved SIFT. In *2010 International Conference on Innovative Computing and Communication and 2010 Asia-Pacific Conference on Information Technology and Ocean Engineering* (pp. 98–101). IEEE.

86. Liu, Z., Zhu, Y. S., Fan, Y., Sun, Z. Q., & Lin, H. C. (2016, August). A SIFT-based robust watermarking scheme in DWT-SVD domain using majority voting mechanism. In *Eighth*

International Conference on Digital Image Processing (ICDIP 2016) (Vol. 10033, pp. 569–574). SPIE.

87. Bhatnagar, G., Wu, Q. J., & Liu, Z. (2013). Directive contrast based multimodal medical image fusion in NSCT domain. *IEEE Transactions on Multimedia*, 15(5), 1014–1024.

88. Wen, Q., SUN, T. F., & Wang, S. X. (2003). Concept and application of zero-watermark. *Acta Electonica Sinica*, 31(2), 214.

89. Cedillo-Hernandez, M., Cedillo-Hernandez, A., Nakano-Miyatake, M., & Perez-Meana, H. (2020). Improving the management of medical imaging by using robust and secure dual watermarking. *Biomedical Signal Processing and Control*, 56, 101695.

90. Mardanpour, M., & Chahooki, M. A. Z. (2016). Robust transparent image watermarking with Shearlet transform and bidiagonal singular value decomposition. *AEU-International Journal of Electronics and Communications*, 70(6), 790–798.

91. Shinde, G., & Mulani, A. (2019). A robust digital image watermarking using DWT-PCA. *International Journal of Innovations in Engineering Research and Technology*, 6(4), 1–7.

92. Prathap, I., Natarajan, V., & Anitha, R. (2014). Hybrid robust watermarking for color images. *Computers & Electrical Engineering*, 40(3), 920–930.

93. Fan, D., Li, Y., Gao, S., Chi, W., & Lv, C. (2022). A novel zero watermark optimization algorithm based on Gabor transform and discrete cosine transform. *Concurrency and Computation: Practice and Experience*, 34(14), e5689.

94. Xiyao, L., Zhang, Y., Du, S., Zhang, J., Jiang, M., & Fang, H. (2022). DIBR Zero-watermarking based on Invariant Feature and Geometric Rectification. *IEEE MultiMedia*.

95. Zheng, Q., Liu, N., Cao, B., Wang, F., & Yang, Y. (2020). Zero-watermarking algorithm in transform domain based on RGB channel and voting strategy. *Journal of Information Processing Systems*, 16(6), 1391–1406.

96. Lu, W., Sun, W., & Lu, H. (2012). Novel robust image watermarking based on sub-sampling and DWT. *Multimedia Tools and Applications*, 60(1), 31–46.

97. Sadreazami, H., & Amini, M. (2012). A robust spread spectrum based image watermarking in ridgelet domain. *AEU-International Journal of Electronics and Communications*, 66(5), 364–371.

98. Zhao, J., Zhang, N., Jia, J., & Wang, H. (2015). Digital watermarking algorithm based on scale-invariant feature regions in non-subsampled contourlet transform domain. *Journal of Systems Engineering and Electronics*, 26(6), 1309–1314.

99. Bazargani, M., Ebrahimi, H., & Dianat, R. (2012). Digital image watermarking in wavelet, contourlet and curvelet domains. *Journal of Basic and Applied Scientific Research*, 2(11), 11296–11308.

100. Sadreazami, H., Ahmad, M. O., & Swamy, M. N. S. (2015). Multiplicative watermark decoder in contourlet domain using the normal inverse Gaussian distribution. *IEEE Transactions on Multimedia*, 18(2), 196–207.

101. Li, Y. M., Wei, D., & Zhang, L. (2021). Double-encrypted watermarking algorithm based on cosine transform and fractional Fourier transform in invariant wavelet domain. *Information Sciences*, 551, 205–227.

102. Nana, Z. (2016, July). Watermarking algorithm of spatial domain image based on SVD. In *2016 International Conference on Audio, Language and Image Processing (ICALIP)* (pp. 361–365). IEEE.

103. Su, Q., & Chen, B. (2018). Robust color image watermarking technique in the spatial domain. *Soft Computing*, 22(1), 91–106.

104. Chan, H. T., Hwang, W. J., & Cheng, C. J. (2015). Digital hologram authentication using a hadamard-based reversible fragile watermarking algorithm. *Journal of Display Technology*, 11(2), 193–203.

105. Luo, Y., Li, L., Liu, J., Tang, S., Cao, L., Zhang, S., ... & Cao, Y. (2021). A multi-scale image watermarking based on integer wavelet transform and singular value decomposition. *Expert Systems with Applications*, 168, 114272.

106. Chrysochos, E., Fotopoulos, V., & Skodras, A. N. (2008, August). Robust watermarking of digital images based on chaotic mapping and DCT. In *2008 16th European Signal Processing Conference* (pp. 1–5). IEEE.

107. Rawat, S., & Raman, B. (2011). A chaotic system based fragile watermarking scheme for image tamper detection. *AEU-International Journal of Electronics and Communications*, 65(10), 840–847.

108. Arora, S. M. (2018). A DWT-SVD based robust digital watermarking for digital images. *Procedia Computer Science*, 132, 1441–1448.

109. Yamasaki, T., & Aizawa, K. (2013). SIFT-based non-blind watermarking robust to non-linear geometrical distortions. *IEICE Transactions on Information And Systems*, 96(6), 1368–1375.

110. Huang, T., Xu, J., Yang, Y., Tu, S., & Han, B. (2021, October). Zero-Watermarking Algorithm for Medical Images Based on Nonsubsampled Contourlet Transform and Double Singular Value Decomposition. In *2021 5th Asian Conference on Artificial Intelligence Technology (ACAIT)* (pp. 65–76). IEEE.

111. Kang, X. B., Lin, G. F., Chen, Y. J., Zhao, F., Zhang, E. H., & Jing, C. N. (2020). Robust and secure zero-watermarking algorithm for color images based on majority voting pattern and hyper-chaotic encryption. *Multimedia Tools and Applications*, 79(1), 1169–1202.

112. Lee, S., Yoo, C. D., & Kalker, T. (2007). Reversible image watermarking based on integer-to-integer wavelet transform. *IEEE Transactions on Information Forensics and Security*, 2(3), 321–330.

113. Fontani, M., De Rosa, A., Caldelli, R., Filippini, F., Piva, A., Consalvo, M., & Cappellini, V. (2010, September). Reversible watermarking for image integrity verification in hierarchical pacs. In *Proceedings of the 12th ACM workshop on Multimedia and security* (pp. 161–168).

114. Mostafa, S. A., El-Sheimy, N., Tolba, A. S., Abdelkader, F. M., & Elhindy, H. M. (2010). Wavelet packets-based blind watermarking for medical image management. *The Open Biomedical Engineering Journal*, 4, 93.

115. Al-Qershi, O. M., & Khoo, B. E. (2011). Authentication and data hiding using a hybrid ROI-based watermarking scheme for DICOM images. *Journal of digital Imaging*, 24(1), 114–125.

116. Al-Qershi, O. M., & Khoo, B. E. (2011). High capacity data hiding schemes for medical images based on difference expansion. *Journal of Systems and Software*, 84(1), 105–112.

117. Memon, N. A., & Gilani, S. A. M. (2011). Watermarking of chest CT scan medical images for content authentication. *International Journal of Computer Mathematics*, 88(2), 265–280.

118. Nambakhsh, M. S., Ahmadian, A., & Zaidi, H. (2011). A contextual based double watermarking of PET images by patient ID and ECG signal. *Computer Methods and Programs in Biomedicine*, 104(3), 418–425.

119. Tan, C. K., Ng, J. C., Xu, X., Poh, C. L., Guan, Y. L., & Sheah, K. (2011). Security protection of DICOM medical images using dual-layer reversible watermarking with tamper detection capability. *Journal of Digital Imaging*, 24(3), 528–540.

120. BW, T. A., & Permana, F. P. (2012, July). Medical image watermarking with tamper detection and recovery using reversible watermarking with LSB modification and run length encoding (RLE) compression. In *2012 IEEE International Conference on Communication, Networks and Satellite (ComNetSat)* (pp. 167–171). IEEE.

121. Roček, A., Slavíček, K., Dostál, O., & Javorník, M. (2016). A new approach to fully-reversible watermarking in medical imaging with breakthrough visibility parameters. *Biomedical Signal Processing and Control*, 29, 44–52.

122. Lee, J. E., Kang, J. W., Kim, W. S., Kim, J. K., Seo, Y. H., & Kim, D. W. (2021). Digital image watermarking processor based on deep learning. *Electronics*, 10(10), 1183.

123. Kumar, S., & Dutta, A. (2016, May). A study on robustness of block entropy based digital image watermarking techniques with respect to various attacks. In *2016 IEEE International conference on recent trends in electronics, information & communication technology (RTEICT)* (pp. 1802–1806). IEEE.

124. Song, C., Sudirman, S., Merabti, M., & Llewellyn-Jones, D. (2010, January). Analysis of digital image watermark attacks. In *2010 7th IEEE Consumer Communications and Networking Conference* (pp. 1–5). IEEE.

125. Voloshynovskiy, S., Pereira, S., Pun, T., Eggers, J. J., & Su, J. K. (2001). Attacks on digital watermarks: Classification, estimation based attacks, and benchmarks. *IEEE Communications Magazine*, 39(8), 118–126.

126. Duan, G., Ho, A. T., & Zhao, X. (2008). A Novel non-redundant contourlet transform for robust image watermarking against non-geometrical and geometrical attacks.

127. Hussein, E., & Belal, M. A. (2012). Digital watermarking techniques, applications and attacks applied to digital media: A survey. *Threshold*, 5, 6.

128. Katzenbeisser, S., & Veith, H. (2002, April). Securing symmetric watermarking schemes against protocol attacks. In *Security and Watermarking of Multimedia Contents IV* (Vol. 4675, pp. 260–268). SPIE.

129. Evsutin, O., & Dzhanashia, K. (2022). Watermarking schemes for digital images: Robustness overview. *Signal Processing: Image Communication*, 100, 116523.

130. Ali, M., Ahn, C. W., Pant, M., Kumar, S., Singh, M. K., & Saini, D. (2020). An optimized digital watermarking scheme based on invariant DC coefficients in spatial domain. *Electronics*, 9(9), 1428.

131. Abraham, J., & Paul, V. (2019). An imperceptible spatial domain color image watermarking scheme. *Journal of King Saud University-Computer and Information Sciences*, 31(1), 125–133.

132. Hosny, K. M., Darwish, M. M., Li, K., & Salah, A. (2018). Parallel multi-core CPU and GPU for fast and robust medical image watermarking. *IEEE Access*, 6, 77212–77225.

133. PVSSR, C. M. (2017). A robust semi-blind watermarking for color images based on multiple decompositions. *Multimedia Tools and Applications*, 76(24), 25623–25656.

134. Yuan, Z., Liu, D., Zhang, X., & Su, Q. (2020). New image blind watermarking method based on two-dimensional discrete cosine transform. *Optik*, 204, 164152.

135. Liu, S., Pan, Z., & Song, H. (2017). Digital image watermarking method based on DCT and fractal encoding. *IET image Processing*, 11(10), 815–821.

136. Savakar, D. G., & Ghuli, A. (2019). Robust invisible digital image watermarking using hybrid scheme. *Arabian Journal for Science and Engineering*, 44(4), 3995–4008.

137. Wu, J., Shi, H., Zhang, S., Lei, Z., Yang, Y., & Li, S. Z. (2018, February). De-Mark GAN: Removing dense watermark with generative adversarial network. In *2018 International Conference on Biometrics (ICB)* (pp. 69–74). IEEE.

138. Cu, V. L., Burie, J. C., Ogier, J. M., & Liu, C. L. (2019, September). A robust data hiding scheme using generated content for securing genuine documents. In *2019 International Conference on Document Analysis and Recognition (ICDAR)* (pp. 787–792). IEEE.

139. Wei, Q., Wang, H., & Zhang, G. (2020). A robust image watermarking approach using cycle variational autoencoder. *Security and Communication Networks*, 2020, 1–9.

140. Zhang, L., Li, W., & Ye, H. (2021, October). A blind watermarking system based on deep learning model. In *2021 IEEE 20th International Conference on Trust, Security and Privacy in Computing and Communications (TrustCom)* (pp. 1208–1213). IEEE.

141. Fan, B., Li, Z., & Gao, J. (2022). DwiMark: A multiscale robust deep watermarking framework for diffusion-weighted imaging images. *Multimedia Systems*, 28(1), 295–310.

142. Fang, H., Jia, Z., Zhou, H., Ma, Z., & Zhang, W. (2022). Encoded feature enhancement in watermarking network for distortion in real scenes. *IEEE Transactions on Multimedia*.

143. Neekhara, P., Hussain, S., Zhang, X., Huang, K., McAuley, J., & Koushanfar, F. (2022). FaceSigns: Semi-fragile neural watermarks for media authentication and countering deepfakes. arXiv preprint arXiv:2204.01960.

144. Siddaraju, P. M., Jayadevappa, D., & Ezhilarasan, K. (2015). Application of fractional wave packet transform for robust watermarking of mammograms. *International Journal of Telemedicine and Applications*, 2015.

145. Thakkar, F. N., & Srivastava, V. K. (2017). A blind medical image watermarking: DWT-SVD based robust and secure approach for telemedicine applications. *Multimedia Tools and Applications*, 76(3), 3669–3697.

146. Jiang, L., Xu, Z., & Xu, Y. (2013, July). A new comprehensive security protection for remote sensing image based on the integration of encryption and watermarking. In *2013 IEEE International Geoscience and Remote Sensing Symposium-IGARSS* (pp. 2577–2580). IEEE.

147. Li, D., Che, X., Luo, W., Hu, Y., Wang, Y., Yu, Z., & Yuan, L. (2019). Digital watermarking scheme for colour remote sensing image based on quaternion wavelet transform and tensor decomposition. *Mathematical Methods in the Applied Sciences*, 42(14), 4664–4678.

148. Tong, D., Ren, N., & Zhu, C. (2019). Secure and robust watermarking algorithm for remote sensing images based on compressive sensing. *Multimedia Tools and Applications*, 78(12), 16053–16076.

149. Fei, P., Li, C., & Min, L. (2013). A reversible watermark scheme for 2D vector map based on reversible contrast mapping. *Security and Communication Networks*, 6(9), 1117–1125.

150. Ray, A., & Roy, S. (2020). Recent trends in image watermarking techniques for copyright protection: a survey. *International Journal of Multimedia Information Retrieval*, 9(4), 249–270.

151. Nin, J., & Ricciardi, S. (2013, March). Digital watermarking techniques and security issues in the information and communication society. In *2013 27th International Conference on Advanced Information Networking and Applications Workshops* (pp. 1553–1558). IEEE.

152. Hu, M. C., Lou, D. C., & Chang, M. C. (2007). Dual-wrapped digital watermarking scheme for image copyright protection. *Computers & Security*, 26(4), 319–330.

153. Zhong, J., & Gan, Y. (2013). Copy-paste forgery image blind detection algorithm based on histogram invariant moments. *Sensors & Transducers*, 161(12), 92.

154. Jahnke, T., & Seitz, J. (2004). An introduction in digital watermarking: Applications, principles, and problems. In *E-Commerce and M-Commerce Technologies* (pp. 117–141). IGI Global.

155. Cox, I., Miller, M., Bloom, J., Fridrich, J., & Kalker, T. (2007). Digital watermarking and steganography. *Morgan kaufmann*.

156. Agarwal, N., Singh, A. K., & Singh, P. K. (2019). Survey of robust and imperceptible watermarking. *Multimedia Tools and Applications*, 78(7), 8603–8633.

157. Pizzolante, R., Castiglione, A., Carpentieri, B., De Santis, A., & Castiglione, A. (2014, September). Protection of microscopy images through digital watermarking techniques. In *2014 international conference on intelligent networking and collaborative systems* (pp. 65–72). IEEE.

158. Zhang, Y., Wang, C., Wang, X., & Wang, M. (2017). Feature-based image watermarking algorithm using SVD and APBT for copyright protection. *Future Internet*, 9(2), 13.

159. Zear, A., Singh, A. K., & Kumar, P. (2018). A proposed secure multiple watermarking technique based on DWT, DCT and SVD for application in medicine. *Multimedia Tools and Applications*, 77(4), 4863–4882.

160. Zhou, X., Zhang, H., & Wang, C. (2018). A robust image watermarking technique based on DWT, APDCBT, and SVD. *Symmetry*, 10(3), 77.

161. Singh, A. K., Kumar, B., Singh, S. K., Ghrera, S. P., & Mohan, A. (2018). Multiple watermarking technique for securing online social network contents using back propagation neural network. *Future Generation Computer Systems*, 86, 926–939.

162. Takore, T. T., Kumar, P. R., & Devi, G. L. (2018). A new robust and imperceptible image watermarking scheme based on hybrid transform and PSO. *International Journal of Intelligent Systems and Applications*, 10(11), 50.

163. Phadikar, A., Jana, P., & Mandal, H. (2019). Reversible data hiding for DICOM image using lifting and companding. *Cryptography*, 3(3), 21.

CHAPTER 9

Image Fusion Techniques and Applications for Remote Sensing and Medical Images

Emadalden Alhatami[1], MengXing Huang[1], and Uzair Aslam Bhatti[1,2]

[1]*School of Information and Communication Engineering, Hainan University, Haikou, China*
[2]*State Key Laboratory of Marine Resource Utilization in the South China Sea, Hainan University, Haikou, China*

9.1 INTRODUCTION

Image fusion refers to the merging of two or more images of the same scene to create a single image that contains greater information and detail than any individual image. After the process of image fusion, the resulting image incorporates the most significant features and characteristics of each input image, thereby improving the overall visual perception of the scene [1].

Image fusion applications are numerous in practical fields such as remote sensing, medical imaging, and military surveillance [2]. In remote sensing, image fusion is used to combine data from different sensors to create high-resolution digital maps of land cover and surface features. In medical imaging, image fusion is used to combine data from different imaging modalities such as MRI and CT scans, to create more accurate and comprehensive images of internal structures. In military surveillance, image fusion is used to combine data from multiple sources such as radar, infrared cameras, and optical cameras to create a fused image that provides a more comprehensive view of the battlefield [3].

The image fusion process usually involves a series of spatial and spectral transformations that are applied to the input images. Spatial transformations involve aligning the images based on their position and orientation and combining them using a weighted average or other mathematical operations to create a new image. Spectral transformations involve combining the pixel values of input images based on their spectral content or color composition [4,5].

Pixel-level, feature-level, and decision-level fusion are among the various techniques that can be employed for image fusion. Pixel-level fusion involves combining the pixel

DOI: 10.1201/9781003427674-9

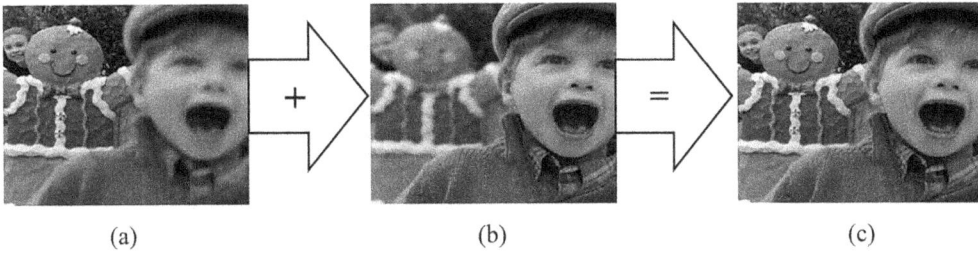

FIGURE 9.1 (a) Input image 1, (b) input image 2, and (c) fused image.

intensities of the input images at each pixel location, resulting in a fused image that contains more spatial and spectral detail. Feature-level fusion involves extracting the salient features from each input image and combining them to create a fused image that highlights the most important features of the scene. Decision-level fusion involves using machine learning algorithms to make decisions about which features to include in the fused image based on a set of pre-defined rules [6]. The results are very promising thereby, providing motivation to employ different sensors or cameras for collecting data about a single scene. In the recent past, serious experimentation with multiple sensors and intelligent machines has been of interest to the research communities. Fusion continues to occupy a prominent position in image processing algorithms of its success in addressing the needs of surveillance, robotics, navigation, clinical diagnosis, and remote sensing [7].

Image fusion combines the details from multiple images and provides the end user with one informative image [8]. A fusion algorithm is preferred when it is free of any artifacts in the final delivered image, the process of image fusion is presented as a simple illustration in Figure 9.1.

It is easy to note that the information from each input image is integrated in an effective manner [6].

Multiple images containing complementary and redundant information can undergo processing and fusion to generate a new image that is richer in information, accuracy, and reliability. The resulting image is more suitable for human vision as well as for computer-based tasks such as detection, recognition, classification, and understanding [9]. Several mathematical techniques have been devised for image fusion, including but not limited to average and weighted averaging, color mapping, nonlinear approaches, optimization techniques, Markov random fields, simulated annealing, artificial neural networks, image pyramids, and wavelet transform [10,11].

9.2 RULE OF IMAGE FUSION

The concept of the image fusion rule pertains to the criteria or guidelines that are utilized to assess the quality of a merged image. These criteria are based on several factors, including the level of detail, contrast, and spatial resolution of the fused image, as well as its overall visual quality and applicability to the specific task or application [12]. In Table 9.1, the sum of common rules of image fusion is shown.

TABLE 9.1   Rule of Image Fusion

Rule	Progress
Spatial registration	Spatial registration of input images is imperative to ensure that the corresponding pixels in each image represent the same location within the scene. This can be achieved by using image alignment techniques such as feature-based registration, intensity-based registration, or geometric transformation.
Spectral compatibility	The input images should be spectrally compatible, meaning that they should have similar spectral characteristics and cover the same spectral range. This ensures that the information from each image can be effectively combined to create a single, informative image.
Quality image	The quality of the input images plays a crucial role in the process of image fusion. Higher quality images, with good resolution, contrast, and signal-to-noise ratio, will result in a better-fused image.
Fusion method	Various image fusion methods exist, and each has its own set of advantages and limitations. The selection of a specific fusion method is dependent on the application and the properties of the input images.
Evaluation	Assessing the quality of the merged image is crucial to ensure that it meets the requirements of the intended application. This can be done using objective measures such as entropy, mutual information, or structural similarity, as well as subjective measures such as visual inspection.

9.3 LEVELS OF IMAGE FUSION

Image fusion can be categorized into various levels, based on the complexity and characteristics of the fusion process. Pixel-level fusion, feature-level fusion, and decision-level fusion are the widely recognized levels of image fusion. Image fusion techniques frequently employ a combination of these levels to attain the desired result [6,8] (Figure 9.2).

9.3.1 Pixel-Level Image Fusion

Pixel-level image fusion involves merging several images of a scene to create a single image that contains all of the relevant information from the original images. Pixel-level image fusion aims to produce a merged image that is more informative and visually appealing than any of the individual input images.

There are many different algorithms for pixel-level image fusion, including methods based on multi-resolution analysis, wavelet transforms, and sparse representation. These

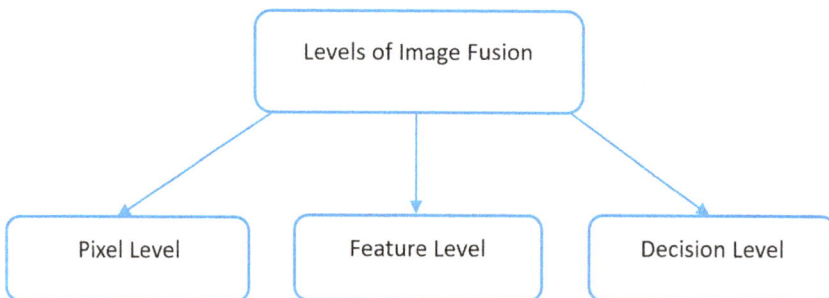

FIGURE 9.2   Levels of image fusion.

algorithms typically involve decomposing the input images into different frequency bands, combining the frequency bands in some way, and then reconstructing the fused image. The selection of an algorithm is dependent on the particular application and the properties of the input images.

Pixel-level image fusion is often used in fields such as remote sensing, surveillance, and medical imaging. In remote sensing, for example, pixel-level image fusion can be used to combine images from different sensors to create a single image that contains more information about the target area. In surveillance, pixel-level image fusion can be used to combine images from different cameras to create a single image that provides a more complete view of the scene. In medical imaging, pixel-level image fusion can be used to combine images from different modalities to create a single image that provides more detailed information about the patient's condition [6,17].

9.3.2 Feature-Level Image Fusion

Feature-level image fusion involves extracting relevant features from multiple input images and then combining them into a single fused image. The aim of feature-level image fusion is to create a final image that contains the most important information from each of the input images while minimizing redundancy and artifacts.

Feature-level image fusion is often used in computer vision applications, including object recognition, tracking, and classification, as well as in medical imaging for diagnosis and treatment planning.

There are several approaches to feature-level image fusion, including transform-based methods, statistical methods, and deep learning–based methods. These techniques have accomplished exceptional performance in numerous computer vision tasks. The selection of a method is contingent on the particular application and the properties of the input images [13].

9.3.3 Decision-Level Image Fusion

Decision-level image fusion entails merging the output of multiple classifiers or decision-making algorithms that have been applied to the input images. The goal of decision-level image fusion is to improve the performance of the final decision by taking into account the strengths and weaknesses of each individual classifier or algorithm.

Decision-level image fusion is often used in applications such as object recognition, target detection, and classification, where multiple classifiers are used to make a final decision. For example, in object recognition, multiple classifiers may be used to detect different features of an object, such as its shape, texture, and colors. The output of each classifier can then be combined using a fusion rule to make a final decision about the object's identity.

There are several approaches to decision-level image fusion, including majority voting, weighted voting, and Bayesian inference. The output of each individual classifier or algorithm is used to update the posterior probability distribution over the final decision. The selection of the approach depends on the specific implementation and the properties of the individual classifiers or algorithms [6,14].

9.4 IMAGE FUSION METHODS

There are three types of image fusion methods, which are

- Spatial domain.

- Frequency domain (transform domain).

- Deep learning–based methods (Figure 9.3).

Spatial domain combination primarily involves combining the pixels of the original images. It fuses the entire image by utilizing local spatial features including gradient, spatial variance, and local standard deviation. On the other hand, frequency domain combination involves transforming the images into the frequency domain. This method involves the projection of the source images onto localized bases that are crafted to capture the edges and sharpness of an image. The transformed coefficients obtained through this process are utilized to extract relevant features from the input images, which are then used to generate the merged images.

9.4.1 Spatial Domain Fusion Methods

The spatial domain in image fusion involves combining the images in the spatial domain by directly processing the pixel values of the input images. These methods are based on mathematical operations.

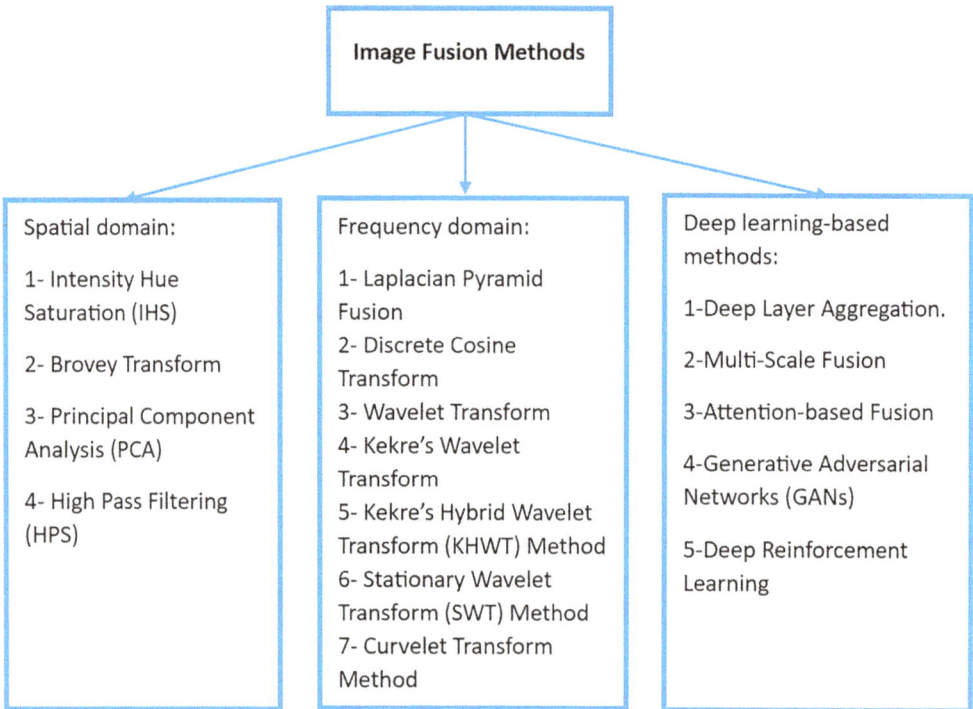

Image Fusion Methods

Spatial domain:

1- Intensity Hue Saturation (IHS)

2- Brovey Transform

3- Principal Component Analysis (PCA)

4- High Pass Filtering (HPS)

Frequency domain:

1- Laplacian Pyramid Fusion
2- Discrete Cosine Transform
3- Wavelet Transform
4- Kekre's Wavelet Transform
5- Kekre's Hybrid Wavelet Transform (KHWT) Method
6- Stationary Wavelet Transform (SWT) Method
7- Curvelet Transform Method

Deep learning-based methods:

1-Deep Layer Aggregation.

2-Multi-Scale Fusion

3-Attention-based Fusion

4-Generative Adversarial Networks (GANs)

5-Deep Reinforcement Learning

FIGURE 9.3 Image fusion methods.

9.4.1.1 Intensity Hue Saturation

The intensity hue saturation (IHS) method of image fusion is a popular technique used to integrate images from different sensors or sources. The process involves converting the images into three separate components, which are then processed individually and merged to create the final fused image [34].

The intensity component represents the brightness values in the image, while the hue and saturation components represent the color information. To perform IHS fusion, the intensity component is extracted from the high-resolution image, while the hue and saturation components are extracted from the low-resolution image.

The intensity component is then combined with the hue and saturation components from the low-resolution image to create the fused image. This method allows for the preservation of high-resolution details while incorporating color information from the low-resolution image.

IHS fusion is commonly used in remote sensing applications, where it is used to fuse data from multiple detectors, such as optical and microwave sensors, to create a comprehensive and informative image. It is also used in medical imaging, where it is used to combine images from multiple modalities, such as CT and MRI scans [34].

9.4.1.2 Brovey Transform

Brovey Transform is a commonly used technique for fusing multi-spectral and pan-chromatic images. This technique aims to improve the spatial resolution of multi-spectral images to match that of the high-resolution panchromatic image. The merged image produced via the Brovey Transform technique retains the spectral information of the original multi-spectral image and exhibits improved spatial details from the high-resolution panchromatic image [15].

The Brovey Transform is applied by dividing the panchromatic image by the sum of individual bands of the multispectral image. The resulting fractions are then multiplied by each band of the multispectral image to obtain the fused image. Mathematically, the Brovey Transform can be expressed as:

$$\text{Fused image} = (P \times (R + G + B))/(R + G + B)$$

where P is the panchromatic image; R, G, and B are the individual bands of the multi-spectral image; and the resulting fused image combines the spatial information from the panchromatic image along with the spectral information from the multispectral image.

The Brovey Transform is a straightforward yet effective method for merging satellite images and is widely employed in remote sensing applications including change detection, land cover classification, and vegetation analysis [15].

9.4.1.3 Principal Component Analysis (PCA)

The PCA method is the principal component analysis method. The fusion principle of this transformation is to register the multi-spectral images, calculate the eigenvalues and eigenvectors of the principal component transformation matrix, and calculate the principal components after sorting them in order from high to small. After the histogram matching, the first principal component is replaced with the panchromatic image, and the inverse principal component transformation is completed at last to obtain the fused image [16].

The design of the PCA image fusion applet is the same as that of the wavelet image fusion applet, and its GUI interface.

9.4.1.4 High-Pass Filtering

High-pass filtering is a type of image filtering that enhances the high-frequency components of an image while reducing the low-frequency components. High-frequency components are those that vary rapidly in space, such as edges and fine details, while low-frequency components are those that vary slowly, such as smooth regions and large structures.

High-pass filtering can be used for several purposes, such as edge detection, noise reduction, and sharpening. The most common high-pass filter used in image processing is the Laplacian filter, which highlights edges and other high-frequency features in an image.

The Laplacian filter works by convolving the image with a 3×3 or 5×5 kernel that approximates the Laplacian operator. The Laplacian operator is a differential operator of the second order that gauges the intensity variation of an image among its neighboring pixels. By applying the Laplacian filter to an image, the high-frequency components are enhanced, while the low-frequency components are suppressed. However, high-pass filtering can also enhance noise in an image, particularly in regions of high contrast. Therefore, it is important to carefully choose the type and parameters of the high-pass filter used in a given application and to apply additional processing steps, such as smoothing or thresholding, to reduce the noise and artifacts in the filtered image [17].

9.4.2 Frequency Domain Fusion Methods

Frequency domain fusion methods involve the use of Fourier transforms to convert the pixel values of images into frequency domain representations. These representations can then be processed in the frequency domain by applying different mathematical operations that are later inverse-transformed to generate a fused image.

9.4.2.1 Laplacian Pyramid Fusion Technique

The Laplacian Pyramid Fusion Technique is a method used for merging several images or videos into a combined output image or video. It involves creating a Laplacian pyramid of each input image or video, which is a hierarchical set of images with different levels of detail. The original image constitutes the bottommost level of the pyramid, and each subsequent level represents an increasingly blurred version of the image [18–53].

Next, the Laplacian pyramid of each input image or video is combined using a weight map that determines the contribution of each pixel to the output image. The weight map can either be manually created by the user or it can be automatically generated based on image characteristics such as contrast or brightness. The combined Laplacian pyramid is reconstructed into a single output image or video. The Laplacian Pyramid Fusion Technique is commonly used in image and video processing applications such as panoramic stitching, image blending, and dynamic scene reconstruction [35].

9.4.2.2 Discrete Cosine Transform

The discrete cosine transform (DCT) is a method that transforms a signal or image into its frequency domain. It has been widely used in image processing applications, including image fusion. In image fusion, the DCT can be used to convert images into their frequency domain representations. The DCT coefficients represent the information that is used to construct the final fused image. The specific image fusion method being used determines whether the DCT is utilized as a pre-processing or post-processing step [18].

One of the advantages of using the DCT for image fusion is its ability to compress the information in an image. This makes it possible to extract the most relevant information from each input image while minimizing the amount of noise and redundancy. Some of the popular image fusion methods that use the DCT include PCA-based image fusion, principal frequency selection, and wavelet-based image fusion. These methods use various algorithms to combine the DCT coefficients from the input images into a final fused image [18]. The DCT is a useful tool for image fusion as it allows for the extraction and selective combination of key features from multiple input images.

9.4.2.3 Wavelet Transform

The wavelet transform (WT) is a signal processing technique that breaks down a signal or image into various frequency components. It is widely used in image fusion to merge information from different sources or to combine different modalities.

WT can effectively represent the spatial and spectral information of an image in a concise manner. The decomposition can be done iteratively to obtain multiple levels of decomposition. Each level of decomposition provides different frequency bands that capture different types of features in the image [19].

The fusion process involves the selection of appropriate frequency bands from multiple images and combining them to obtain a new image. The selection of appropriate frequency bands is based on factors such as image quality, target application, and user preferences.

WT has several advantages over other image fusion techniques. It helps to preserve relevant features in the fused image and reduces visual artifacts. It also allows for flexible and adaptable fusion schemes that can be tailored to specific applications.

The use of the WT in image fusion has become widespread due to its ability to effectively integrate information from different sources or modalities while preserving important features in the fused image [19].

9.4.2.4 Kekre's Wavelet Transform

The Kekre's wavelet transform (KWT) method is a transformation technique used to analyze signals and images to improve the performance and efficiency of the traditional wavelet transform method.

The KWT method employs both the DWT and DCT to extract features from the image or signal. In the KWT method, the DWT is utilized to analyze the signal into its frequency sub-bands, and the DCT is employed to extract features from each of these sub-bands.

The KWT method is known for its ability to reduce the size of the signal and achieve better compression ratios compared to other techniques. It also offers high accuracy when applied to image processing applications including edge detection, image compression, image fusion, feature extraction, and image recognition. The KWT firstly is computationally efficient, making it suitable for real-time applications. Secondly, it allows for the precise localization of features in an image.

KWT is a potent tool for extracting features in various image-processing applications. Its ability to accurately locate features and its computational efficiency make it an attractive option for many real-world applications [20].

9.4.2.5 Kekre's Hybrid Wavelet Transform

The Kekre's hybrid wavelet transform (KHWT) method is a mathematical algorithm used for image compression and feature extraction. The creation of this method involves merging the discrete wavelet transform (DWT) and singular value decomposition (SVD) techniques.

The KHWT method first applies the DWT technique, which breaks down an image into multiple sub-bands that represent different frequency components. These sub-bands are then further decomposed using the SVD technique, which applies a mathematical operation to extract the most significant information and reduce the dimensionality of the data.

The result of the KHWT method is a compressed image that contains only the essential information required for the reconstruction of the original image. This method has been shown to provide high compression ratios while maintaining image quality, making it useful in many applications, such as digital imaging, robotics, and signal processing.

Overall, the KHWT method is a powerful image-processing tool that combines two efficient techniques to provide enhanced results [21].

9.4.2.6 Stationary Wavelet Transform

The stationary wavelet transform (SWT) is a signal processing technique that employs wavelets to decompose signals into high-frequency and low-frequency components.

The SWT method differs from the standard wavelet transform in that it utilizes a fixed-length window and uses a convolution-based approach to avoid signal expansion due to downsampling. This makes the SWT more suitable for analyzing nonstationary signals, which are signals whose frequency content changes over time. The SWT method applies a series of high-pass and low-pass filters to the signal, creating a set of wavelet coefficients at different scales and positions. These coefficients can be utilized to restore the initial signal or to analyze the signal for characteristics such as local frequency content and time-varying features. The advantages of using the SWT method include the ability to analyze nonstationary signals and to accurately capture high-frequency components. Additionally, the SWT method has better time localization and higher frequency resolution than traditional Fourier analysis, making it more suitable for applications such as signal processing, data compression, and noise reduction. The SWT method has applications in a variety of fields including image processing, biomedical engineering, geophysics, and financial analysis [22].

9.4.2.7 Curvelet Transform

The curvelet transform is based on the idea that signals can be represented by a collection of curves of varying sizes and orientations. These curves are then analyzed using Fourier analysis or wavelet analysis, depending on their frequency content.

The curvelet transform is particularly effective at analyzing signals that contain edges, such as images with sharp boundaries between different regions. This is because the curves in the curvelet transform are designed to match the edges in the signal. By using curvelets to analyze the signal, it is possible to separate out the edges from the smoother regions, which can be useful for tasks such as image compression or noise reduction.

The curvelet transform has been applied to a wide variety of signal-processing problems, including image processing, seismic data analysis, and medical imaging. It has been shown to outperform other methods such as wavelet transforms in many cases, particularly when dealing with signals containing curved edges or textures [23].

9.4.3 Deep Learning Methods

Deep learning image fusion with the advancements made in deep learning methods, image fusion techniques have improved significantly. Some of the deep learning methods for image fusion include the following:

Deep Layer Aggregation: The deep layer aggregation method uses a convolutional neural network (CNN) to learn a set of filters that can extract spatiotemporal features from multiple input images. The network then aggregates the learned features into a single fused image.

Multi-Scale Fusion: The multiscale fusion method uses a multiscale CNN to learn features at different scales of the input images. The network then combines the learned features to create a fused image that contains information from all scales.

Attention-Based Fusion: The attention-based fusion method uses a CNN to learn a set of attention maps that highlight the important regions in each input image. The attention maps are then used to selectively fuse the features from the input images, resulting in a fused image that emphasizes important regions [24].

Generative Adversarial Networks (GANs): GANs are a type of deep learning method that capability to use for image fusion. The generative network of the GAN is trained to generate a fused image that is indistinguishable from the real fused image. The discriminative network of the GAN is trained to distinguish between real and fake fused images.

Deep Reinforcement Learning (DRL): The DRL method uses a reinforcement learning algorithm to learn a policy for the fusion process. The DRL agent receives multiple input images and learns to select the optimal fusion strategy to maximize a reward signal. All of these deep learning–based methods have shown improved performance and accuracy on image fusion tasks compared to traditional methods [25,26].

9.5 TECHNIQUES FOR THE ASSESSMENT OF IMAGE FUSION QUALITY

The evaluation of fused images can be performed through two standard methods: quantitative analysis and qualitative analysis.

In qualitative analysis, the fused image is visually compared to the input images. Various statistical criteria such as graphical patterns, spatial detail, color, object size, and spectral information are used to assess the accuracy of the fused image. This analysis relies on human judgment and interpretation. On the other hand, the quantitative analysis aims to overcome the potential impact of human visual judgment by providing objective assessments of image fusion efficacy through mathematical metrics. There are two main approaches to quantitative analysis [27–31].

The first approach involves metric measurements using a reference image. Metrics such as root mean square error (RMSE), signal-to-noise ratio (SNR), peak signal-to-noise ratio (PSNR), cross-correlation (CC), mutual information (MI), and structural similarity index (SSIM) are commonly used to compare the fused image with a reference image. The second approach employs metric measurements that do not rely on references. Metrics like fusion quality index, entropy, and standard deviation are utilized to evaluate the quality and characteristics of the fused image without relying on a specific reference image. Qualitative and quantitative analysis methods contribute to the comprehensive evaluation of fused images, providing insights into their visual accuracy and performance based on statistical measures and human perception [27–31]. Table 9.2 presents the commonly utilized performance measure for various types of image fusion techniques.

TABLE 9.2   Measures Used to Evaluate the Performance of Image Fusion

Name of metric	Mathematical formula	Optimal value
RMSE [27–31]	$\text{RMSE} = \sqrt{\frac{1}{MN}\sum_{i=1}^{M}\sum_{j=1}^{N}\left(Ir\,(i,\,j) - if\,(i,\,j)^2\right)}$	Lower (Close to Zero)
SNR [27–31]	$\text{SNR} = 10\log_{10}\left[\frac{\sum_{i=1}^{M}\sum_{j=1}^{N}(I_r(i,j))^2}{\sum_{i=1}^{M}\sum_{j=1}^{N}(I_r(i,j)-I_f(i,j))^2}\right]$	Higher value
PSNR [27–31]	$\text{PSNR} = 20\log_{10}\left[\frac{L^2}{\sum_{i=1}^{M}\sum_{j=1}^{N}(I_r(i,j)-I_f(i,j))^2}\right]$	Higher value
CC [27–31]	$\text{CC} = \frac{2C_{rf}}{C_r + C_f}$	Higher value (CloseTtoT+1)
MI [27–31]	$\text{MI} = \sum_{i=1}^{M}\sum_{j=1}^{N} hI_r I_f\,(i,\,j)\log_2\left[\frac{I_r I_f\,(i,j)}{I_r(i,j)I_f(i,j)}\right]$	Higher value
Spatial Frequency	$SF = \left(\sqrt{(RF)^2 + (CF)^2}\right)$	Higher value
Column Frequency	$CF = \sqrt{\sum_{i=1}^{M}\sum_{j=1}^{N}(F\,(i,\,j) - F\,(i-1,\,j))^2}$	Higher value
Row Frequency	$RF = \sqrt{\sum_{i=1}^{M}\sum_{j=1}^{N}(F\,(i,\,j) - F\,(i,\,j-1))^2}$	Higher value

9.6 IMAGE FUSION CATEGORIZATION

Image fusion involves merging a source image with reference images to create a single integrated image. Several authors have proposed varying techniques to accomplish the intended objectives of image fusion. These techniques can be categorized into several major classes, including single-sensor fusion, multi-sensor fusion, multi-modal fusion, multi-view fusion, multi-focus fusion, and multi-temporal fusion. Each of these classes encompasses different methods that will be further discussed below.

9.6.1 Single Sensor

This fusion method involves merging multi-modal data, which entails collecting and processing data from different modalities, and then integrating them to generate a single composite image.

SSIF techniques are widely employed in diverse fields including medical imaging, robotics, surveillance, and remote sensing. These techniques make it possible to produce images with enhanced contrast, higher spatial resolution, and greater dynamic range. The different types of SSIF techniques include intensity-based fusion, feature-based fusion, decision-based fusion, and pixel-level fusion.

The goal of SSIF is to combine the information from multiple images into a single image that represents the most accurate and complete information possible [32].

9.6.2 Multi-Sensors

Multi-sensor image fusion overcomes the restrictions of single-sensor image fusion by merging images from various sensors to generate a composite image. This technique involves the integration of images captured by an infrared camera and a digital camera to produce the final fused image. The infrared camera is particularly effective in low-light conditions, while the digital camera excels in capturing images under daylight conditions. Multi-sensor image fusion finds applications in various fields such as machine vision, medical imaging, robotics, and object detection, particularly in military contexts. Its primary purpose is to enhance the overall information content by resolving and con-solidating relevant information from multiple images [32].

9.6.3 Multiview Fusion

Multiview image fusion refers to the process of integrating multiple images captured by different cameras or from different viewpoints into a single, more comprehensive image. This technique is commonly used in computer vision, remote sensing, and medical imaging applications.

The goal of multiview image fusion is to combine the complementary information from different images to improve the overall image quality, enhance the visual percep-tion, and enable better analysis and interpretation of the scene.

There are different techniques for multiview image fusion, including pixel methods, feature methods, and deep learning methods. Each approach has its strengths and weaknesses and must be selected based on the specific application and requirements [32,33].

Multiview image fusion is a valuable technique for improving the quality and richness of visual information gathered from multiple sources.

9.6.4 Multimodal Fusion

Multimodal image fusion technique is commonly used in medical imaging, remote sensing, and other applications where multiple modalities provide complementary information that can aid in detection, diagnosis, and analysis. There are many different methods for multimodal image fusion, including intensity-based fusion, feature-based fusion, and deep learning–based fusion, each with its own trade-offs in terms of accuracy, computational complexity, and robustness.

9.6.5 Multi-Focus Fusion

Multi-focus image fusion is a technique that combines multiple images with different focuses to perform multi-focus image fusion; a set of images is first captured with different focus points. Then, the images are aligned and their corresponding regions are extracted. Next, the quality of each region is evaluated using various measures such as sharpness, contrast, and energy. Finally, the regions are blended together to create a fused image with optimal quality [32,33].

There are different methods to perform multi-focus image fusion, including traditional methods such as weighted average and Laplacian pyramid, as well as advanced methods such as deep learning–based approaches.

9.6.6 Multi-Temporal Fusion

Multi-temporal fusion involves capturing images of the same scene at different time intervals. This technique is crucial for monitoring and estimating changes that occur over time on the ground. By acquiring long-term and short-term observations, it becomes possible to analyze and predict the occurrence of changes in a given area. Remote-sensing satellites with revisit capabilities enable the acquisition of multi-temporal images for diverse time periods. These images play a vital role in detecting and monitoring land surface variations across large geographical areas [33].

9.7 IMAGE FUSION APPLICATIONS

Image fusion is likely to find even more applications in the future; image fusion is a widely used technique with numerous applications in various fields. In remote sensing, image fusion is used to combine data from different sensors such as optical, radar, and thermal sensors to create a more comprehensive and accurate representation of the earth's surface. Applications of this technology are numerous and varied, encompassing land use/cover mapping, disaster management, and agriculture.

In medical imaging, image fusion can combine data from multiple imaging modalities such as CT and MRI to improve diagnosis and treatment planning. This can be used for applications such as tum or detection, surgical planning, and disease.

In surveillance image fusion can combine data from multiple cameras, or from different modalities such as visible and infrared, to enhance object detection and tracking. This can be used for applications such as security monitoring and crowd management.

In computer vision, image fusion can be used for feature extraction, object recognition, and image classification. This can be used for applications such as autonomous vehicles, robotics, and machine learning.

In military and defined applications, image fusion can be used to enhance situational awareness by combining data from different sensors such as cameras, radar, and lidar. This technology can be utilized in diverse fields, such as target detection, reconnaissance, surveillance, and identification.

In the multimedia and entertainment applications, the image fusion can be used for video enhancement, special effects, and virtual reality. This can be used for applications such as movie-making, gaming, and augmented reality. These are just a few examples of the many applications of image fusion diagnosis [2,3,7,34].

9.7.1 Medical Image Fusion

Medical imaging is a technology employed to obtain a visual description of human organs inside the body. Various imaging systems like X-ray, CT scan, ultrasound, MRI, PET, SPECT, etc., are developed to meet different clinical requirements. The images captured using medical imaging techniques are helpful in the early diagnosis of severe diseases thereby decreasing the mortality risk (Figure 9.4).

9.7.1.1 Types of Medical Images

Medical imaging assists in viewing tissues and organs inside different parts of the human body [34]. Physicians will be interested in understanding the functional and anatomical conditions of various human organs. Different perspectives of an organ can be captured based on the choice of imaging technique [34–37]. According to the nature captured by the medical image, they are categorized into two types:

1. Structural imaging modalities: It provides the structural details of the human body at various locations like limbs, lungs, head, etc. Bone fractures and dislocations can be identified. The medical imaging modality scanners that give the structural information are X-rays, CT, and MRI.

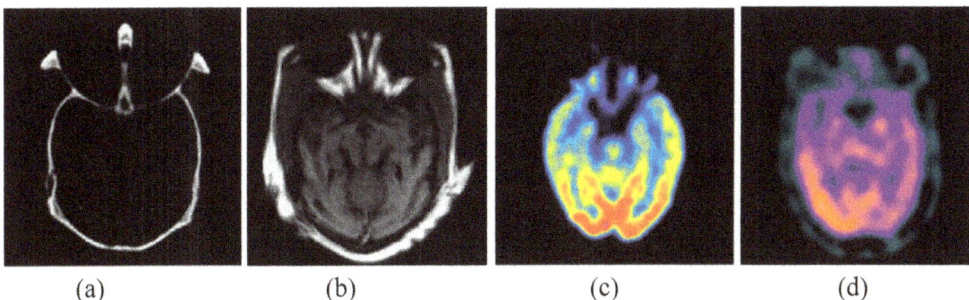

FIGURE 9.4   (a) CT images, (b) MRI images, (c) PET image, and (d) SPECT image.

2. Functional imaging modalities: Functional behavior like blood flow in an organ is of interest in this case. The activity of the brain can be imaged. The type of functional imaging modalities is positron emission tomography (PET) and single photon emission computerized tomography (SPECT).

9.7.1.2 Combining of Imaging Modalities

Medical imaging gives various visual representations of interior body parts, diseased tissues, and damaged organs. It plays a crucial role in public health support by offering medical analysis, disease monitoring, and treatment. Each imaging modality is capable of capturing a piece of unique information based on the sensor employed. If we are interested in developing a single imaging system that can capture multimodal information, there is a risk of exposure to different radiations for a long duration. It may end up in additional medical complications. In addition to this, the multimodal image-capturing device will be very expensive. Hence, a simple and cost-efficient method of developing images with multimodal details is required. It is attempted by developing various image fusion algorithms [35].

This section introduces the fusion of different medical image modalities with some illustrations:

1. Fusion of MRI-CT: It is the integration of information contained in an MRI and CT into a single image, as shown in Figure 9.5.

2. Fusion of MRI-PET: It is defined as the merging of structural and anatomical details of MRI and functional details and blood flow activity of PET [36]. The process is illustrated in Figure 9.6.

3. Fusion of MRI-SPECT: Typically, MRI yields high-resolution images, whereas SPECT generates low-resolution images. For an effective diagnosis, we required the information obtained from both imaging modalities since MRI is having the capability of identifying the soft tissues while SPECT efficiently provides the blood flow and molecular details. This helps to analyze the organs' information [37]. The brain-fused image can be seen in Figure 9.7.

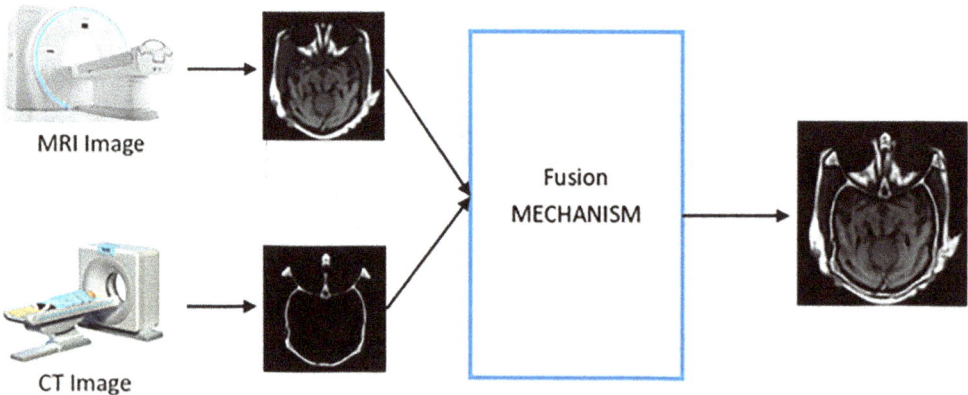

FIGURE 9.5 Fusion of MRI-CT.

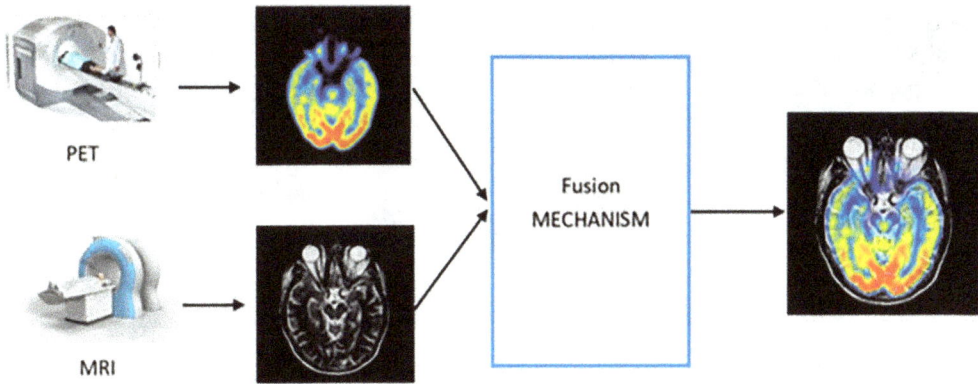

FIGURE 9.6   Fusion of PET-MRI.

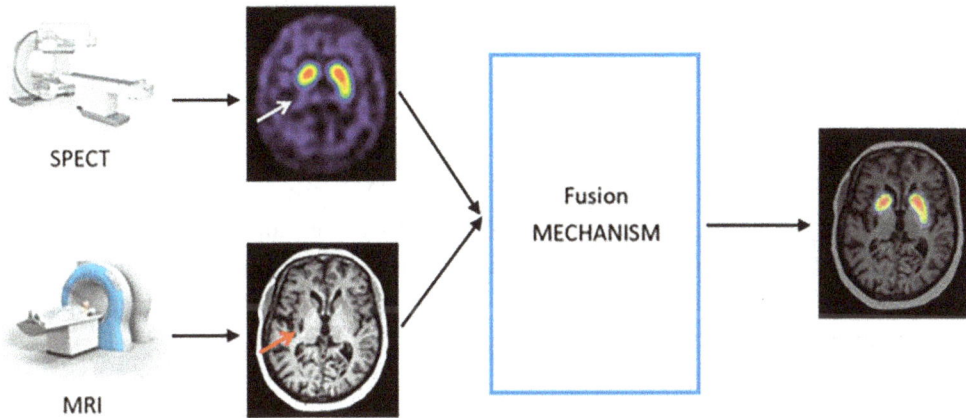

FIGURE 9.7   Fusion of SPECT-MRI.

9.7.2 Remote-Sensing Image Fusion

In remote-sensing image fusion, two different satellite images, termed *panchromatic* and *multispectral images,* are considered. Panchromatic images give information in simple white and black colors [38]. A multispectral image contains one or more spectra or wavelengths in it. The fused image of their different resolution images gives useful information and it is also used in remote-sensing applications (Figure 9.8).

9.7.2.1 Types of Satellite Images

Aerial and satellite images of various kinds enable precise mapping of land cover and facilitate comprehension of landscape characteristics on both regional and global scales. Satellite sensors collect images above the surface of the Earth from hundreds of miles whereas aircraft fly from a mere thousand feet up at altitudes. The height from which an image is attained and the sensor's physical properties determine the area covered and the amount of detail shown in the image. In general, the photographs taken from low altitudes that cover small areas show more detail, while satellite images

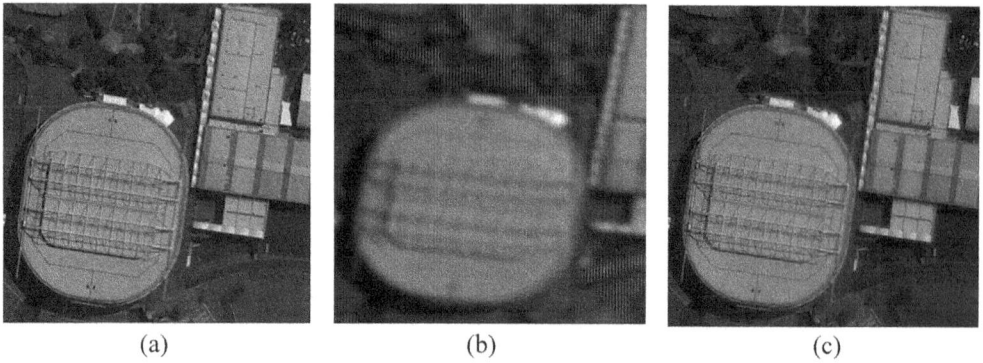

(a)   (b)   (c)

FIGURE 9.8   Remote-sensing image fusion. (a) MS image, (b) PAN image, and (c) fused image.

cover relatively larger areas but show less detail. By comparing images taken at different times, environmental changes on the earth's surface over a period can be studied. The most commonly available satellite imagery for land surface monitoring is ALOS, Aster, Digital Globe, EO-1, Geo Eye, IKONOS, Landsat, MODIS, Rapid Eye, RESOURCESAT, and SPOT [39].

9.7.2.1.1 Single-Band Images   Panchromatic images (PAN) are single-band images that are generally displayed in grayscale with high spatial resolution, and can be used for image feature extraction [39]. Since these images do not possess any spectral information, they have limited applications. PAN images are rich in spatial resolution. For example, PAN images acquired by SAR have a wavelength of 1 mm to 1 m.

9.7.2.1.2 Multispectral Images   This is a series in the same scene with many monochrome images. Each of them is acquired with a sensor that is immune to wavelengths. A well-known multi-band or multi-spectral image is an RGB colors image, composed of bands of red, blue, and green. Multi-spectral sensor records multiple images in the visible, thermal infrared, and near-infrared ranges in dissimilar frequency bands with the pixel values representing a reflectance factor within each wavelength band [39]. For instance, seven-band images are produced by Landsat 5 with wavelength ranges ranging from 0.45 to 1.25 μm. These images are very low spatial resolution but have a high spectral resolution. Hence, most of remote-sensing applications use multispectral images [39,40].

9.7.3 Visible-Infrared Fusion

Visible-infrared fusion refers to the process of combining visible light and infrared radiation to enhance the overall image quality of a scene or object. This process is commonly used in imaging technologies such as thermal cameras, where the visible light spectrum is combined with the infrared spectrum to create a more accurate and comprehensive image. The result of the visible infrared fusion is a greater level of detail and clarity in the image, as well as the ability to see objects and details that would not be visible in either spectrum alone [41]. The technique is particularly useful in industries such as the military, surveillance, and medical imaging (Figure 9.9).

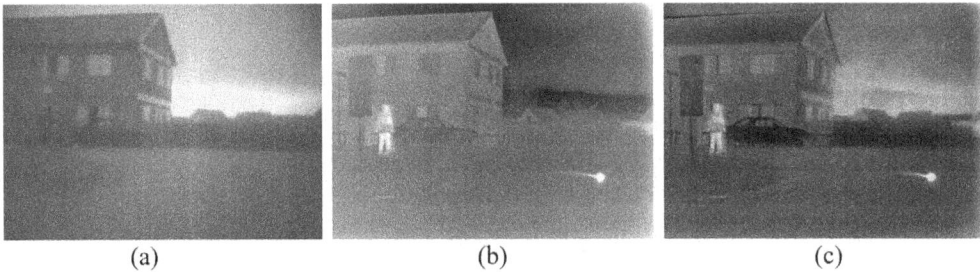

FIGURE 9.9 Examples of image fusion in the surveillance domain. (a) A visible image, (b) an infrared image, and (c) a fused image.

9.7.3.1 Types of Visible-Infrared Images

The type of image produced by visible-infrared fusion depends on the specific method used for fusion. Some common types of visible-infrared fusion images include the following.

9.7.3.1.1 False-Color Composites   In this type of image, the visible and infrared bands are assigned different colors and combined to create a false-color image. For example, the visible bands may be assigned to red, green, and blue channels, while the infrared band is assigned to the red or blue channel to create a composite image that highlights the spatial distribution of vegetation.

9.7.3.1.2 Grayscale Composites   In this type of image, the visible and infrared bands are combined to create a grayscale image. This type of image can be useful for highlighting subtle differences in the reflectance of different materials, such as vegetation, water, and bare soil.

9.7.3.1.3 Merged Images   In this type of image, the visible and infrared bands are merged using a specific fusion method to create a single, integrated image. The Brovey transform method can be used to merge the visible and infrared bands, resulting in an image that combines high spatial resolution with the spectral information from both bands [42].

Visible-infrared fusion images can be useful for a wide range of remote-sensing applications, including vegetation monitoring, land cover classification, and mineral exploration. By combining data from multiple spectral ranges, it is possible to create images that provide a more comprehensive view of the Earth's surface.

9.7.4 Multi-Focus Image Fusion

Multi-focus image fusion involves the combination of images that have different objects in focus. The objective is to generate a merged image that is in focus across the entire scene, even if various parts of the scene were captured at different focus distances. Multi-focus image fusion is commonly used in applications such as microscopy, where it is important to capture detailed images of small objects at different focus distances, and in surveillance systems, where it is important to have a clear image of objects in both the foreground and background [9,23,43].

FIGURE 9.10 Examples of image fusion in the photography domain. (a) Source image 1, (b) source image 2, and (c) fused image.

There are several methods for multi-focus image fusion, including the weighted average method. This technique computes each pixel in the merged image as a weighted average of the corresponding pixels in the input images. The weights assigned to each pixel are determined by a focus measure that quantifies the sharpness of each image.

In the Laplacian pyramid method, the input images are broken down into multiple levels using a Laplacian pyramid, and the fusion is performed by combining the high-frequency details from each level, and in a wavelet transform method the input images are decomposed into multiple levels using a wavelet transform, and the fusion is performed by combining the wavelet coefficients from each level (Figure 9.10).

The output of multi-focus image fusion is a single image that is in focus across the entire scene, with increased visual detail and improved image quality. Multi-focus image fusion is a potent instrument in a range of applications where it is important to capture a clear and detailed image of a scene with varying depths of field.

9.8 CONCLUSION

Image fusion techniques are essential in merging two or more images of the same object or scene, captured from distinct sensors or modalities, to augment the overall information content and quality of the merged image. Medical imaging and remote sensing are two fields where image fusion techniques have become increasingly popular due to the benefits it offers. In medical image fusion, it has been used to provide a more comprehensive diagnostic image, providing a more complete visualization of the area of interest, better precision in targeting specific regions, and improved accuracy in image-guided surgeries. It has also been used to merge different modalities such as CT and MRI to provide a more comprehensive view of the human anatomy. On the other hand, remote-sensing image fusion helps improve the analytical capability of multispectral, hyperspectral, and radar data by enhancing the spatial and spectral resolution of the resultant fused image. It also enables easy distinction and interpretation of different features on the ground or other objects in space, enhancing land use mapping, environmental studies, crop monitoring, and resource management. Image fusion techniques have made significant contributions to the field of medical imaging and remote sensing and, with further advancements, can lead to improved diagnosis, treatment, and analysis of data in both fields.

REFERENCES

1. Hu, Z, Tang, J, Wang, Z, Zhang, K, Zhang, L & Sun, Q. (2018). Deep learning for image-based cancer detection and Diagnosis—A survey. *Pattern Recognit*, 83: 134–149.
2. Anitha, S, Subhashini, T & Kamaraju, M. (2015). A novel multi-modal medical image fusion approach based on phase congruency and directive contrast in NSCT domain. *Int J Comput Appl*, 129(10): 30–35.
3. Bhavana, V & Krishnappa, HK. (2015). Multi-modality medical image fusion using discrete wavelet transform. *Procedia Comput Sci*, 70: 625–631.
4. Sara, D, Mandava, AK, Kumar, A, Duela, S & Jude, A. (2021). Hyperspectral and multi-spectral image fusion techniques for high-resolution applications: A review. *Earth Sci Inf*, 14: 1685–1705.
5. Jeevanand, N, Verma, PA & Saran, S. (2018). Fusion of hyperspectral and multispectral imagery with regression Kriging and the Lulu operators: A comparison. *Int Arch Photogramm Remote Sens Spatial Inf Sci*, 5: 583–588.
6. Kalaivani, K & Asnath, Y. (2016). Analysis of image fusion techniques based on quality assessment metrics. *Indian J Sci Technol*, 9(31): 1–15.
7. Gomathi, PS & Kalaavathi, B. (2016). Multimodal medical image fusion in non-subsampled contourlet transform domain. *Circuits Syst*, 7(8): 1598–1610.
8. Meher, B, Agrawal, S, Panda, R & Abraham, A. (2019). A survey on region-based image fusion methods. *Inf Fus*, 1(48): 119–13 2.
9. Kaur, H, Koundal, D, Kadyan, V, Kaur, N & Polat, K. (2021). Automated multimodal image fusion for brain tumor detection. *J Artif Intell Syst*, 3: 68–82.
10. Liu, K & Kang, G. (2017). Multiview convolutional neural networks for lung nodule classification. *Int J Imaging Syst Technol*, 27(1): 12–22.
11. Das, R, Thepade, S & Ghosh, S. (2015). Content-based image recognition by Information fusion with multiview features. *Int J Inf Technol Comput Sci*, 7(10): 61–73.
12. Nanmaran, R, Srimathi, S, Yamuna, G, Thanigaivel, S, Vickram, AS & Priya, AK. (2022). Investigating the role of image fusion in brain tumor classification models based on machine learning algorithm for personalized medicine. *Comput Math Methods Med*, 1–13.
13. Feng, X, He, L, Cheng, Q, Long, X & Yuan, Y. (2020). Hyperspectral and multispectral remote sensing image fusion based on endmember spatial information. *Remote Sens*, 12: 1009.
14. Jiang, D, Zhuang, D, Huang, Y & Fu, J. (2011). Survey of multispectral image fusion techniques in remote sensing applications in image fusion and its applications. *IntechOpen*, 1–23.
15. Meng, F, Song, M, Guo, B, Shi, R & Shan, D. (2017). Image fusion based on object region detection and non-subsampled contourlet transform. *Comput Elect Eng*, 62: 375–383.
16. Liu, Y, Jin, J, Wang, Q, Shen, Y & Dong, X. (2014). Region level based multi-focus image fusion using quaternion wavelet and normalized cut. *Signal Process*, 97: 9–30.
17. Wei, Q, Bioucas, J, Dobigeon, N & Tourneret, JY. (2015). Hyperspectral and multispectral image fusion based on a sparse representation. *IEEE Trans Geosci Remote Sensing*, 53(7): 3658–3668.
18. Naji, N & Aghagolzadeh, A. (2015). A new multi-focus image fusion technique based on variance in DCT domain. *2nd International Conference on Knowledge-Based Engineering and Innovation (KBEI)*.
19. Dousty, M, Daneshvar, S & Sotero, R. (2016). Multifocus image fusion via the Hartley transform. *IEEE Canadian Conference on Electrical and Computer Engineering (CCECE)*.
20. Kekre, HB, Sarode, T & Dhannawat, R. (2012). Kekre's wavelet transform for image fusion and comparison with other pixel based image fusion techniques. *Int J Comput Sci Inf Secur*, 10(3): 23–31.

21. Udomhunsakul, S, Yamsang, P, Tumthong, S & Borwonwatanadelok, P. (2011). Multiresolution edge fusion using SWT and SFM. *Proc World Congr Eng*, 2: 6–8.

22. Dhannawat, R & Sarode, T. (2013). Kekre's hybrid wavelet transform technique with DCT WALSH HARTLEY and Kekre's transform for image fusion. *Int J Comput Eng Technol (IJCET)*, 4(1): 195–202.

23. Selvapandian, A & Manivannan, K. (2018). Fusion based Glioma brain tumordetection and segmentation using ANFIS classification. *Comput Methods Programs Biomed*, 166: 33–38.

24. Liu, Y, Chen, X, Wang, Z, Wang, ZJ, Ward, RK & Wang, X. (2018). Deep learning for pixel-level image fusion: recent advances and future prospects. *Inf Fus*, 1(42): 158–173.

25. Liu, Z, Chai, Y, Yin, H, Zhou, J & Zhu, Z. (2017). A novel multi-focus image fusion approach based on image decomposition. *Inf Fus*, 1(35): 102–116.

26. Ma, J, Ma, Y & Li, C. (2019). Infrared and visible image fusion methods and applications: a survey. *Inf Fus*, 1(45): 153–178.

27. Patil, V, Sale, D & Joshi, MA. (2013). Image fusion methods and quality assessment parameters. *Asian J Eng Appl Technol*, 2(1): 40–46.

28. Kosesoy, I, Cetin, M & Tepecik, A. (2015). A toolbox for teaching image fusion in Matlab. *Procedia-Soc Behav Sci*, 25(197): 525–530.

29. Paramanandham, N & Rajendiran, K. (2018). Multi sensor image fusion for surveillance applications using hybrid image fusion algorithm. *Multimedia Tools Appl*, 77(10): 12405–12436.

30. Jin, X, Jiang, Q, Yao, S, Zhou, D, Nie, R, Hai, J & He, K. (2017). A survey of infrared and visual image fusion methods. *Infrared Phys Technol*, 1(85): 478–501.

31. Dogra, A, Goyal, B & Agrawal, S. (2017). From multi-scale decomposition to non-multi-scale decomposition methods: a comprehensive survey of image fusion techniques and its applications. *IEEE Access*, 5: 16040–16067.

32. Sharma, M. (2016). A review: image fusion techniques and applications. *Int J Comput Sci Inf Technol*, 7(3): 1082–1085.

33. Jeevanand, N, Verma, PA & Saran, S. (2018). Fusion of hyperspectral and multispectral imagery with regression Kriging and the Lulu operators. A comparison. *Int Arch Photogramm Remote Sens Spatial Inf Sci*, 5: 583–588.

34. Liu, X, Yu, A, Wei, X, Pan, Z & Tang, J. (2020). Multimodal MR image synthesis using gradient prior and adversarial learning. *IEEE J Sel Top Signal Process*, 14: 1176–1188.

35. Amin, J, Sharif, M, Gul, N, Yasmin, M & Shad, SA. (2020). Brain tumor classification based on DWT fusion of MRI sequences using convolutional neural network. *Pattern Recognit Lett*, 129: 115–122.

36. Preethi, S & Aishwarya, P. (2021). An efficient wavelet-based image fusion for brain tumor detection and segmentation over PET and MRI image. *Multimed Tools Appl*, 80: 14789–14806.

37. Umri, BK, Akhyari, MW & Kusrini, K. (2020). Detection of covid-19 in chest X-ray image using CLAHE and convolutional neural network. *2nd International Conference on Cybernetics and Intelligent System (ICORIS)*, IEEE, Manado, Indonesia, 1–5.

38. Deng, LJ, Feng, M & Tai, XC. (2019). The fusion of panchromatic and multispectral remote sensing images via tensor-based sparse modeling and hyper-Laplacian prior. *Information Fusion*, 52, 76–89.

39. Liu, Q, Wang, Y & Zhang, Z. (2014). Pan-sharpening based on geometric clustered neighbor embedding. *Opt Eng*, 53(9): 2014.

40. Yuan, Q, Wei, Y, Meng, X, Shen, H & Zhang, L. (2018). A multiscale and multi-depthconvolutional neural network for remote sensing imagery pan-sharpening. *IEEE J Sel Top Appl Earth Observ Remote Sens*, 11(3): 978–989.

41. Adu, J, Wang, M, Wu, Z & Hu, J. (2012). Infrared image and visible light image fusion based on nonsubsampled contourlet transform and the gradient of uniformity. *Int J Adv Comput Technol*, 4(5): 114–121.

42. Liu, S, Piao, Y & Tahir, M. (2016). Research on fusion technology based on low-light visibleimage and infrared image. *Opt Eng*, 55(12): 123104.
43. Bhat, S & Koundal, D. (2021). Multi-focus image fusion techniques: a survey. *Artif Intell Rev*, 54: 5735–5787.
44. Li, D, Li, J, Bhatti, UA, Nawaz, SA, Liu, J, Chen, YW & Cao, L. (2023). Hybrid encrypted watermarking algorithm for medical images based on DCT and improved DarkNet53. *Electronics*, 12(7): 1554.
45. Bhatti, UA, Tang, H, Wu, G, Marjan, S & Hussain, A. (2023). Deep learning with graph convolutional networks: an overview and latest applications in computational intelligence. *Int J Intell Syst*, 2023: 1–28.
46. Sheng, M, Li, J, Bhatti, UA, Liu, J, Huang, M & Chen, YW. (2023). Zero watermarking algorithm for medical image based on Resnet50-DCT. *CMC-Comput Mater Continua*, 75(1): 293–309.
47. Liu, J, Li, J, Ma, J, Sadiq, N, Bhatti, UA & Ai, Y. (2019). A robust multi-watermarking algorithm for medical images based on DTCWT-DCT and Henon map. *Appl Sci*, 9(4): 700.
48. Fan, Y, Li, J, Bhatti, UA, Shao, C, Gong, C, Cheng, J & Chen, Y. (2023). A multi-watermarking algorithm for medical images using Inception V3 and DCT. *CMC-Comput Mater Continua*, 74(1): 1279–1302.
49. Li, T, Li, J, Liu, J, Huang, M, Chen, YW & Bhatti, UA. (2022). Robust watermarking algorithm for medical images based on log-polar transform. *EURASIP J Wireless Commun Networking*, 2022(1): 1–11.
50. Bhatti, UA, Huang, M, Wu, D, Zhang, Y, Mehmood, A & Han, H. (2019). Recommendation system using feature extraction and pattern recognition in clinical care systems. *Enterprise Information Syst*, 13(3): 329–351.
51. Bhatti, UA, Yu, Z, Chanussot, J, Zeeshan, Z, Yuan, L, Luo, W, … & Mehmood, A. (2021). Local similarity-based spatial–spectral fusion hyperspectral image classification with deep CNN and Gabor filtering. *IEEE Trans Geosci Remote Sensing*, 60: 1–15.
52. Bhatti, UA, Yu, Z, Li, J, Nawaz, SA, Mehmood, A, Zhang, K & Yuan, L. (2020). Hybrid watermarking algorithm using Clifford algebra with Arnold scrambling and chaotic encryption. *IEEE Access*, 8: 76386–76398.
53. Liu, J, Li, J, Zhang, K, Bhatti, UA & Ai, Y. (2019). Zero-watermarking algorithm for medical images based on dual-tree complex wavelet transform and discrete cosine transform. *J Med Imaging Health Inform*, 9(1): 188–194.

CHAPTER 10

Detecting Phishing URLs through Deep Learning Models

Shah Noor[1], Sibghat Ullah Bazai[2], Saima Tareen[1], and Shafi Ullah[2]

[1]*Department of Computer Engineering, Balochistan University of Information Technology, Engineering, and Management Sciences (BUITEMS), Quetta, Pakistan*
[2]*Department of Computer Science, Balochistan University of Information Technology, Engineering, and Management Sciences (BUITEMS), Quetta, Pakistan*

10.1 INTRODUCTION

The development of communication technologies and digitalization has made life faster and more accessible particularly during the pandemic lockdown when all of the transactions and life necessities had to be obtained online such as shopping and payments, as opposed to physically doing so. To accomplish daily tasks, people just need to open a smart electronic device; enter a URL; or search websites such as shopping stores, bookstores, educational platforms, and pharmacy websites. Opposite of this, the expansion of e-services raises the possibility for attackers to steal, get, and exploit user data including phone numbers, names, credit card information, and identification. Thus, users encounter various cyber-attacks and online threats every day. Therefore, hackers have begun to deceive online web customers by delivering a website, message, or email that resembles a shopping or banking website. Additionally, if a person clicks on a malicious link or email unintentionally and performs transactions, the hacker has the ability to get highly sensitive data and even install malware applications into the user system.

Phishing is a cybercrime that has been defined as the most known approach for hackers to attain a user's confidential data, including credit card numbers, passwords, and user names (Paliath, Qbeitah, & Aldwairi, 2020). DNS-based phishing, web-based delivery, phone phishing, email/spam, malware-based phishing, and spoofing are a few well-known examples of phishing attempts. Phishing attacks come in various forms and generally include different types of communication channels including social media,

DOI: 10.1201/9781003427674-10

email, instant chats, and quick response (QR) codes (Geng, Yan, Zeng, & Jin, 2018). Frequently attackers imitate famous banks, companies, and credit card websites to persuade a user to log in or sign up for phishing websites and submit their credentials records (Jahangeer et al., 2023). In recent years, the number of phishing attacks has increased exponentially according to a report published in the Anti-Phishing Working Group (APWG); around 68,000 and 94,000 phishing attack cases have been recorded per month since the early 2020s; 12,70883 total phishing attack cases were observed in the third quarter of 2022.

In the literature, various approaches exist for identifying and resolving phishing URL problems as illustrated in Figure 10.1 (Yang, Zheng, Wu, Wu, & Wang, 2021). The blacklist method is used to store malicious or phishing URLs. Whenever a client tries to open a URL that exists in the blacklist, the system will prevent the user from opening it (Prakash, Kumar, Kompella, & Gupta, 2010). Then, several rule-based techniques have been employed to identify phishing websites. Unfortunately, all these methods have limited URL data features and require regular updates.

Artificial intelligence (AI) is rapidly evolving intelligent approaches like machine learning (ML) and deep learning (DL), which are effective in managing cybersecurity and providing protection for computer activities (Basit et al., 2021). In classical ML approaches,

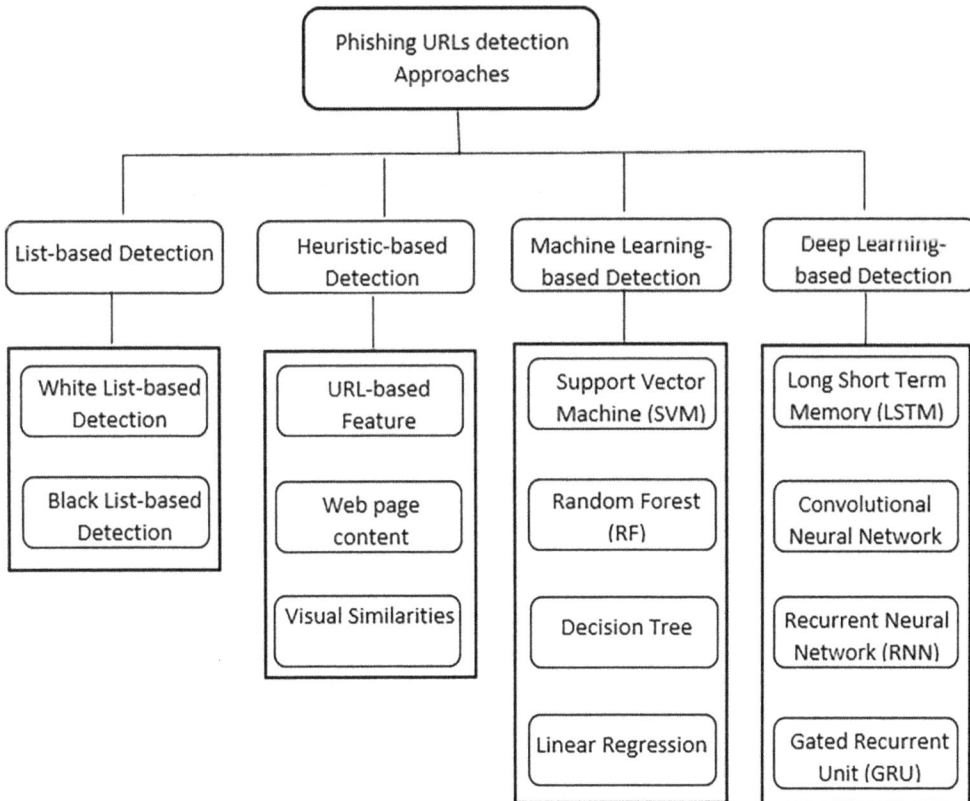

FIGURE 10.1 Phishing URL detection methods overview.

human knowledge is required for selection and feature extraction. Classification and feature selection tasks are separated (Ghafoor, Roomi, Aqeel, Sadiq, & Bazai, 2021). In contrast to ML, DL models reduce the requirement for manual attribute extracting and depend on third-party services due to automatic feature extracting and learning (Alsariera, Adeyemo, Balogun, & Alazzawi, 2020; Alsariera, Elijah, & Balogun, 2020).

The main benefit of DL over traditional ML techniques includes great performance and end-to-end problem solutions, particularly in situations involving big data sets like image classification, speech recognition, and phishing detection (Qazi et al., 2017). As opposed to ML and DL algorithms in several investigations, authors concluded that DL-based approaches outperformed ML-based approaches in terms of accuracy for detecting phishing websites (Bagui, Nandi, Bagui, & White, 2021). DL models train a system to imitate human brains by learning them using sample data sets. Using the DL-based method, a computer algorithm may also learn directly how to carry out detection works utilizing large data sets that include images, voice, and text. Data must be preprocessed before being fed into DL models since they require the input data to be transformed into numerical vectors. Because of the large number of training parameters, a neural network model's training time represents a considerable overhead. Despite a few drawbacks of DL models, their capacity to accurately classify the input data and automatically extract higher-level features from the URL makes them ideal for detecting phishing URLs (Tareen et al., 2022).

The chapter will go through DL models and benchmarks mostly used for training and evaluating models to detect cybersecurity threats. This chapter also seeks to provide a comprehensive overview of previous works that used different DL-based models for detecting phishing URLs and highlight data sets, features, and DL-based architecture that are used for detecting cybersecurity threats, specifically phishing URLs.

The rest of this chapter is organized as follows: section 10.2 represents a brief description of DL-based models that are mostly used for cybersecurity classification problems. Section 10.3 overviews matrices that are frequently used for model evaluation purposes. Section 10.4 discusses numerous cybersecurity threats and previous work related to them. Existing work related to phishing URL detection is presented in section 10.5. Section 10.6 concludes the chapter by providing a short summary of the overall chapter.

10.2 DL MODELS USED IN CYBERSECURITY

10.2.1 Convolutional Neural Network

Convolutional neural network (CNN) is a DL-based model designed to process input from arrays. CNN is mostly used for image processing tasks (2D array of the pixel) and three-dimensional (3D) arrays, such as volumetric photos and videos (Aamir et al., 2019). CNN is also used with a one-dimensional array (1D) of signals although less frequently. Regardless of dimensionality, CNNs are used when there is spatial or temporal ordering (Mosavi, Ardabili, & Varkonyi-Koczy, 2020).

CNN consists of a three-layer pooling layer, a convolutional layer, and a fully connected layer, as illustrated in Figure 10.2. The training process has two stages: the feed-forward

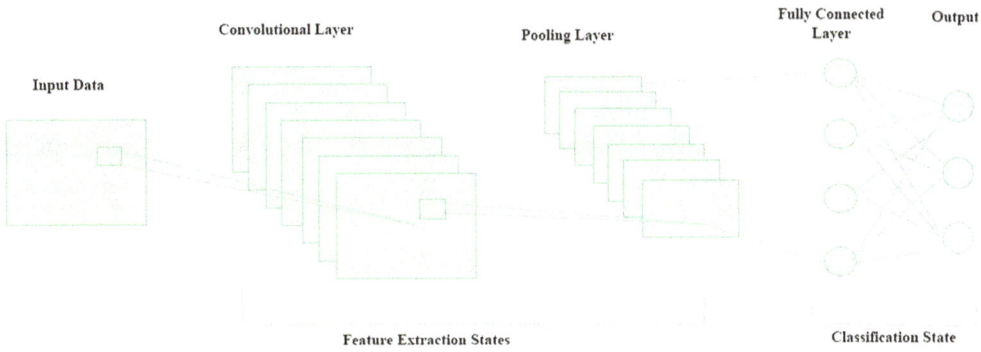

FIGURE 10.2   CNN architecture.

stage and the back-propagation stage in each CNN. The convolutional layer is used to extract numerous features from given input data. This layer is primarily responsible for computation as convolution mathematical operation is executed among input data and particular size filters. Performing required computation tasks and reducing the representation's spatial size is the responsibility of the pooling layer where the fully connected layer is used to link between the prior and recent layers.

10.2.2 Recurrent Neural Networks

A recurrent neural network (RNN) extends the potential of a traditional neural network by handling input sequences of variable length, which can only accept inputs of fixed length. A RNN's primary duty is to recognize patterns and sequences in input data, which can include speech, handwriting, text, and other applications. In a RNN architecture, the outputs from earlier states are fed into the current state. Remembering data is the capability of hidden layers. The hidden state is updated based on the results generated in the previous state. Figure 10.3 represents the complete architecture of RNN.

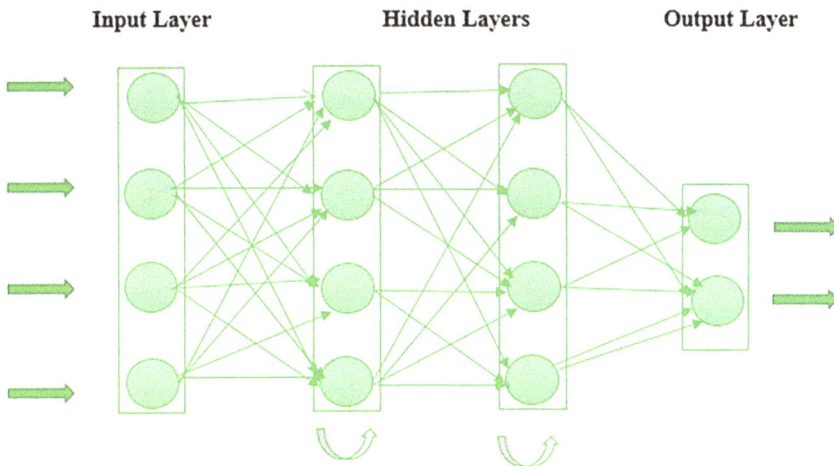

FIGURE 10.3   RNN architecture.

10.2.3 Long Short-Term Memory

Long short-term memory (LSTM) is a sequential neural network that is able to learn long-term dependencies from input data. LSTM is used for image processing applications and sequence and pattern detection tasks. LSTM consists of three units named input gate, output gate, and forget gate. The input gate is in charge of acquiring new information from the input data, and the output gate is utilized to transmit the updated data to the next timestamp. The forget gate determines whether irrelevant information from the previous state should be remembered or not.

10.2.4 Deep Belief Networks

One of the most trustworthy deep learning techniques, deep belief network (DBN) has outstanding accuracy and computational efficiency. As a result, there are many different application domains, including fascinating applications in a variety of scientific and engineering challenges. DBN consists of different layers, except for connections among units within each layer, this approach has many connections between the layers. As a hybrid, multi-layered neural network with both directed and undirected connections, DBNs can be considered. In order to regenerate the inputs, DBN models are often trained by individually altering the weights in each hidden layer.

10.2.5 Multi-Layer Perceptron

As shown in Figure 10.4, the multi-layer perceptron (MLP) is a neural network with three layers: the input layer, hidden layer, and output layer. The input layer is where the signal is received for processing. The essential activities, including categorization and prediction, are carried out by the output layer. An arbitrary number of hidden layers responsible for real computation are placed between the input and output layers. MLP neurons use backpropagation learning algorithms for the training state. MPL architecture is mostly used for pattern prediction, classification, and identification.

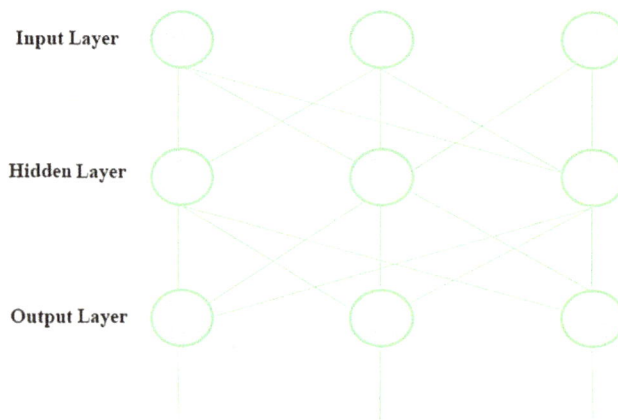

FIGURE 10.4 Multi-layer perceptron architecture.

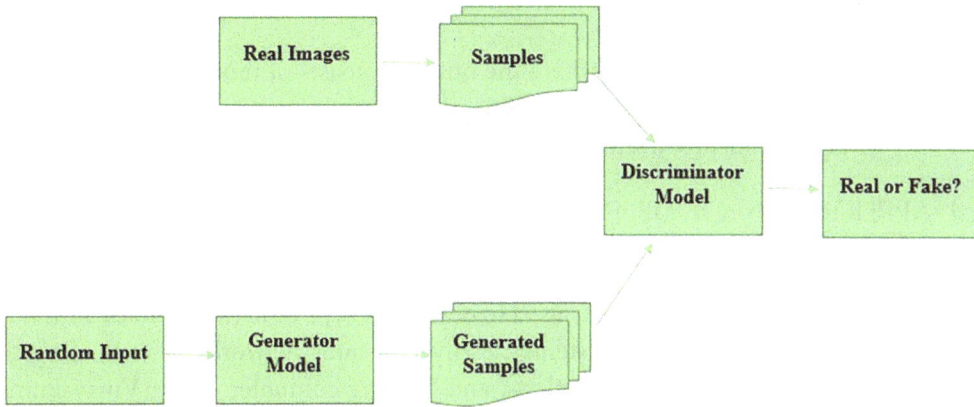

FIGURE 10.5   GAN architecture.

10.2.6 Generative Adversarial Network

Generative adversarial network (GAN) is an unsupervised DL-based model that automatically learns patterns and sequences from the data set. Two networks are arranged in opposition to one another in GAN architecture where each network plays a unique role. New data instances are generated by one neural network, known as the generator, where the discriminator neural network is used to discriminate generated samples as real or fake. Usually, GAN training time is huge. GAN architecture is illustrated in Figure 10.5.

10.3 METRICS

Metrics are used in ML and DL to evaluate model performance, compare different models, and identify the best-performing model. This section represents numerous benchmarks mostly used by the author cited in the next section.

10.3.1 Accuracy

Accuracy can be defined as the ratio of samples with accurate classification to all items. When data set classes are unbalanced, accuracy usefulness is lower.

10.3.2 Precision

Precision explains the proportion of correctly predicted scenarios that actually resulted in a positive outcome. Precision is helpful when false positives are more concerned than false negative samples.

10.3.3 Recall (Sensitivity)

Recall indicates the proportion of successful predictions made by the model. The recall is a more valuable matrix when a false negative is more significant than a false positive.

10.3.4 F1 Score

F1 score measures proposed model accuracy by combining precision and recall scores. The F1 score is mostly used for binary classification problems.

10.3.5 Confusion Matrix

For the DL classification problems where the output consists of more than one class, the confusion matrix is performance evaluation. A confusion matrix is basically a table consisting of actual and predicted values.

10.4 APPLICATION OF DEEP LEARNING IN CYBERSECURITY USE CASES

10.4.1 Intrusion Detection System

An intrusion detection system (IDS) is used to detect any type of network attack within the environment. Intrusion detection systems classify malicious networks by evaluating collected packets, blocking attack connections, and alerting computer users (Vinayakumar, Soman, & Poornachandran, 2017). For network security, IDS connects with the firewall as a fundamental tool. Generally, intrusion detection techniques are categorized into two types: anomaly detection and signature-based detection. Anomaly detection–based approaches are more effective as they detect new and unknown abnormal traffic or attack, whereas signature-based approaches detect known attacks and their patterns are already available (Khraisat, Gondal, Vamplew, & Kamruzzaman, 2019). Signature-based methods fail to detect unknown attacks.

The DL-based model shows remarkable results in resolving intrusion detection problems. In the paper (Su, Sun, Zhu, Wang, & Li, 2020), the authors propose a BAT-MC algorithm for intrusion detection where the model is composed of BLSTM and an attention mechanism. BLSTM mechanism is utilized to excerpt traffic attributes from the traffic bytes of each packet and attention is used to perform features engineering tasks on the network flow vector. The proposed methodology achieves 84% accuracy. For resolving network intrusion detection issues, authors (Imrana, Xiang, Ali, & Abdul-Rauf, 2021) use the bidirectional long short-term memory (BLSTM) model. The proposed methodology uses NSL-KDD for training and evaluating the model.

10.4.2 Malware Detection

Malware is a program that minimizes performance and vulnerability by damaging files or data in the system. It may result in the complete corruption of a system or server in some circumstances (Saxe & Berlin, 2015). Using unauthorized software, malware is easily spread through multiple environments. Trojans, viruses, worms, rootkits, adware, and ransomware are involved in malware. In recent years, malware detection has become a more challenging and difficult task (Xu et al., 2019). To deceive detectors and prevent "pattern matching" detection, malware writers employ a variety of techniques, including polymorphism and metamorphism. In order to quickly develop malware programs with a limitless number of versions that may readily avoid conventional pattern-matching detections, malware attackers have built a number of automated malware-generating toolkits.

To resolve these issues, recently researchers use DL-based models such as authors (Elayan & Mustafa, 2021) use GRU deep learning–based model for detecting malware in Android os applications. The author uses the CICAndMal2017 data set for training the GRU model. The author performs a comparative analysis between the proposed model and the traditional machine learning model where the GRU model achieves 98.2% higher accuracy compared to machine learning approaches (Aamir et al., 2023). In the paper (Akhtar & Feng, 2022), the authors use a hybrid CNN-LSTM model for malware detection, where the authors achieved 98% accuracy using 43,867 data samples.

10.4.3 Botnet Detection

A botnet is a network that has been infected with malware and is managed by a single attacker known as the bot-herder. Using many channels, including emails and spam, a cyber-attacker targets numerous users with their malware script and takes over the command and control (CC) of the victim device (Vormayr, Zseby, & Fabini, 2017). Attackers use these affected devices for designing a network and perform several destructive tasks like executing distributed denial-of-service (DDoS) attacks and sending spam emails. Botnet attacks are the most serious cyber threat for all inter-connected systems, as finding bot-herders and the compromised network requires a complex investigation by cybersecurity analysts (Milosevic, Dehghantanha, & Choo, 2017).

Anomaly-based and signature-based methods were used for botnet detection, but they required regular updates. ML-based methods are also used for botnet detection, but these algorithms depend on third-party services for feature engineering. To overcome these issues, recently DL methods were used for botnet detection. In the paper (Shi & Sun, 2020), the author proposes a DL-based method to detect botnets by analyzing time-based attributes of network traffic. LSTM, RNN, and hybrid models (RNN-LSTM) are used for training purposes, where the author achieved 95%, 97%, and 99% accuracy for RNN, LSTM, and LSTM-RNN models. For detecting zero-day botnet network traffic, in the paper (Nugraha, Nambiar, & Bauschert, 2020), authors use four DL models, named CNN, LSTM, CNN-LSTM, and multilayer perceptron, where the authors use a CTU-13 data set for training and evaluating the model.

10.4.4 Network Traffic Identification

For detecting potential security threats and enhancing the quality of service, network traffic detection is one of the effective methods that enables the prediction of traffic on a network (Ullah, Bazai, Aslam, & Shah, 2023). Because of this reason, network traffic classification has become an essential part of network management for governmental organizations, Internet service providers (ISPs), and major corporations. Network traffic detection is used in numerous areas, such as network security, network management, and network design (Zaland, Bazai, Marjan, & Ashraf, 2021).

Nowadays, DL-based methods are used for network traffic identification as in the paper (Izadi, Ahmadi, & Rajabzadeh, 2022); the authors propose deep learning and a data diffusion–based approach for classifying network traffic with a higher accuracy rate. The authors use three algorithms (CNN, MLP, and deep belief network [DBN]) to extract features and classify network traffic. Then a Bayesian fusion approach is used to combine the result of the three algorithms and classification. The authors achieved 97% accuracy.

10.4.5 Credit Card Fraud Detection

One of the key detections that banks are concerned with is credit card fraud, which comprises of detecting credit card payments and stolen cards. Due to the popularity and enhancement of online banking, credit card fraud detection is increasing. Detecting fraudulent credit card transactions has become a major issue for Internet shoppers. Various effective methods are used for protecting credit card transactions such as tokenization and credit card data encryption (Iwasokun, Omomule, & Akinyede, 2018). Unfortunately, these methods do not completely protect credit card transactions from fraud. The DL-based method shows more effective results compared to other traditional methods, as in the paper (Pillai, Hashem, Brohi, Kaur, & Marjani, 2018); the authors used a multi-layer perceptron model for detecting credit card fraud detection problems. The model is trained or evaluated using 284,807 transaction records, taken from the data set that had been collected through the research cooperation of Machine Learning Group and Wordline. The proposed model achieved 82% sensitivity. In 2019, the authors (Shenvi, Samant, Kumar, & Kulkarni, 2019) proposed a deep neural network model for credit card fraud detection, where the authors used oversampling and under-sampling methods for enhancing accuracy.

For resolving cybersecurity issues, DL can provide new methods. DL-based models show remarkable performance as opposed to traditional rule-based, signature-based, and classical ML-based methodologies. Table 10.1 lists the DL papers applied to solving various cybersecurity problems. This table lists the deep learning–based algorithm, data sets, and benchmarks that are used by the author for training and evaluating the model.

Several researchers used various matrices and different data sets, as highlighted in Table 10.1, so it becomes difficult to make conclusions regarding the effectiveness of any one technique. However, most authors used 1D CNN, LSTM, and CNN-LSTM models for detection purposes.

10.5 EXISTING WORK RELATED TO PHISHING URL DETECTION USING DL MODELS

The phishing URL issue is challenging and complex in itself since there is no specific approach to effectively eliminate all the risks. DL-based phishing URL detection methods have emerged to identify phishing URLs. Moreover, the DL-based models have become more efficient in cybersecurity.

In the paper (Singh, Singh, & Pandey, 2020), the author uses the CNN model to detect phishing URLs. The proposed technique works on the basis of real-time analysis of malicious websites. The data set consists of 73,575 URLs. For model training, the authors first tokenize the URL data, pass tokenized data into the embedding layer, and then use CNN for classification purposes. The author (Al-Ahmadi, 2020) uses two CNNs for website phishing detection. The author used both image and URL data sets for training and evaluating the model, where one CNN is used for extracting URL features and detecting whether the URL is benign, and another CNN is used for image data to extract and classify images as benign or malicious. After combining both models' results, the outcome is reported as benign or malicious web.

TABLE 10.1 DL-Based Methods Used in Different Cybersecurity Threats

Year	Cybersecurity area	Algorithm	Data set	Data set size	Features	Benchmark
2018 (Pillai et al., 2018)	Credit Card Fraud Detection	MLP	Machine Learning Group and Wordline Collaboration data	284,807 records	28 Features	Sensitivity, Accuracy, F1-score Precision
2019 (Shenvi et al., 2019)	Credit Card Fraud Detection	Neural Network	Aforementioned data set	2,84,315 Transactions	Features (Automatic)	Accuracy, Precision, Recall, Confusion Matrix
2019 (Gharib, Mohammadi, Dastgerdi, & Sabokrou, 2019)	Intrusion Detection	Auto-encoder	NSLKDD	125973	Features (Automatic)	Accuracy, Precision, Recall
2019 (Chen, Wang, Wen, Lai, & Sun, 2019)	Malware Detection	CNN	GUN Open Source Project, Honeypot, NCHC Windows System Files, Github, Malware Knowledge Base	22,000 source and binary code	Features (Automatic)	Recall, Accuracy, Precision, Confusion Matrix
2020 (Cheng et al., 2020)	Credit Card Fraud Detection	3D CNN, Attention	Real-World Data Set	236,706	Features (Automatic)	Accuracy, Precision, Recall, F1-score
2020 (Su et al., 2020)	Intrusion Detection	Hybrid Model (LSTM, Attention)	NSL-KDD	125,973 Instances	Features (Automatic)	Accuracy, Confusion Matrix
2020 (Shi & Sun, 2020)	Botnet Detection	RNN, LSRM, Hybrid Model (LSTM-RNN)	Malware Capture Facility Project (MCFP)	Not Mentioned	35 network features	Accuracy, Precision, Recall, F1-score
2020 (Nugraha et al., 2020)	Botnet Detection	CNN, LSTM, CNN-LSTM, MLP	CTU-13 Data Set	Use three scenarios. Each scenario has different instances	15 Features	Sensitivity, Accuracy, Specificity, F1-score Precision

(Continued)

TABLE 10.1 (Continued) DL-Based Methods Used in Different Cybersecurity Threats

Year	Cybersecurity area	Algorithm	Data set	Data set size	Features	Benchmark
2021 (Imrana et al., 2021)	Intrusion Detection	BLSTM	NSL-KDD	22,544 Traffic Instances	Features (Automatic)	Accuracy, False Alarm Rate, Recall, Precision, F-score
2021 (Alkahtani & Aldhyani, 2021)	Botnet Detection	CNN-LSTM	N-BaIoT	65,179	Features (Automatic)	Accuracy, Precision, Recall
2021 (Coleman & Hwang, 2021)	Android Malware	1D CNN-LSTM	Debin, APK Pure, Amazon	20,000 Application Packages	16 Features (Automatic)	Accuracy, recall, F1-score, Precision
2021 (Elayan & Mustafa, 2021)	Android Malware	GRU	CICAndMal2017	712 samples	Two static features(Android applications permissions, API calls)	Recall, Accuracy, Precision
2022 (Akhtar & Feng, 2022)	Malware Detection	Hybrid Model (CNN-LSTM)	Kaggle	43867 Samples	Features (Automatic)	Recall, Accuracy, Precision, F1-score
2022 (Izadi et al., 2022)	Network Traffic Identification	CNN, MLP, DBN	ISCX VPN-nonVPN	28 GB data	Features (Automatic)	Recall, Accuracy, Precision

The paper (Ariyadasa, Fernando, & Fernando, 2020) proposes CNN- and LSTM-based models for phishing URL detection. URLs and HTMLs are used as the data set for training and evaluating the model. LSTM and 1D convolutional are used for learning URLs and other 1D convolutional work remembering the HTML features. Both networks were trained separately while combined through the sigmoid layer. The proposed model achieved 97% accuracy. In (Yerima & Alzaylaee, 2020), the author proposed the CNN model for phishing URL detection. The model is trained and evaluated using 6,157 benign and 4,898 benchmark phishing URL data sets. (Yang et al., 2021) proposed multi-level phishing URL detection using random forest and a CNN model. For converting URLs into a fixed-size input matrix, a character embedding method is used. CNN is used for extracting attributes at various levels, whereas random forest is used for multi-level attribute classification. In the last for output prediction, the author used the winner-take-all method. To solve the phishing URL problem, the authors (Zhang, Bu, Chen, Zhang, & Lu, 2021) proposed the CNN-BiLST algorithm. CNN-BiLST automatically extracts URL attributes using a hybrid neural network. Sensitive word segmentation is used for word segmentation processing on URL data. After converting the URL into a feature vector matrix, CNN automatically extracts local features and BiLSTM is used to obtain bidirectional long-distance dependent attributes. In the last, multi-level attributes are fed into the full connection layer. The Softmax function is used to classify data using collected features.

(Assefa & Katarya, 2022) proposed an auto-encoder-based model for resolving phishing URL problems. In the autoencoder, the outlier analysis method is used to classify and discriminate website URLs as benign or malicious. (Aljabri & Mirza, 2022) used an ML-based method for phishing URL detection. The author used two data sets for evaluating and training the model, where the first data set consisted of 11,055 URLs and the second data set comprised 14,093 URLs. Three main features were found in both data sets: content-based, domain-based, and URL lexical-based attributes. The authors use multiple algorithms, such as random forest, SVM, logistic regression, and CNN to perform comparative analysis among both data sets. In (Almousa, Zhang, Sarrafzadeh, & Anwar, 2022), the authors proposed a systematic solution for URL phishing classification using DL and a hyperparameter optimization algorithm. The proposed methodology utilized three categories of website attributes: content-based, URL-based, and hybrid. The authors performed a comparative analysis among fully connected DNN, CNN, LSTM, and two optimization algorithms.

In (Alshingiti et al., 2023), the authors proposed CNN, LSTM, and LSTM-CNN models for phishing URL classification. The authors used the SelectKBest method for preprocessing data and then applied a classification algorithm, where LSTM-CNN performs better than other proposed approaches.

A brief summary of the literature review is represented in Table 10.2. Five different matrices are used in the table, where the algorithm column represents all models that are used by the author for detecting phishing URLs, data size shows the total number of URLs used for the training model, and the number of features used for the detection of phishing URLs and benchmarks represent the evaluating matrices used for testing the model.

TABLE 10.2 List of Studies Based on Phishing URL Detection Using Deep Learning Model

Year	Algorithm	Data set source	Data set size	Features	Benchmark
2020 (Ariyadasa et al., 2020)	Hybrid Model based on LSTM and CNN	Github, Phishtank	40,000 Instances	Automatic (Abstract-Level Features)	Accuracy, Precision, Confusion Matrix, Recall
2020 (Rasymas & Dovydaitis, 2020)	Hybrid Model based on CNN and LSTM	Kaggle, Phishtank	2,585,146	Automatic (Word-Level Embedding Feature, Lexical Feature, Character-Level Embedding Feature)	Accuracy, Precision, ROC Curve, F1-Score
2020 (Aljofey, Jiang, Qu, Huang, & Niyigena, 2020)	CNN	Alexa, Openphish, Spamhaus, Techhelplist	318,642 URLs	Automatic Character-Level Embedding Features	Accuracy, F1-Score, Precision, Recall, Training And Test Time, AUC Value
2020 (Wei et al., 2020)	CNN	Common Crawl, PhishTank	21,208 URLs	Automatic	Accuracy, ROC Curve, Precision
	Multilayer Perceptron (MPL)	UCI, Kaggle	13,511 URLs	31 Features	Accuracy, Confusion Matrix Precision, F-Measure, Recall
2020 (Somesha, Pais, Rao, & Rathour, 2020)	Deep Neural Network (DNN), LSTM, CNN	PhishTank, Alexa	3,526 Samples	10 Features	Accuracy, Error Rate
2020 (Saha et al., 2020)	MLP	Kaggle	10,000	10 Features	Accuracy, Confusion Matrix
2021 (Pham, Pham, Hoang, & Ta, 2021)	GAN, LSTM, Gated recurrent units (GRU)	Common Crawl, Common Crawl 2	100,000	Automatic (Character Embedding Features)	Accuracy, F1 Value, Precision
2021 (Dutta, 2021)	LSTM	PhishTank, Alexa	13,700 URLs	Automatic	Learning Rate, Accuracy, F1 Score
2021 (Dilhara, 2021)	Three non-hybrid DL methods such as GRU, CNN (1D), LSTM, and four hybrid DL methods named BI (LSTM)-LSTM, GRU-LSTM, BI(GRU)-LSTM, LSTM-LSTM	Mendeley Data	88,647 URLs	Automatic	Accuracy, Recall Rate, F1 Value, Confusion Matrix, Precision

Year (Reference)	Model	Dataset	Data Size	Features	Metrics
2021 (Zhang et al., 2021)	Hybrid CNN-BiLSTM Model	PhishTank, MalwarePatrol, DMOZ, Alexa	206,200 Tagged URLs	Automatic (Multi-Layer Features)	Accuracy, Recall Rate, F1 Value
2021 (Yang et al., 2021)	Hybrid model based on CNN and Random Forest	ALEXA, PhishTank	47,210 URLs	Automatic (Multi-Layer Features)	Accuracy, Precision, Recall, F-Measure
2021 (Catal, Donmez, & Senturk, 2021)	Hybrid DL-based on LSTM and DNN algorithm	Ebbu2017, PhishTank	73,575, 26,000	40 (Nlp Features), Character Embedding	Confusion Matrix. Accuracy, Area Under Roc Curve (Auc), F1-Score
2022 (Fujita, 2022)	CNN, BiGRU, BiLSTM	Ebbu2017 Phishing Dataset	73,575	Automatic (Lexical Features)	Accuracy, Precision, Receiver Operating Characteristic Curve (ROC)
2022 (Alanzi & Uliyan, 2022)	LSTM	PhishTank	194,798 URLs	10 Features	Accuracy, Precision, Recall, F-Measure
2022 (Assefa & Katarya, 2022)	Autoencoder	PhishTank	10,000 URLs	15 Features	Accuracy, Confusion Matrix, F-Score
2022 (Siddiq, Arifuzzaman, & Islam, 2022)	Neural network, CNN	UCI Machine Learning Repository	10,055 Samples	30 Categorical Features	Accuracy, Precision
2022 (Almousa et al., 2022)	LSTM, Fully Connected DNN, CNN	Tan data set, Kumar data set, UCI data set, AZA data set	10,000 Webpages, 11,055 Webpages, 1353 Websites, 50 300 Webpages	29 URL-Based Features, 19 Content-Based Features	Accuracy, Precision
2023 (Alshingiti et al., 2023)	CNN, LSTM, LSTM-CNN	URL 2016	20,000 Records	30 Features	Accuracy, Precision, Recall, F-Score
2023 (Dhanavanthini & Chakkravarthy, 2023)	RNN	PhishTank, OpenPhish, Common Crawl	46,839 Instances	20 Lexical Features	Accuracy, Precision, Recall, F-Score, and Inference Time

Numerous DL-based methods are used for phishing URL detection, as illustrated in Table 10.2, where CNN, LSTM, RNN, CNN-LSTM, and GRU models show more promising results. PhishTank, ALEXA, and Common Crawl data sets are mostly used by authors for training models. For evaluating the model, the authors mostly used accuracy, precision, recall, and F1-score evaluation matrices.

10.6 CONCLUSION

Phishing attacks are the most effective and critical attacks in cybersecurity threats. Different traditional or ML-based methods are used for phishing URL detection purposes. Unfortunately, these methods require regular updates or another third-party service for feature extraction. Recently, DL-based methods are used for detecting purposes in various cybersecurity threats. This chapter aims to survey those articles, which use DL-based methods for cybersecurity threats and phishing URL detection specifically. This paper highlights benchmarks and data sets that they use for training and evaluating purposes.

REFERENCES

Aamir, M., Bazai, S. U., Bhatti, U. A., Dayo, Z. A., Liu, J., & Zhang, K. (2023). Applications of machine learning in medicine: Current trends and prospects. *2023 Global Conference on Wireless and Optical Technologies (GCWOT)*, (pp. 1–4).

Aamir, M., Rahman, Z., Ahmed Abro, W., Tahir, M., & Mustajar Ahmed, S. (2019). An optimized architecture of image classification using convolutional neural network. *International Journal of Image, Graphics and Signal Processing*, 11, 30–39 10.5815/ijigsp.2019.10.05.

Akhtar, M. S., & Feng, T. (2022). Detection of malware by deep learning as CNN-LSTM machine learning techniques in real time. *Symmetry*, 14(11), 2308.

Al-Ahmadi, S. (2020). A deep learning technique for web phishing detection combined URL features and visual similarity. *International Journal of Computer Networks & Communications (IJCNC)*, 12, 41–54.

Alanzi, B. M., & Uliyan, D. M. (2022). Detection of phishing websites by investigating their URLs using LSTM algorithm. *IJCSNS*, 22(5), 419.

Aljabri, M., & Mirza, S. (2022). Phishing attacks detection using machine learning and deep learning models. Paper presented at the 2022 7th International Conference on Data Science and Machine Learning Applications (CDMA).

Aljofey, A., Jiang, Q., Qu, Q., Huang, M., & Niyigena, J.-P. (2020). An effective phishing detection model based on character level convolutional neural network from URL. *Electronics*, 9(9), 1514.

Alkahtani, H., & Aldhyani, T. H. (2021). Botnet attack detection by using CNN-LSTM model for Internet of Things applications. *Security and Communication Networks*, 2021, 1–23.

Almousa, M., Zhang, T., Sarrafzadeh, A., & Anwar, M. (2022). Phishing website detection: How effective are deep learning-based models and hyperparameter optimization? *Security and Privacy*, 5(6), e256.

Alsariera, Y. A., Adeyemo, V. E., Balogun, A. O., & Alazzawi, A. K. (2020). AI meta-learners and extra-trees algorithm for the detection of phishing websites. *IEEE Access*, 8, 142532–142542.

Alsariera, Y. A., Elijah, A. V., & Balogun, A. O. (2020). Phishing website detection: Forest by penalizing attributes algorithm and its enhanced variations. *Arabian Journal for Science and Engineering*, 45, 10459–10470.

Alshingiti, Z., Alaqel, R., Al-Muhtadi, J., Haq, Q. E. U., Saleem, K., & Faheem, M. H. (2023). A deep learning-based phishing detection system using CNN, LSTM, and LSTM-CNN. *Electronics*, 12(1), 232.

Ariyadasa, S., Fernando, S., & Fernando, S. (2020). Detecting phishing attacks using a combined model of LSTM and CNN. *International Journal of Advanced and Applied Sciences*, 7(7), 56–67.

Assefa, A., & Katarya, R. (2022). Intelligent phishing website detection using deep learning. Paper presented at the 2022 8th International Conference on Advanced Computing and Communication Systems (ICACCS).

Bagui, S., Nandi, D., Bagui, S., & White, R. J. (2021). Machine learning and deep learning for phishing email classification using one-hot encoding. *Journal of Computer Science*, 17(7), 610–623.

Basit, A., Zafar, M., Liu, X., Javed, A. R., Jalil, Z., & Kifayat, K. (2021). A comprehensive survey of AI-enabled phishing attacks detection techniques. *Telecommunication Systems*, 76, 139–154.

Chen, C.-M., Wang, S.-H., Wen, D.-W., Lai, G.-H., & Sun, M.-K. (2019). Applying convolutional neural network for malware detection. Paper presented at the 2019 IEEE 10th International Conference on Awareness Science and Technology (ICAST).

Cheng, D., Xiang, S., Shang, C., Zhang, Y., Yang, F., & Zhang, L. (2020). Spatio-temporal attention-based neural network for credit card fraud detection. Paper presented at the Proceedings of the AAAI Conference on Artificial Intelligence.

Coleman, S.-P. W., & Hwang, Y.-S. (2021). Malware detection by merging 1D CNN and bi-directional LSTM utilizing sequential data. Paper presented at the Information Science and Applications: Proceedings of ICISA 2020.

Dhanavanthini, P., & Chakkravarthy, S. S. (2023). Phish-armour: Phishing detection using deep recurrent neural networks. *Soft Computing*, 1–13.

Dilhara, B. (2021). Phishing URL detection: A novel hybrid approach using long short-term memory and gated recurrent units. *International Journal of Computer Applications*, 975, 8887.

Dutta, A. K. (2021). Detecting phishing websites using machine learning technique. *PloS One*, 16(10), e0258361.

Elayan, O. N., & Mustafa, A. M. (2021). Android malware detection using deep learning. *Procedia Computer Science*, 184, 847–852.

Fujita, H. (2022). Malicious URL detection with distributed representation and deep learning. In New Trends in Intelligent Software Methodologies, Tools and Techniques: Proceedings of the 21st International Conference on New Trends in Intelligent Software Methodologies, Tools and Techniques (SoMeT_22) (Vol. 355, p. 171). IOS Press.

Geng, G.-G., Yan, Z.-W., Zeng, Y., & Jin, X.-B. (2018). RRPhish: Anti-phishing via mining brand resources request. Paper presented at the 2018 IEEE International Conference on Consumer Electronics (ICCE).

Ghafoor, M. I., Roomi, M. S., Aqeel, M., Sadiq, U., & Bazai, S. U. (2021). Multi-features classi-fication of SMD screen in smart cities using randomised machine learning algorithms. 2021 2nd International Informatics and Software Engineering Conference (IISEC), (pp. 1–5).

Gharib, M., Mohammadi, B., Dastgerdi, S. H., & Sabokrou, M. (2019). Autoids: auto-encoder based method for intrusion detection system. *arXiv preprint arXiv:1911.03306*.

Imrana, Y., Xiang, Y., Ali, L., & Abdul-Rauf, Z. (2021). A bidirectional LSTM deep learning approach for intrusion detection. *Expert Systems with Applications*, 185, 115524.

Iwasokun, G. B., Omomule, T. G., & Akinyede, R. O. (2018). Encryption and tokenization-based system for credit card information security. *International Journal of Cyber Security and Digital Forensics*, 7(3), 283–293.

Izadi, S., Ahmadi, M., & Rajabzadeh, A. (2022). Network traffic classification using deep learning networks and Bayesian data fusion. *Journal of Network and Systems Management*, 30(2), 25.

Jahangeer, A., Bazai, S. U., Aslam, S., Marjan, S., Anas, M., & Hashemi, S. H. (2023). A review on the security of IoT networks: From network layer's perspective. *IEEE Access*, 11, 71073–71087.

Khraisat, A., Gondal, I., Vamplew, P., & Kamruzzaman, J. (2019). Survey of intrusion detection systems: Techniques, data sets and challenges. *Cybersecurity*, 2(1), 1–22.

Milosevic, N., Dehghantanha, A., & Choo, K.-K. R. (2017). Machine learning aided Android malware classification. *Computers & Electrical Engineering*, 61, 266–274.

Mosavi, A., Ardabili, S., & Varkonyi-Koczy, A. R. (2020). List of deep learning models. Paper presented at the Engineering for Sustainable Future: Selected papers of the 18th International Conference on Global Research and Education Inter-Academia–2019.

Nugraha, B., Nambiar, A., & Bauschert, T. (2020). Performance evaluation of botnet detection using deep learning techniques. Paper presented at the 2020 11th International Conference on Network of the Future (NoF).

Ozcan, A., Catal, C., Donmez, E., & Senturk, B. (2021). A hybrid DNN–LSTM model for detecting phishing URLs. *Neural Computing and Applications*, 35, 4957–4973.

Paliath, S., Qbeitah, M. A., & Aldwairi, M. (2020). PhishOut: Effective phishing detection using selected features. Paper presented at the 2020 27th International Conference on Telecommunications (ICT).

Pham, T. D., Pham, T. T. T., Hoang, S. T., & Ta, V. C. (2021). Exploring efficiency of GAN-based generated URLs for phishing URL detection. Paper presented at the 2021 International Conference on Multimedia Analysis and Pattern Recognition (MAPR).

Pillai, T. R., Hashem, I. A. T., Brohi, S. N., Kaur, S., & Marjani, M. (2018). Credit card fraud detection using deep learning technique. Paper presented at the 2018 Fourth International Conference on Advances in Computing, Communication & Automation (ICACCA).

Prakash, P., Kumar, M., Kompella, R. R., & Gupta, M. (2010). Phishnet: Predictive blacklisting to detect phishing attacks. Paper presented at the 2010 Proceedings IEEE INFOCOM.

Qazi, E.-u.-H., Hussain, M., Aboalsamh, H., Malik, A. S., Amin, H. U., & Bamatraf, S. (2017). Single trial EEG patterns for the prediction of individual differences in fluid intelligence. *Frontiers in Human Neuroscience*, 10, 687.

Rasymas, T., & Dovydaitis, L. (2020). Detection of phishing URLs by using deep learning approach and multiple features combinations. *Baltic Journal of Modern Computing*, 8(3), 471–483.

Saha, I., Sarma, D., Chakma, R. J., Alam, M. N., Sultana, A., & Hossain, S. (2020). Phishing attacks detection using deep learning approach. Paper presented at the 2020 Third International Conference on Smart Systems and Inventive Technology (ICSSIT).

Saxe, J., & Berlin, K. (2015). Deep neural network based malware detection using two dimensional binary program features. Paper presented at the 2015 10th International Conference on Malicious and Unwanted Software (MALWARE).

Shenvi, P., Samant, N., Kumar, S., & Kulkarni, V. (2019). Credit card fraud detection using deep learning. Paper presented at the 2019 IEEE 5th International Conference for Convergence in Technology (I2CT).

Shi, W.-C., & Sun, H.-M. (2020). DeepBot: A time-based botnet detection with deep learning. *Soft Computing*, 24, 16605–16616.

Siddiq, M. A. A., Arifuzzaman, M., & Islam, M. (2022). Phishing website detection using deep learning. Paper presented at the Proceedings of the 2nd International Conference on Computing Advancements.

Singh, S., Singh, M., & Pandey, R. (2020). Phishing detection from URLs using deep learning approach. Paper presented at the 2020 5th International Conference on Computing, Communication and Security (ICCCS).

Somesha, M., Pais, A. R., Rao, R. S., & Rathour, V. S. (2020). Efficient deep learning techniques for the detection of phishing websites. *Sādhanā*, 45, 1–18.

Su, T., Sun, H., Zhu, J., Wang, S., & Li, Y. (2020). BAT: Deep learning methods on network intrusion detection using NSL-KDD data set. *IEEE Access*, 8, 29575–29585.

Tareen, S., Bazai, S. U., Ullah, S., Ullah, R., Marjan, S., & Ghafoor, M. I. (2022). Phishing and intrusion attacks: An overview of classification mechanisms. 2022 3rd International Informatics and Software Engineering Conference (IISEC), (pp. 1–5).

Ullah, R., Bazai, S. U., Aslam, U., & Shah, S. A. (2023). Utilizing blockchain technology to enhance smart home security and privacy. Proceedings of International Conference on Information Technology and Applications: ICITA 2022, (pp. 491–498).

Vinayakumar, R., Soman, K., & Poornachandran, P. (2017). Evaluating effectiveness of shallow and deep networks to intrusion detection system. Paper presented at the 2017 International Conference on Advances in Computing, Communications and Informatics (ICACCI).

Vormayr, G., Zseby, T., & Fabini, J. (2017). Botnet communication patterns. *IEEE Communications Surveys & Tutorials*, 19(4), 2768–2796.

Wei, W., Ke, Q., Nowak, J., Korytkowski, M., Scherer, R., & Woźniak, M. (2020). Accurate and fast URL phishing detector: A convolutional neural network approach. *Computer Networks*, 178, 107275.

Xu, X., Liu, Q., Zhang, X., Zhang, J., Qi, L., & Dou, W. (2019). A blockchain-powered crowd-sourcing method with privacy preservation in mobile environment. *IEEE Transactions on Computational Social Systems*, 6(6), 1407–1419.

Yang, R., Zheng, K., Wu, B., Wu, C., & Wang, X. (2021). Phishing website detection based on deep convolutional neural network and random forest ensemble learning. *Sensors*, 21(24), 8281.

Yerima, S. Y., & Alzaylaee, M. K. (2020). High accuracy phishing detection based on convolutional neural networks. Paper presented at the 2020 3rd International Conference on Computer Applications & Information Security (ICCAIS).

Zaland, Z., Bazai, S. U., Marjan, S., & Ashraf, M. (2021). Three-tier password security algorithm for online databases. 2021 2nd International Informatics and Software Engineering Conference (IISEC), (pp. 1–6).

Zhang, Q., Bu, Y., Chen, B., Zhang, S., & Lu, X. (2021). Research on phishing webpage detection technology based on CNN-BiLSTM algorithm. Paper presented at the Journal of Physics: Conference Series.

CHAPTER 11

Augmenting Multimedia Analysis

A Fusion of Deep Learning with Differential Privacy

Iqra Tabassum and Sibghat Ullah Bazai

Balochistan University of Information Technology, Engineering, and Management Sciences (BUITEMS), Quetta, Pakistan

11.1 INTRODUCTION

Multimedia data analysis in crowd sensing refers to the process of collecting, analyzing, and interpreting multimedia data such as images, videos, and audio from multiple crowdsourced sources, typically from mobile devices and social media platforms to gain insights or to solve problems. This data can be used to extract information from videos and images, such as identifying objects and events in real time, and can also be used to analyze social media content to monitor public sentiment or track the spread of information (AbualSaud et al., 2018). The process of multimedia data analysis in crowd sensing typically involves the following steps:

1. *Data Collection*: Crowdsensing leverages the collective intelligence and participation of a large number of individuals to collect data from diverse sources, typically through their mobile phones. In multimedia data analysis, the data collected can include images, videos, audio recordings, and other types of multimedia. For example, in a traffic monitoring application, individuals may use their mobile devices to capture images or videos to capture traffic congestion and road accidents. This data is then collected and used for further analysis.

2. *Data Preprocessing*: The collected data is preprocessed to remove any noise and irrelevant data. This step ensures that the data is clean and ready for analysis. In multimedia analysis, preprocessing can involve removing duplicates or low-quality

DOI: 10.1201/9781003427674-11

data, correcting errors in the data, and normalizing the data to ensure consistency across different sources.

3. *Data Analysis*: The preprocessed data is analyzed using various data analysis techniques, such as ML, computer vision, and natural NLP. In multimedia data analysis, these techniques can be used to identify patterns, trends, and insights in the data. For example, machine learning algorithms can be used to classify images or videos based on their content, such as identifying objects or events in the images. Natural language processing can be used to analyze audio recordings, such as speech recognition or sentiment analysis.

4. *Data Interpretation*: For a deeper understanding of the problem or topic under study, the data analysis findings are interpreted. This step involves making decisions or taking actions based on the insights gained from the data. For example, in a traffic monitoring application, the insights gained from analyzing the collected data may be used to optimize traffic flow or improve road safety.

Due to its inexpensive deployment costs and wide geographic coverage, crowdsensing is recognized as a potential data collection technique (L. Wang et al., 2017). Multimedia data analysis is a technique that can be utilized in crowdsensing to obtain valuable information from the data that is gathered by persons and sensors, such as identifying traffic congestion patterns from traffic camera images or detecting anomalies in air quality data from sensors. In a case to monitor traffic congestion, data is collected from the GPS sensors in individuals' mobile devices, which is then analyzed to generate real-time information on traffic flow and congestion.

DL techniques, such as CNNs and RNNs, have been particularly effective in analyzing multimedia data in crowdsensing. For example, CNNs can be used to automatically detect and classify objects in images (Aamir et al., 2023), while RNNs can be used to analyze speech and language data (Aamir et al., 2019).

However, privacy issues are significant in multimedia analysis, especially when working with personal information in images and videos. For example, when data is sensed in front-end devices, the sensed data might include some private information about mobile users. In addition, data can be manipulated with, or faked by, outside muggers or even contaminated by inside attackers (i.e., rogue users). Cloud servers may be inquiring enough to deduce the private/classified information of the query users and owners of data for the purpose of storing and processing data in the back-end devices, or even malevolent enough to change some query results (L. Wang et al., 2017). In these situations, it is important to consider a wide range of parameters, including data confidentiality, user privacy, data reliability, data quality, and access control. The privacy of individuals must be maintained while still allowing for data analysis; hence, methods like differential privacy and other privacy-preserving algorithms must be employed (Table 11.1).

TABLE 11.1   A List of the Chapter's Acronyms

DL	Deep Learning
ML	Machine Learning
NLP	Natural Language Processing
DNN	Deep Neural Network
CNNs	Convolutional Neural Networks
RNN	Recurrent Neural Network
NN	Neural Network
DP	Differential Privacy
CDP	Central Differential Privacy
LDP	Local Differential Privacy
FL	Federated Learning
HE	Homomorphic Encryption
SMC	Secure Multi-Party Computation
DP-SGD	Differential Private Stochastic Gradient Descent
SGD	Stochastic Gradient Descent
PII	Personally Identifiable Information
MLaaS	Machine Learning-as-a-Service

11.2 MULTIMEDIA DATA AND CROWDSENSING PRIVACY CONCERNS

Crowdsensing, which involves the collection of data from individuals through their mobile devices, can pose several privacy problems (AbualSaud et al., 2018; Waheed et al., 2020). Some of them are the following:

1. *Location Privacy:* Location data is often collected in crowdsensing applications, which can reveal a person's movements and activities. This information can be sensitive and can be used to track a person's whereabouts, creating a potential privacy risk.

2. *Personally Identifiable Information (PII):* Crowdsensing may collect PII, such as a person's name, email address, phone number, or other sensitive information. This data can be used to identify an individual and can be misused if it falls into the wrong hands.

3. *Inference Attacks:* Inference attacks involve using seemingly innocuous data to infer sensitive information about an individual. For example, data collected on a person's daily routine can reveal their medical data.

4. *Data Misuse:* The data collected in crowdsensing can be misused, either intentionally or unintentionally, by the researchers or other parties with access to the data. This can result in harm to individuals, such as discrimination, identity theft, or targeted advertising.

5. *Lack of User Control:* Crowdsensing often involves collecting data from individuals without their explicit consent or knowledge. This lack of control over personal data can make individuals uncomfortable and can create privacy concerns.

To address these privacy problems, it is important to implement strong privacy policies and data protection measures. This entails using privacy-preserving methods like encryption, anonymization, and DP as well as offering users more control over their data and maintaining transparency in data collection and use.

11.3 DEEP LEARNING AND PRIVACY RISKS

Deep neural networks serve as the cornerstone of the ML discipline known as deep learning. Many research fields, including computer vision, NLP, medical research, voice recognition, and many more, have adopted DL technologies to achieve cutting-edge outcomes in their quest to uncover complex patterns in high-dimensional data. Neural network layers serve as the building elements of the DL model. The model's parameters are trained by optimizing the output after the input data has been processed. The objective is to establish a connection among the input samples and the prediction, which improves the classification accuracy on large-scale, highly organized databases dramatically. Since DL has strong capabilities for data abstraction, representation learning is carried out using a variety of non-linear transformations. The connection is seen as a function:

$$F: X \rightarrow Y$$

Internet titans like Google, Amazon AWS ML, Microsoft Azure ML, etc. offer outsourced learning such as MLaaS, a black box API that offers DL service. Users can submit data sets, run clustering or regression tasks, and obtain predictive results on MLaaS. The ML API uses the features of input tests, which contain sensitive information, to produce predictions. Here, the idea of privacy leaks in ML and DL arises primarily as a result of overfitting, which causes the model to implicitly retain some details about training data. It is also linked to the structure and type of the model itself (Zhao et al., 2019). Threats to privacy in DL models can be categorized based on the training and prediction phases.

1. *Training Phase:* Finding a relation F through the minimization of the objective function L, also known as the cost function or loss function, is the aim of the training phase. In an effort to improve the loss function, the gradients are computed and updated with a learning rate using back propagation and SGD (Zhang et al., 2022). The DL deployment structure and privacy threats are tightly associated during the training phase.

2. *Prediction/Inference Phase:* Model F is used in this phase to anticipate yet-to-be-observed data, which is also known as the testing phase or the deployment phase. The same distribution's input data are used as input by the DL model, which predicts the results (Zhang et al., 2022).

Additionally, there are two distinct learning models, as per the ML/DL architecture.

1. *Centralized Learning:* A server that holds the training data is also where the models are trained and hosted by the central organization, and it is the default system

option for DL models. High efficiency and accuracy are achieved by having all the data available in a centralized learning environment, but user privacy may be compromised as sensitive data is directly accessible from the centralized server (Liu et al., 2021). In this system, the model gains a lot of accuracy by gathering a lot of data for training. The centralized server is put under a lot of strain as a result, and once an attack occurs, all personal data is vulnerable. Furthermore, the level of the adversary's access to the system hosting the model and data determines how private the data and the model are, not how the model is used (Zhao et al., 2019). It is depicted in Figure 11.1. For a variety of reasons, it is not always a wise choice:

- The amount of data is too enormous to fit into a single system.

- Users are cautious to disclose unprocessed data.

- To improve prediction accuracy, users seek to train their neural networks using a variety of instances.

2. *Distributed Learning:* A networked system where each component is located on a distinct computer yet connected as a whole. There is typically one server and several clients. Data is stored by the active clients, and the system server distributes the data features and builds the model (Liu et al., 2021). Distributed learning comes in a variety of forms:

- Collaborative learning

- Federated learning

- Split learning

(Shokri et al., 2017) presented collaborative learning, a version of distributed learning, wherein local users and centralized servers participate in the training tasks, but only share

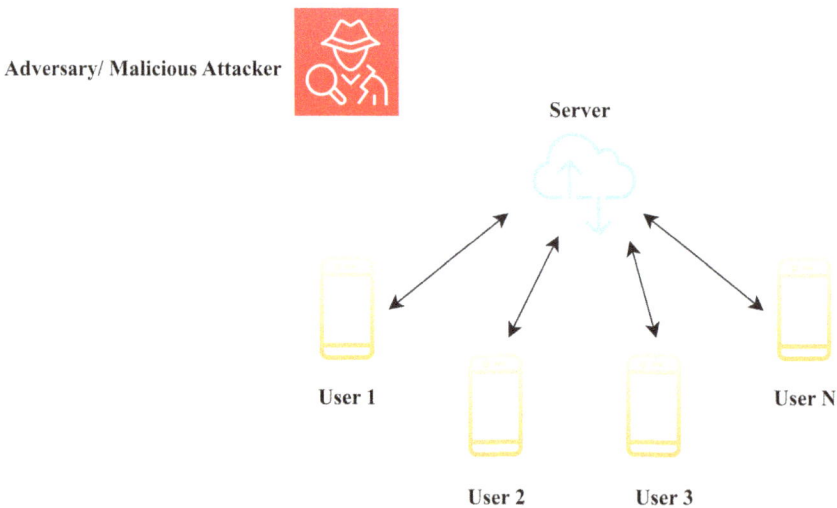

FIGURE 11.1 Centralized learning.

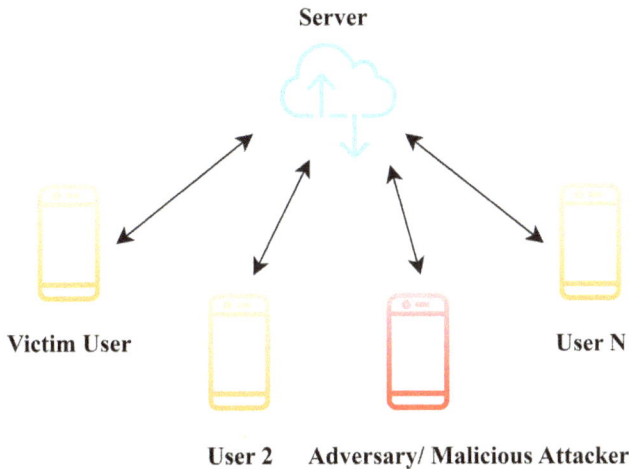

FIGURE 11.2   Collaborative learning.

a portion of the parameters. However, if there are any fraudulent participants, they can teach GANs to be both a friend and a foe in the fight against information theft, which is depicted in Figure 11.2.

The malicious user convinces the victim to disclose their personal information while it is still in the training phase.

11.3.1 Privacy Attacks in Deep Learning Pipeline

Deep learning systems are susceptible to a variety of threats, and there are two categories of privacy:

- *Privacy at the Data Set Level*: This refers to the confidential information contained within the training data set.

- *Privacy at the Model Level*: Referring to the model's structure, parameters, or hyperparameters.

To extract sensitive data, one might attack both the data set and the model. These attack techniques include:

1. *Membership Inference Attacks:* The goal of the attack is to identify either a particular sample of data was used during the training phase of the model. Despite not having direct access to the training data set or the trained model, the attacker can still query and use the trained model for predictions. This implies that the trained model's output is utilized to determine if the training data samples were added in the training data set (Zhang et al., 2022).

   (Shokri et al., 2017) proposed this attack, which operates in a "black-box" environment, meaning the victim model was accessible to the attacker and could use "shadow training" to mimic the victim model's behavior in their queries to gain a confidence score. The author applies supervised training to these models, explicitly instructing them to distinguish between outputs that correspond to members and non-members of the training data set (with the labels "in" and "out"). When a data sample was included in the training data set of the targeted model, it would result in a strong level of certainty for the targeted model's prediction. They also discovered that the primary elements that make a model susceptible to membership inference attacks include overfitting, type, and model structure.

2. *Model Inversion Attacks:* Model inversion attacks primarily operate in white-box models, while they can also employ less effective black-box techniques. The attack adjusts the weights and obtains the features of all classes in the network by tracking the gradient in a trained network to reverse-engineer (get f^{-1}) the network's weights. The classes can still replicate the prototype example even without prior knowledge (Liu et al., 2021).

   Fredrikson et al. conducted a white-box attack that could "acquire sensitive genomic information about individuals". The core idea was to populate the target feature vector "with every conceivable value and then calculate a weighted probability estimate that it is the correct value", based on a linear regression model f. In addition Fredrikson et al., expand the attack to facial recognition models, targeting two distinct objectives: the reconstruction attack, which generates "an image of the individual represented by a certain label", and the deblurred attack, which creates the deblurred image of the specified person that is provided "a face that is blurred in an image". The aim of such attacks is to "utilize GD to minimize a cost function involving f".

3. *Model Extraction Attacks:* Model extraction attacks, also known as model reverse engineering or model stealing attacks, seek to access private data from DL models. In this type of assault, the adversary only has black-box access and is unaware of the training data or the model parameter beforehand. The attack is made to obtain the model's parameters that were developed using private information. The attacker's aim is to generate a model that execute similarly to the target model in verifying data during prediction. Since the model's parameters are highly dependent on the training set, the confidentiality of the training set will also be compromised if the black-box model's parameters are revealed. By giving black-box model samples constantly and storing the prediction vector, it is possible to create a model f' that is comparable to the target model f (Zhao et al., 2019).

Model extraction attacks have been the subject of many works. To extrapolate additional model features, such as the NN architecture, Oh et al. created meta-models. Papernot et al. created an attack to take the ML model's hyperparameters. A hyperparameter is employed to maintain a balance between the regularization term and loss function within the objective function. The training data and model can be used by the opponent to obtain this value. Figure 11.3 shows the threat model in different stages of DL/ML.

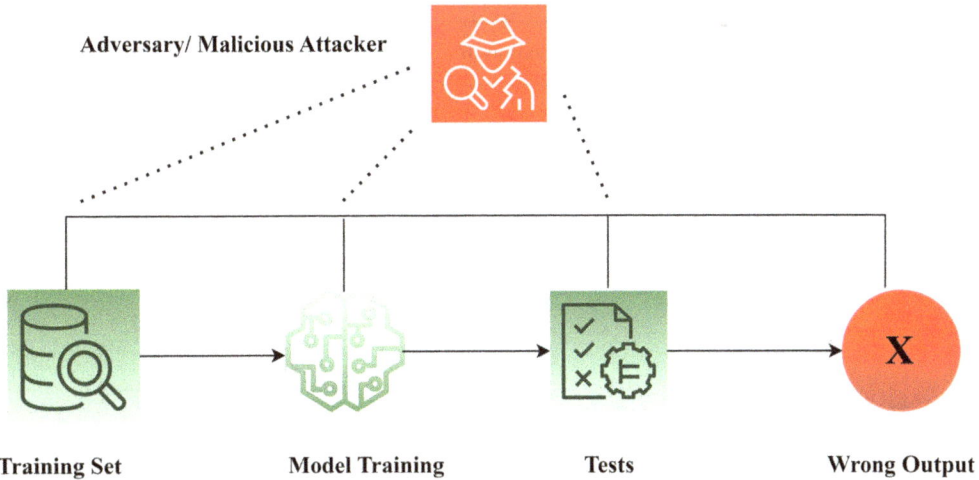

FIGURE 11.3   Deep learning threat model.

11.4 ALGORITHMS FOR PRESERVING PRIVACY

Due to the tendency for people to share high-quality photos and videos and the potential for the collection of sensitive personal information from those taking part in the sensing, privacy is a major problem in crowdsensing applications. The following algorithms are used to preserve the privacy of multimedia data:

1. *Differential Privacy (DP):* When using the data for statistical analysis, this technique adds random noise to the data to protect privacy. Differential privacy ensures that an individual's contribution to the overall data set remains anonymous. It increases the amount of noise in the data or DL model until the attacker is unable to tell apart a specific data sample from the data set or unable to retrieve particular sensitive data from the model. It ensures that a particular data set's privacy will be maintained (Zhang et al., 2022). In a study paper published in 2006, Cynthia Dwork, Frank McSherry, Kobbi Nissim, and Adam Smith presented the idea of DP.

In order to process bulk databases, differential privacy offers robust privacy guarantees. It is explained in relation to two databases that are close to one another but have different data records. The training data set for a DNN has neighboring databases that contain many labels for image pairs. A single picture pair label that is present in one database but not in the other indicates that two databases are adjacent (Zhao et al., 2019).

Differential privacy is defined as follows: For any data set D, there is a neighboring data set D' with a randomized algorithm M, which provides (ε, δ)-differential privacy for every set of outputs Ω if M meets certain conditions.

$$Pr\left[M\left(D\right) \in \Omega\right] \le e^{\varepsilon}. \; Pr\left[M\left(D'\right) \in \Omega\right] + \delta$$

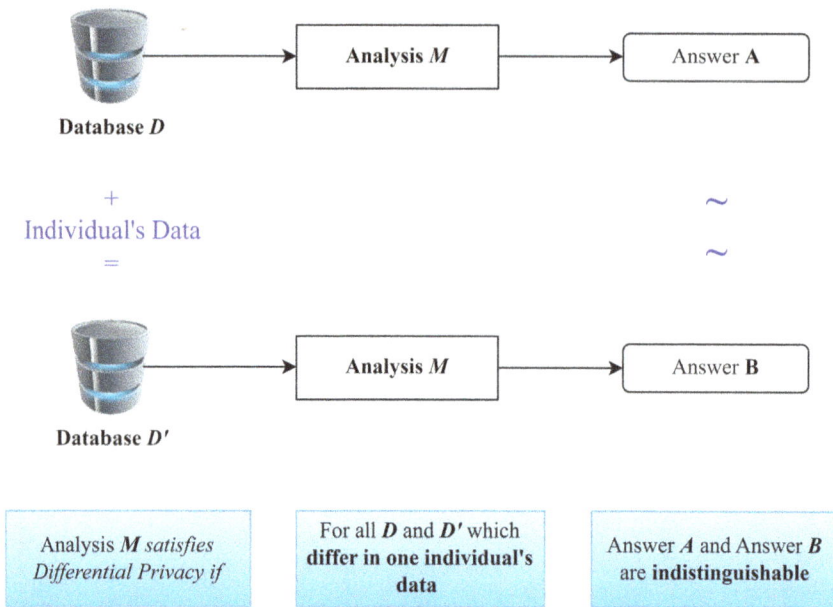

FIGURE 11.4   Differential privacy.

In other words, the likelihood that the algorithm will produce a specific result on the original data set is nearly identical to the likelihood that the method would provide the same result on any nearby data set. It is displayed in Figure 11.4.

ε is the privacy budget parameter that determines the level of privacy. A smaller level of epsilon indicates a higher level of protection, and δ loosens the bound of the error means allowing for more noise to be added to protect privacy. Central differential privacy and local differential privacy are the two categories into which DP applications can be separated:

- Central Differential Privacy: Users must have trust in the database owners (i.e., the data curators) to safeguard their privacy as a means to use CDP. CDP involves gathering data from individuals and then introducing random noise. The original data set or the result of queries executed on the original data set both now include random noise (Ouadrhiri & Abdelhadi, 2022).

- Local Differential Privacy: Before transferring the data to a service provider who is not trusted for data gathering and analysis, each user adjusts it locally. Under a measurable and exacting mechanism, LDP achieves reasonable denial for each individual. The collection of statistics while protecting privacy is intensively researched in the literature.

Preliminary of LDP: By introducing local data to noise that is controlled by a predetermined parameter, LDP broadens the concept of differential privacy. Simply described, a perturbation algorithm A probabilistically transforms a local raw value v_i to another value in the same range of potential outputs k. Subsequently, the server receives the modified value. Even after the server has collected all altered values, a learning task on the statistical properties (such as mean and frequency) of such data preserves a certain level of accuracy. A credible privacy guarantee that is constrained by a privacy budget ε is nevertheless possible for each individual.

Traditionally, the perturbation method satisfies the ε-LDP principle if and only if we have the inputs v_i and v_j of any two individuals:

$$Pr\left[A\left(vi\right) = s\right] \le e^{\varepsilon}.\ Pr\left[A\left(vj\right) = s\right]$$

where $s \in k$. With a larger user population and privacy budget ε, perturbed data is undoubtedly closer to the original data. As the noises are applied directly to the data set, if the budget is low, this tactic might significantly affect how well a model performs (Zheng et al., 2020).

2. *Federated Learning (FL):* A ML approach called FL enables data to be trained across a distributed network of devices without the requirement for centralized data storage. In the context of crowdsensing, this means that data can be analyzed on users' devices without requiring it to be transmitted to a central server. This approach preserves privacy by keeping user data local and avoiding the need to share sensitive information with third-party servers.

3. *Homomorphic Encryption (HE):* This technique allows data to be encrypted before it is collected so that it can be processed without decrypting. This ensures the data remains private, even if it is intercepted by unauthorized parties. The ability to train a model on an encrypted data set makes HE a powerful privacy protection tool. HE achieves the same accuracy as if the training were done on the original data set, which is the unencrypted version of the data set (Ouadrhiri & Abdelhadi, 2022). HE has potential application in many areas such as personalized medicine, where sensitive health data must be protected, or in financial modeling, where private financial data must be kept secure. However, HE is computationally expensive, and there are currently limitations on the size and complexity of computations that can be performed using this technique.

4. *Secure Multi-Party Computation (SMC):* The technique that protects users' privacy during the training phase is a domain of cryptography that permits methods where multiple parties can compute a function together without revealing their individual inputs either to each other or to the central server. This ensures sensitive data remains private while still allowing for meaningful analysis. Therefore, SMC does not require a reliable third party (Ouadrhiri & Abdelhadi, 2022). If an SMC protocol satisfies the following characteristics, it is considered secure:

   • Privacy: A client shouldn't be able to discover any information about another client in the network unless it comes from his or her own input and output.

- Correctness: Every participant should obtain accurate results.

- Independence of Input: The inputs of honest clients should not be affected by the inputs of malicious clients.

- Guarantee of Output: The delivery of their outputs to legitimate clients shouldn't be blocked by malicious clients.

- Fairness: Malicious clients can only benefit from outputs if and when genuine participants also benefit from them.

5. *Anonymization:* One way to escalate privacy is through data anonymization. This involves removing or obscuring any detail that could be used to identify individuals in the data. This technique involves removing or modifying any personally identifiable information (PII) from the data before it is analyzed. For example, personal identifiers such as names or addresses can be removed, or data can be aggregated to ensure that individual contributions cannot be distinguished. Anonymization can help to protect the privacy of individuals, but it can also make it more difficult to perform meaningful analysis.

6. *Blockchain-Based Approaches:* These strategies protect the data's security and privacy using distributed ledger technology. Blockchain technology allows data to be stored and processed in a decentralized manner, which can provide enhanced privacy and security.

11.5 THE DIFFERENTIAL PRIVACY DISTRIBUTIONS

Differential privacy algorithms aim to add randomness to the data, or the queries made on the data in a way that preserves the privacy of individual data points despite allowing useful insights to be extracted from the data. DL commonly uses the Laplace mechanism and the exponential mechanism (Zhao et al., 2019) to ensure DP. Several DP distributions include the following:

1. *Laplace Mechanism:* The Laplace distribution's noise is added to the query's output using this approach. The amount of noise added depends on the sensitivity of the query, which is a measure of how much the output would change if a single individual's data were removed. This mechanism is widely used for statistical queries.

2. *Exponential Mechanism*: This approach applies noise from the exponential distribution to a query's output. The amount of noise added depends on the utility of the result, which is a measure of how useful the result is in achieving analysis goals. It is widely used for complex queries.

3. *Approximate Gaussian Mechanism:* This algorithm is similar to the Laplace mechanism, but uses a Gaussian distribution instead of a Laplace distribution to add noise to the output of a query. It can be more efficient than the Laplace mechanism for certain types of queries.

4. *Randomized Response*: The randomized response mechanism is used when collecting data about sensitive topics or asking questions. It adds noise to the individual response in a way that makes it difficult to determine the true answer.

5. *PrivBayes:* This algorithm is used for differentially private synthetic data generation. It creates a synthetic data set that has similar properties to the original data set but with added noise to protect the privacy of individuals.

6. *Private Multiplicative Weights*: The private multiplicative weights algorithm is used for online learning and data analysis. It allows for efficient updates of a model while preserving the privacy of the data.

7. *Private Information Retrieval*: The private information retrieval algorithm is used for privacy-preserving data storage and retrieval. It allows a user to retrieve data from a database without revealing the specific data point they are accessing.

8. *Local Differential Privacy (LDP)*: LDP adds noise to individual data points before being transmitted to the primary server for analysis. This ensures that even if the server is compromised, individual data remains private.

9. *Global Differential Privacy*: Global differential privacy adds noise to the final output of analysis to ensure privacy.

10. *Smooth Sensitivity*: This algorithm is used for queries that involve multiple inputs, such as linear regression. It involves adding noise to the output of the query in a way that depends on the sensitivity of each input.

11. *Zero-Knowledge Proofs*: Zero-knowledge proofs allow data to be analyzed without being revealed, using cryptographic techniques.

12. *Differential Private Database*: A differential private database is a type of database that is designed to provide strong privacy assurance for individual records in the database.

13. *Sparse Vector Technique*: This algorithm is used for queries that involve sparse data, where most of the data points are zero. It involves adding noise to a subset of the data points and then using a sparse recovery algorithm to recover the full result.

14. *Private Set Intersection*: This technique computes the intersection of two sets of data without disclosing any details about the individual data points that make up the intersection. It involves increasing noise while maintaining privacy.

The algorithm selected will depend on the particular goals and specifications of the study as well as the type of data being examined.

11.6 HOW DIFFERENTIAL PRIVACY FUSES WITH DEEP LEARNING

It has been proven that attacks that take advantage of model information leakage are capable of compromising DL algorithms, such as neural networks. Differential privacy can be fused with DL in various ways to protect the privacy of sensitive data while still allowing for accurate and effective analysis. Here are some of these approaches by adding noise to the input data or the gradients during training, or to the model's output during inference, or using hashing functions.

1. *Adding Noise to the Training Data*: One of the simplest ways to integrate DP with DL is to inject noise into the training data before it is used to train a DL model. This ensures that individual data points cannot be identified, while allowing the model to still gain knowledge from the data and make accurate predictions. Laplacian noise or Gaussian noise are different types of noise that can be added, and the portion of noise injected can be controlled to balance privacy and accuracy.

2. *Using Differentially Private Algorithms*: Another way to integrate DP with DL is to use differentially private algorithms for training DL models. These algorithms have been specifically designed to provide privacy guarantees while still maintaining the privacy of the model. For example, the DL algorithm called deep learning with differential privacy (DP-SGD) uses a technique called "stochastic gradient descent" with added noise to ensure differential privacy.

3. *Encrypted Deep Learning*: Encrypted DL is an approach that allows data to be encrypted while still allowing for computation on the encrypted data. This approach can be used to integrate DP with DL by allowing data to be encrypted prior to being utilized for training a DL model. The model can then be trained on the encrypted data without decrypting it, preserving the privacy of the data. This technique is called *homomorphic encryption*.

By incorporating DP with DL, it is possible to achieve accurate and effective analysis of sensitive data while still protecting individual privacy. For example, a DL model trained with DP could be used to analyze data collected from devices to identify patterns in human mobility. The DP guarantees would help ensure that sensitive information about individuals' locations or movements is not exposed. However, it is important to carefully balance privacy and accuracy to ensure that the resulting model is both useful and reliable.

11.7 METHODOLOGY: EXPLORING THE INTERSECTION OF MULTIMEDIA DATA WITH DEEP LEARNING AND PRIVACY IN LITERATURE

In many fields, DNNs have produced futuristic outcomes and put in place themselves as leaders. However, deep learning models require a lot of data. The General Data Protection Regulation (GDPR) of the European Union and the Health Insurance Portability and Accountability Act (HIPPA) of the United States set rules and guidelines for the storage and exchange of personally identifiable information and health information. Additionally, the capacity to maintain total control and secrecy over one's personal information is something else that ethical principles support (Adnan et al., 2022).

In order to build policies for assuring the responsible and moral use of personal data, this work attempts to offer a more thorough knowledge of the intersection of multimedia data privacy (pictures and videos) in deep learning. It might also aid in the creation of best practices for the incorporation of multimedia data into DL models with an emphasis on advancing privacy and safeguarding individual data.

11.7.1 Preserving-Privacy Image Analysis

The Renyi-Differentially Private GAN approach was proposed by (Ma et al., 2023) and uses one hot encoding to divide the original data samples into a group of 0 and 1 before training. Instead of using parameters, they achieve DP within the GAN framework by deliberately introducing random noise to the value of the loss function. They note that performance loss is comparably minor and injecting noise to the loss numbers is simple to achieve.

The authors of the research (Adnan et al., 2022) identify the issue of histopathology images, which are unable to be gathered and shared in significant amount due to rules as well as due to data capacity limitations for their high resolution and gigapixel nature. They investigate FL as a model for collaboration that enables models to be trained across several institutions without explicitly sharing patient data by using a privacy reservation mechanism. When compared to training without collaboration, Renyi differential private accountant can enhance the performance of histopathology images.

Due to the high model complexity of GANs, which makes it easy for them to remember the training samples, (Xu et al., 2019) presented GANobfuscator, a differentially private GAN to attenuate information. It can achieve DP by introducing noise throughout the learning process. Theoretically, they demonstrated that GANobfuscator satisfies (ε, δ)-DP.

It is quite difficult to train big neural networks with adequate accuracy and privacy. First, train the model using DP on the ImageNet data set (Kurakin et al., 2022), and research the different DP-SGD training methods and parameters that can affect accuracy. When these discoveries are combined, train the ResNet-18 using ImageNet and discover that it can obtain a very high value of epsilon, which indicates a loss of privacy.

The same pipeline as Kurakin et al.'s work is used by (Bu et al., 2022); however, they demonstrated that DP training can be effective and have no negative effects on accuracy when used with big CNNs and ViT with convolutional layers (Table 11.2).

11.7.2 Preserving-Privacy Video Analysis

Yaqub et al., 2022 suggested using an image hashing approach for online student proctoring. It makes use of an in-house data set that has been manually classified as either a normal pose or an anomaly pose. It was hashed using dHashing-based imaging, MediaPipe, Dlib, Gaussian Blurring, Single White Masking, and Gaussian Blurring. The outcomes of the experiments are encouraging.

Giorgi et al., 2022 use DL autoencoders with DP to conduct privacy-preserving anomaly detection in video frames. They established that the use of autoencoders to anonymize and reconstruct video frames can slightly impair the ability to detect anomalies.

To create private videos for any private analysis, H. Wang et al., 2020 developed a sampling-based differentially private mechanism. Additionally, it gives unreliable analysts a flexible platform to privately conduct any analysis over privately generated videos (Table 11.3).

TABLE 11.2 Some Techniques and Methods for Ensuring DP Privacy When Using DL for Images

Ref	Objective	Privacy-preserving algorithm	Deep learning algorithm	Data set	Data set size	Benchmarks
(Ma et al., 2023)	The Renyi-Differential Private-GAN (RDP-GAN) is a suggestion that accomplishes DP by introducing noise to the value of the exchanged loss function during training.	Random Gaussian noise	DC-GAN, DP-GAN, PATE-GAN	Adult having Individual attributes. MNIST handwritten image digits	Adult 40K has 14 attributes (8 selected), 20K for training, 9K for testing, 1K for validation MNIST= 70K of 28×28, 60K for training, 9K for testing, 1K for validation	Accuracy, Privacy Budget
(Adnan et al., 2022)	To provide clients and patients strict boundaries on the level of privacy, use FL in conjunction with DP.	DP-SGD	Pre-trained DenseNet for feature extractor, MEM model an attention based Multiple Instance Learning (MIL) algorithm.	The Cancer Genome Atlas (TCGA)	2 TB of data, 2580 hematoxylin and eosin (H&E). 1806 for training, 774 for testing samples of WSIs	DP hyperparameters: Gradient Clipping, Noise Multiplier, Privacy Budget, Test Accuracy, External Accuracy
(Ho et al., 2022)	To preserve the robustness of FL, a system with DP was built employing chest X-ray images and symptom information.	DP-SGD	3×3 CNN, CNN-SPP, 5×5 CNN, ResNet18, ResNet50, 1DCNN, ANN, LSTM	Two distinct Covid-19 data sets: Chest X-ray images, symptom data	Chest x-ray images 15153. 14553 for training, 600 for testing. 3616 Covid-19 +ves, 10192 normal, 1345 viral pneumonia. Symptom data set: 5434 samples with 21 cols. 4890 for training, 544 for testing.	Q model constant, Noise, Accuracy

Reference	Objective	Method	Model	Dataset	Dataset Description	Metrics
(Kurakin et al., 2022)	Train large neural networks in real-life with both reasonable privacy and accuracy.	DP-SGD, Gaussian noise distribution	Family of ResNet-v2 i.e, ResNet-18, ResNet-50	ImageNet CIFAR-10 Places365	Places365 1084 images of various scenes with 365 labels of landscapes and buildings	Accuracy, Privacy budget loss bound
(Bu et al., 2022)	Proposed Mixed Ghost Clipping, for ease of private training in large CNNs models.	DP-ViT	CNN, ResNet, VGG, Wide-ResNet, DenseNet	CIFAR-10, CIFAR-100, ImageNet	Not mentioned	Accuracy, Max Batch size, Min Time/ Epoch, Memory (GB)
(Ghazi et al., 2021)	Introduced RRWithPrior algorithm to improve the traditional RR mechanism and applied to the LabelDP problems.	Randomized Response with Prior (RRWithPrior) + Label Differential Privacy (LabelDP)	ResNet18	CIFAR-10+ CIFAR-100+ MovieLens-1M	10 and 100 Class Image classification+ 1M Anonymous ratings of 3,900 movies by 6,040 movie users	Accuracy
(Xu et al., 2019)	Proposed GANobfuscator, could inject noise to the learning process to obtain DP under GAN.	Gaussian noise with zero mean and multiple value of Standard deviation	GAN, DCGAN	MNIST handwritten digits images, LSUN, CelebA face images	MNIST=70K of 28×28, 60K for training, 10K for testing LSUN 1M labelled images of 64×64, 10 scene category. CelebA 200K of 48×48, 40 attribute annotations	Accuracy, Precision, Privacy Budget, Inception Scores, Jensen-Shannan Scores

TABLE 11.3 Some Techniques and Methods for Ensuring Privacy for Videos

Ref	Objective	Privacy-preserving algorithm/ technique	Deep learning algorithm/ technique	Data set	Data set size	Benchmarks
(Yaqub et al., 2022)	Used Image-Hashing Anomaly Detection to proctor online while protecting privacy	Gaussian Blurring, Single White Masking, Image Hashing	DLib, MediaPipe	Inhouse data set from five different Asian origin, 25–30 yrs HELEN, UTK Face, CelebA, RF, LFW	2–3 min long, 1280×720 resolution, 25–30 FPS	Accuracy, Precision, Recall
(Giorgi et al., 2022)	Proposed a privacy-preserving surveillance video streams anomaly detection in real-life environments	Autoencoder Neural Network for privacy protection, DP embedded in autoencoders as an optimization mechanism (DP Adam optimizer, DP Adagrad optimizer).	ResNet-50 to extract features, YOLOv4 as object detector	UCF-Crime data set	128 hours of untrimmed surveillance videos, 13 types of anomalies, 210 videos for training, 90 for testing, with 7 classes	Area Under the Curve (ROC) True +ve rate, False +ve rate
(H. Wang et al., 2020)	Proposed a Differentially Private video analytics platform (VideoDP) with rigorous privacy guarantee.	Pixel sampling, Laplace Pixel interpolation	CNN based attacks to check rigoursy of videoDP against DL.	MOT, 15 videos, Sensitive Visual Elements pedestrians and vehicles UAD anomaly detection data set, sensitive VE pedestrians BVD at 5 different scenes	MOT 846 FPS, 1920×1080 UAD 180 FPS, 740×440 BVD 1200 FPS, 200,000 sequential images taken as videos with 2464×2056 resolution.	KL Divergence, Mean Squared Error

11.7.3 Preserving-Privacy With Other Methods

Luo et al., 2019 released a real-world image data set to test FL object detection algorithms using YOLO and R-CNN with predictable benchmarks.

The Color-NeuraCrypt method for a random private neural network for protecting privacy was put forth by Qi et al., 2023 and enables data owners to encrypt medical data before it is uploaded. Owners of the data can use encrypted data to test and train the model remotely. They also noticed that NeuraCrypt's performance suffers when color images are used; hence, they proposed Color-NeuraCrypt.

Xiong et al., 2021 created two algorithms called ADGAN-I and ADGAN-II that use GAN to produce privacy-preserving images and videos by concealing side channel information in order to avert location privacy leakage from the camera data of self-driving vehicles in offline applications (Table 11.4).

11.8 DISCUSSION

The fusion of deep learning with differential privacy has several implications and opportunities for future research. In this part, we go over a few of these aspects and point out some areas of concern.

Firstly, the use of DP has the potential to increase the trust and transparency of DL systems. By providing a strong guarantee of privacy, DP can increase the accountability of these systems and ensure that the use of sensitive data is transparent and ethical.

Secondly, the overfitting characteristic of DL models, the distribution of data between member and non-member samples, and the impact of data augmentations are the key causes of membership inference attacks and model inversion attacks in DL. There are other factors involved in privacy risks besides overfitting as well. Additionally, the output of the victim model may reveal training data. Then, using model gradients in FL, a series of model inversion attacks are motivated by this factor to reconstruct training data. The parameters, hyperparameters, and DL structure are crucial in model extraction attacks since they affect the model's performance. This attack aims to either completely copy the victim model and all of its parameters for malicious use or to completely destroy the victim model already in place to reduce performance. To undermine model integrity and accuracy are the primary goals.

Thirdly, privacy-preserving techniques have some limitations that needs to be considered when selecting the algorithm. Federated learning requires constant communication between the central server and the edge devices, which can lead to high communication and computation overheads in scenarios where there is a significant number of devices. Federated learning assumes that the data distributed across the edge devices is homogenous, but in practice, the data may vary in terms of quality, quantity, and distribution which can impact the accuracy and generalization of the model. Due to its computational complexity, homomorphic encryption is costly to process and ineffective in real-world applications, particularly when the training data set is too big to fit in the computer memory. When the model is already

TABLE 11.4 Some Other Techniques and Methods for Ensuring Privacy

Ref	Objective	Privacy-preserving algorithm/technique	Deep learning algorithm/technique	Data set	Data set size	Benchmarks
(Luo et al., 2019)	Designed a high-quality real-world labelled Image data set for Federated Learning	Federated Learning	YOLOv3, Faster R-CNN	Street data set of images and videos	26 street monitoring 704×576 resolution, Total 2544 images after removing night and similar scenes legible images 956, 785 for training, 191 for testing	Intersection over Union, Mean Average, Precision
(Qi et al., 2023)	Proposed color-NeuraCrypt a private neural network for preserving privacy	Encryption Color-NeuraCrypt	ViT finetuned ViTB_16	MNIST handwritten digits with 10 classes CIFAR-10	MNIST 70,000 grayscale images, $1 \times 28 \times 28$, 60,000 for training, 10,000 for testing CIFAR-10 60,000 color images, $1 \times 28 \times 28$ resolution, 50,000 for training 10,000 for testing	Accuracy
(Xiong et al., 2021)	ADGAN: Preserves the data's usefulness for further uses while safeguarding individual location privacy.	Privacy based on GAN	ADGAN-I ADGAN-II	Cityscapes (A data set of images and videos) Google Street View images data set	Google Street View 10343 place marking in USA Downscale into 512×512 due to H/W constraints	Pixel Accuracy (PA), Intersection over Union (IoU), FCN Scores

trained and available for use, MLaaS is appropriate. For training complex models over large data sets involving numerous clients, secure multi-party computation is not appropriate. However, SMC is expensive in terms of communication overhead and computational complexity. Anonymization is not a guarantee that personal information will remain private because re-identification attacks may take place when anonymized data is combined with information from other sources, including social media or public records. Additionally, it entails removing or hiding data points that may be used to identify specific people, which can reduce the accuracy and quality of the data.

The application of DP in DL can also facilitate the sharing of data between organizations without compromising individual privacy. This can enable organizations to collaborate and develop more accurate and robust models.

The application of DP in DL, however, is not without its difficulties. One of the main challenges is finding the right balance between privacy and accuracy. This requires careful selection of the privacy parameters and the amount of noise added to the data. Additionally, the scalability and efficiency of DP in DL need to be evaluated for large-scale data sets.

Overall, the integration of DL and DP has the promise of enabling the creation of a DL system that protects user privacy and can be used for a variety of tasks.

11.9 CONCLUSION

Incorporating differential privacy with deep learning pipeline is a promising approach for augmenting multimedia analysis that can help to address privacy concerns associated with the use of sensitive data while still enabling the model to learn from the data. It can also result in a more robust and trustworthy model since it has been trained on data that has been anonymized to some degree.

However, there are still some challenges and opportunities for future research. One of the main challenges is finding the right balance between privacy and accuracy, which requires careful selection of the privacy parameters and the amount of noise added to the data. Additionally, more work is needed to evaluate the scalability and efficiency for large-scale multimedia data sets.

REFERENCES

AbualSaud, K., Elfouly, T. M., Khattab, T., Yaacoub, E., Ismail, L. S., Ahmed, M. H., & Guizani, M. (2018). A Survey on Mobile Crowd-Sensing and Its Applications in the IoT Era. *IEEE Access*, 7(c), 3855–3881. 10.1109/ACCESS.2018.2885918

Adnan, M., Kalra, S., Cresswell, J. C., Taylor, G. W., & Tizhoosh, H. R. (2022). Federated Learning and Differential Privacy for Medical Image Analysis. *Scientific Reports*, 12(1), 1–10. 10.1038/s41598-022-05539-7

Aamir, M., Rahman, Z., Ahmed Abro, W., Aslam Bhatti, U., Ahmed Dayo, Z., & Ishfaq, M. (2023). Brain tumor classification utilizing deep features derived from high-quality regions in MRI images. *Biomedical Signal Processing and Control*, 85, 104988 10.1016/j.bspc.2023.104988.

Aamir, M., Rahman, Z., Ahmed Abro, W., Tahir, M., & Mustajar Ahmed, S. (2019). An Optimized Architecture of Image Classification Using Convolutional Neural Network. *International Journal of Image, Graphics and Signal Processing*, 11, 30–39 10.5815/ijigsp.2019.10.05.

Bu, Z., Mao, J., & Xu, S. (2022). *Scalable and Efficient Training of Large Convolutional Neural Networks with Differential Privacy.* 1–22. http://arxiv.org/abs/2205.10683

Ghazi, B., Golowich, N., Kumar, R., Manurangsi, P., & Zhang, C. (2021). Deep Learning with Label Differential Privacy. *Advances in Neural Information Processing Systems*, 32(NeurIPS), 27131–27145.

Giorgi, G., Abbasi, W., & Saracino, A. (2022). Privacy-Preserving Analysis for Remote Video Anomaly Detection in Real Life Environments. *Journal of Wireless Mobile Networks, Ubiquitous Computing, and Dependable Applications*, 13(1), 112–136. 10.22667/JOWUA. 2022.03.31.112

Ho, T. T., Tran, K. D., & Huang, Y. (2022). FedSGDCOVID: Federated SGD COVID-19 Detection under Local Differential Privacy Using Chest X-ray Images and Symptom Information. *Sensors*, 22(10). 10.3390/s22103728

Kurakin, A., Song, S., Chien, S., Geambasu, R., Terzis, A., & Thakurta, A. (2022). *Toward Training at ImageNet Scale with Differential Privacy. Ml*, 1–25. http://arxiv.org/abs/2201.12328

Liu, B., Ding, M., Shaham, S., Rahayu, W., Farokhi, F., & Lin, Z. (2021). When Machine Learning Meets Privacy: A Survey and Outlook. *ACM Computing Surveys*, 54(2). 10.1145/3436755

Luo, J., Wu, X., Luo, Y., Huang, A., Huang, Y., Liu, Y., & Yang, Q. (2019). *Real-World Image Datasets for Federated Learning.* http://arxiv.org/abs/1910.11089

Ma, C., Li, J., Ding, M., Liu, B., Wei, K., Weng, J., & Poor, H. V. (2023). RDP-GAN: A Rényi-Differential Privacy Based Generative Adversarial Network. *IEEE Transactions on Dependable and Secure Computing*, 1–15. 10.1109/TDSC.2022.3233580

Ouadrhiri, A. El, & Abdelhadi, A. (2022). Differential Privacy for Deep and Federated Learning: A Survey. *IEEE Access*, 10, 22359–22380. 10.1109/ACCESS.2022.3151670

Qi, Z., MaungMaung, A., & Kiya, H. (2023). *Color-NeuraCrypt: Privacy-Preserving Color-Image Classification Using Extended Random Neural Networks.* 10–13. http://arxiv.org/abs/2301.04875

Shokri, R., Stronati, M., Song, C., & Shmatikov, V. (2017). Membership Inference Attacks Against Machine Learning Models. *Proceedings - IEEE Symposium on Security and Privacy*, 3–18. 10.1109/SP.2017.41

Waheed, N., He, X., Ikram, M., Usman, M., Hashmi, S. S., & Usman, M. (2020). *Security and Privacy in IoT Using Machine Learning and Blockchain: Threats and Countermeasures*, 53(6). 10.1145/3417987

Wang, H., Xie, S., & Hong, Y. (2020). VideoDP: A Flexible Platform for Video Analytics with Differential Privacy. *Proceedings on Privacy Enhancing Technologies*, 2020(4), 277–296. 10. 2478/popets-2020-0073

Wang, L., Lu, Z., Sun, H., Hou, Y., & Huang, M. (2017). Security and Privacy in Internet of Things with Crowd-Sensing. *Journal of Electrical and Computer Engineering*, 2017. 10.1155/2017/ 2057965

Xiong, Z., Cai, Z., Han, Q., Alrawais, A., & Li, W. (2021). ADGAN: Protect Your Location Privacy in Camera Data of Auto-Driving Vehicles. *IEEE Transactions on Industrial Informatics*, 17(9), 6200–6210. 10.1109/TII.2020.3032352

Xu, C., Ren, J., Zhang, D., Zhang, Y., Qin, Z., & Ren, K. (2019). GANobfuscator: Mitigating Information Leakage Under GAN via Differential Privacy. *IEEE Transactions on Information Forensics and Security*, 14(9), 2358–2371. 10.1109/TIFS.2019.2897874

Yaqub, W., Mohanty, M., & Suleiman, B. (2022). Privacy-Preserving Online Proctoring using Image-Hashing Anomaly Detection. *2022 International Wireless Communications and Mobile Computing, IWCMC*, 2022, 1113–1118. 10.1109/IWCMC55113.2022.9825119

Zhang, G., Liu, B., Zhu, T., Zhou, A., & Zhou, W. (2022). Visual Privacy Attacks and Defenses in Deep Learning: A Survey. In *Artificial Intelligence Review* (Vol. 55, Issue 6). Springer Netherlands. 10.1007/s10462-021-10123-y

Zhao, J., Chen, Y., & Zhang, W. (2019). Differential Privacy Preservation in Deep Learning: Challenges, Opportunities and Solutions. *IEEE Access*, 7, 48901–48911. 10.1109/ACCESS. 2019.2909559

Zheng, H., Hu, H., & Han, Z. (2020). Preserving User Privacy for Machine Learning: Local Differential Privacy or Federated Machine Learning? *IEEE Intelligent Systems*, 35(4), 5–14. 10.1109/MIS.2020.3010335

Multi-classification Deep Learning Models for Detecting Multiple Chest Infection Using Cough and Breath Sounds

Amna Tahir[1], Hassaan Malik[1], and
Muhammad Umar Chaudhry[2]

[1]*Department of Computer Science, National College of Business Administration & Economics Lahore, Multan Sub Campus, Lahore, Pakistan*
[2]*Department of Computer Engineering, Bahauddin Zakariya University, Multan, Pakistan*

12.1 INTRODUCTION

Every year, cough and respiratory infections place a huge strain on the health system's personnel and resources. These infections cause various chest infections that directly affect the human lungs. These chest infections are AST, COVID-19, NOR, PNEU, and TB. In Wuhan, China, numerous patients with pneumonia of undetermined origin were recorded in December 2019 [1]. The majority of the recorded patients were either employed at or resided close to the Huanan [2] local fish commodity market, which also offered live animals for sale. The novel coronavirus (nCoV) has rapidly spread from Wuhan. From the tests of throat swabs of the patients, nCoV was identified. On 7th January, nCoV was identified by this test through the "Chinese Center for Disease Control and Prevention (CCDCP)" in Wuhan City. The virus was discovered in the majority of nations by early spring 2020, and the World Health Organization (WHO) formally declared the newly discovered virus to be an epidemic by the last days of March 2020 [3]. The novel coronavirus or SARS-CoV-2 has become a pandemic globally. Severe acute respiratory syndrome corona virus 2 (SARS-CoV-2) is the family of coronavirus. SARS-CoV-2, the virus causing the illness, is a member of the zoonotic coronavirus

DOI: 10.1201/9781003427674-12

family. The six various coronaviruses were found to target the respiratory system of humans; severe acute respiratory syndrome (SARS-CoV) and Middle East respiratory syndrome (MERS-CoV) caused outbreaks in the past 20 years while the other four only caused minor symptoms [4]. In November 2002, SARS was discovered in Guangdong. Guangdong is a province in southern China. Guangdong suffered from coronavirus that affected more than 8,000 people. Moreover, almost 37 countries reported 800 deaths between 2002 and 2003. In September 2012, MERS-CoV was first detected in Saudi Arabia. After this, WHO reported approximately 2,000 cases of the MERS CoV disease worldwide. MERS-CoV is a viral disease, which implies that it can spread among humans and animals. According to studies, humans can become infected by coming into contact either directly or indirectly with diseased dromedary camels.

Nearly 12 million people were affected by coronavirus or SARS-COV-2 in the world in July 2020, and 562,039 people died as a result of it [5]. Fever, coughing, exhaustion, sore throats, and body pains are typical COVID-19 symptoms, and countless cases of lost sense of taste have been recorded worldwide. Patients may experience respiratory issues, a high fever, shivers, exhaustion, muscular or body pain, or even pass away in rarer but often more serious conditions. In the world, the system of healthcare and successful testing is trying for a better vaccination of COVID-19 on a large scale. Yet many regions of the world are still suffering. They are at risk from more waves or surges of new COVID-19 cases. Reverse transcriptase polymerase chain reaction (RT-PCR) is the most precise testing technique for the detection of COVID-19. But it takes a lot of time and laboratories are used in this test. This is very costly and not easily available in such countries with low incomes [6].

A type of respiratory infection that makes it difficult to breathe and has a negative impact on the lungs is called pneumonia. Alveoli, which are tiny pockets that make up the lungs, are filled with oxygen when a healthy person breathes. Pneumonia causes the alveoli to become clogged with fluid, making breathing difficult and limiting oxygen absorption [7]. The most prevalent and infectious primary cause of death in children globally is pneumonia. The symptoms of pneumonia are sweating, loss of appetite, fast heartbeat, cough, difficulty in breathing, and tiredness. The chest X-ray enables the physician to diagnose pneumonia and determine its location and seriousness [8]. The heart, lungs, blood vessels, airways, and chest bones can be shown in the chest X-ray [9]. To check the probability of pneumonia, chest X-ray images are used. But the method of detecting this disease by using chest X-ray images is so expensive and not easily accessible in each hospital [10].

Tuberculosis (TB) is one of the diseases that has a bad impact on the lungs of humans. TB is mostly found in such countries in which earnings are low; 95% of cases of TB are related to developing countries [11,12]. The latest detecting technologies for this illness are so expensive because they depend on the particular tools and the method of laboratories. The symptoms of TB are loss of hunger, constant cough that lasts three weeks, loss of weight, and inflammation in the neck. Coughing is one of the main signs of respiratory illnesses, including TB and COVID-19 [13,14]. The thirteenth greatest fatal disease in the world is TB. The second-deadliest contagious disease is TB after COVID-19. Almost 1.5 million people have died globally [15].

An ongoing condition affecting the airways is asthma. The bronchi are often known as the airways. Airways are pipes that deliver air to the lungs. When a patient with asthma breathes, then the atmosphere causes bronchospasm and airflow limitation in an asthmatic person. Although there is an inherent tendency to develop asthma, signs only manifest when a patient is exposed to allergies such as smut, dirt, and pet dandruff. Smoking is the primary cause of an increase in respiratory blockage. The symptoms of asthma include wheezing, coughing and chest tightness, rapid heartbeat, loss of consciousness for a short time, difficulty in breathing, and blue lips or fingers. Coughing is a symptom of a wide range of respiratory conditions in children. Coughing is one of the main signs of asthma. Annually, asthma affects over 300 million people globally and results in 250,000 fatalities. As a result, it is requirement to generate a classifier for the rapid detection of respiratory disorders [16].

The objective of this chapter is to develop a model that will be used to distinguish between respiratory and coughing sounds caused by various chest infections. This model uses the sounds from AST, COVID-19, NOR, PNEU, and TB individuals. When various coughing sounds are applied to the suggested model, the model will be able to determine from the patient's sound whether or not the patient is ill. A convolutional neural network (CNN) is applied to the input spectrogram images, which is a brief representation of a sound wave. Furthermore, a CNN architecture handles the processing of images. In this chapter, the cough and respiratory sounds have been converted into images using a spectrogram, and then CNN architecture is employed to identify the patients with AST, COVID-19, NOR, PNEU, and TB cases. The contribution of this study is as follows:

- The five main kinds of chest infections are identified by the novel proposed DMCIC_Net model using cough and breath sounds. The cough and breath sounds are converted into spectrogram images.

- To create a substantial classifier in this chapter, we lower the model's complexity by reducing the total amount of trainable parameters.

- The issue of class imbalance in medical data sets reduces the accuracy of the CNN model. By employing the SMOTE Tomek up-sampling method to obtain multiple samples of the image at each class, we can get around this problem and improve accuracy.

- The proposed model attained the highest results as compared to six baseline models such as Vgg-16, Vgg-19, ResNet-152, MobileNet, Inception-ResNet-v2, and EfficientNet-B1 in terms of numerous evaluation metrics including accuracy, recall, loss, F1-score, precision, and area under the curve (AUC).

- In addition, the proposed model compares current state-of-the-art classifiers in terms of significant outcomes.

The following sections compose this study: a literature review is presented in Section 12.2. Section 12.3 discusses the materials and methods. Section 12.4 presents the experimental results and discussion. This chapter is concluded in Section 12.5.

12.2 LITERATURE REVIEW

This section describes in depth some of the classification models employed by prior researchers in identifying COVID-19 using sound categorization. Table 12.1 shows the summary of the latest literature on various chest infection diagnoses by using DL models.

From the small number of data sets, Oh et al. [32] introduced CNN architecture in which a patch-based system was applied. By selecting the majority of the patched algorithm results, the algorithm has reached the final result. The publicly accessible data set was utilized in their analysis. In this study, 15,043 total images, representing 8,851 normal cases, 6,012 patients with pneumonia, and 180 patients who tested positive for COVID-19, were used; 88.9% accuracy, 83.4% precision, 85.9% recall, 84.4% F1-score, and 96.4% specificity were all included in CNN's excellent results. Hemdan et al. [33] proposed the CNN model, which was evaluated using a few data sets. This experiment made use of two data sets. In the first data set, there were 1,427 images with 224 COVID-19 positives, 700 images of pneumonia, and the other images were from normal patients. Numerous models including Vgg19, MobileNet v2, Inception, Xception, and Inception-ResNet-v2 were employed in the first data set. However, Vgg19 demonstrated the best performance when compared to the other models. VGG19 increased accuracy by 98.75%. The same number of COVID-19 images as in the first database were included in the second, along with patients who had pneumonia and healthy individuals. MobileNetV2 was employed in the second data set, and 96.78% accuracy was attained. This data set was available publically. The data sets were gathered on three accessible websites: Radiological Society of North America (RSNA), Radiopaedia, and the Italian Society of Medical and Interventional Radiology (SIRM).

Sethy et al. [34] applied the various deep attributes of a CNN and SVM that were utilized to estimate the classification of COVID-19. By the utilization of ResNet50's deep feature, the SVM generated good results of 95.33% accuracy. It consisted of three classes in which COVID-19, pneumonia, and normal cases were included. By utilizing deep learning models, Hemdan et al. [35] proposed the COVIDX-Net model, which consisted of seven CNN architectures to recognize COVID-19 from the CXR snaps. VGG19, DenseNet201, InceptionV3, ResNetV2, Inception-ResNet-V2, Xception, and MobileNetV2 were the seven architectures presented. But VGG19 and DenseNet201 performed better; 90% accuracy was attained for both architectures. A public data set of CXR images, given by doctors Joseph Cohen and Adrian Rosebrock, was used in this study to classify patients with negative and positive COVID-19 results. Hilmizen et al. [36] introduced a technique that contained various transfer learning models. These models were used for the classification of CXR snaps and CT scans: 2,500 CT scan images, including 1,257 COVID-19 images and 1,243 non-COVID-19 images. Just like CT scans, 2,500 chest X-ray images were utilized. According to the results, Densenet-121, Mobile net, Xception, Inception-V3, ResNet-50, and VGG-16 were employed to classify the images in COVID-19 negative and COVID-19 positive classes, with ResNet-50 and VGG-16 providing the best accuracy of 99.87%. The public repository provides access to the CT scan images (radiopedia.org). The data set that

TABLE 12.1 Recent Literature on Deep Learning Models Used for the Classification of COVID-19 Using Sounds

Ref	Year	Model	Type	Objective	Data set	Accuracy
[17]	2023	MSCCov19Net	COVID-19 vs non-COVID-19	To identify the COVID-19 and non-COVID-19 cases by the utilization of cough audio	Coswara, Coughvid, NoCoCoDo, and Virufy	90.4%
[18]	2023	SVM, CNN, LSTM	COVID-19 vs non-COVID-19	To diagnose COVID-19 patients by using cough sounds	Online Website (Sorfeh.com)	95.1%
[19]	2022	(LBP) and Haralick's features	COVID-19 vs non-COVID-19	To identify the Binary classes of COVID-19 and non-COVID-19 using cough sound breath and speech	Coswara	98.9%
[20]	2022	MCDM, ML, ERT, SVM, RF, AB, MLP EGB, GB, LR, K-NN, HGB	COVID-19 vs non-COVID-19	To classify the two classes of COVID-19 positive and negative by using cough audio	Cambridge, Coswara, Virufy, and NoCoCoDa	95%
[21]	2022	Light weight CNN	COVID-19 positive, asthma, pertussis, bronchitis, healthy	To classify the multiple diseases of breathing sound	Crowdsourced	92.32%
[22]	2022	CNN	Binary Class	To identify COVID-19 by using cough sounds	Coughvid	94%
[23]	2022	DCNN	COVID-19 positive, asthma, pertussis, bronchitis, healthy	DCNN is applied to diagnose different respiratory diseases by using cough sounds, voice, and breathing	Crowdsourced	95.45%
[24]	2021	DTL, GoogleNet, ResNet18, ResNet50, ResNet101, MobileNetv2, and NasNetmobile	Binary class	To classify COVID-19 and normal cases by the utilization of cough sounds	Coswara and Sarcos	94.9%
[25]	2021	SVM, K-NN, RNN, RF	Binary class	To detect COVID-19 by applying cough audio	Coswera & Virufy	81.25%

Ref	Year	Method	Class	Objective	Dataset	Accuracy
[26]	2021	SVM	COVID-19 normal pneumonia pertussis	To identify pertussis, COVID-19, pneumonia, and normal by using cough sounds	Coswara	94%
[27]	2020	CNN	Binary class	Identify COVID-19 by using respiratory and coughing audio	Crowdsource	80.7%
[28]	2020	MFCC, VGG 16	Binary class	To identify COVID-19 by using MFCC for audio and VGG 16 for image processing	Audio set & ESC-50	70.58%
[29]	2020	SVM	Binary class	To diagnose the binary class by applying cough and respiratory sounds	Crowdsource	82%
[30]	2020	MFCCs method	COVID-19 pertussis bronchitis normal	To identify COVID-19, pertussis, bronchitis, and healthy by applying cough sounds	ESC	95.60%
[31]	2020	CNN	Binary class	To detect COVID-19 by applying cough audio	Virufy, Corswara	97.5%

was utilized in this study was available to the general public to classify and examine COVID-19, common pneumonia, and normal chest X-ray images.

Ulukaya et al. [17] introduced a model named MSCCov19Net. MSCCov19Net is the abbreviation for the proposed multi-branch model that accepts Mel Frequency Cepstral Coefficients (MFCC), Spectrogram, and Chromagram as inputs. This technique is used to diagnose COVID-19 by using only cough sounds. Four data sets were utilized: Coswara, Coughvid, Virfy, and NoCoCoDa. The accuracy achieved from the Coswara and CoughVid data sets was 74.8% and 61.5% from Virfy. NoCoCoDa gave the best accuracy at 90.4%. Nasab et al. [18] introduced a model that identified COVID-19 by the utilization of cough sounds. Sorfeh.com was the online website used to collect the data set. There were many algorithms of supervised learning used: SVM, random forest, CNN, and long short-term memory (LSTM). By implementing all these algorithms, LSTM achieved a higher accuracy of 95.1%.

Three different COVID-19 input formats, including the cough, breath, and voice sounds, were used in Sharma et al.'s [19] model for sound texture processing. The sound texture of COVID-19 was processed using Haralick's features and local binary patterns (LBPs). The data set was acquired from the University of Cambridge in the United Kingdom. The textural evaluation of cough and breath sounds was performed on the two classes. These sounds were recorded from COVID-19 positive and negative users in the sounds data set of COVID-19. Karim et al. [20] introduced an ensemble-based multi-criteria decision-making (MCDM) approach by using the best machine learning (ML) methods for classifying COVID-19. Extremely randomized trees (ERTs), support vector machine (SVM), random forest (RF), adaptive boosting (AB), multilayer perceptron (MLP), extreme gradient boosting (EGB), gradient boosting (GB), logistic regression (LR), K-nearest neighbor (K-NN) [37], and histogram-based gradient boosting (HGB) were examples of machine learning techniques. Four data sets, including Cambridge, Coswara, Virufy, and NoCoCoDa, were used in this study. These data sets were used to identify COVID-19 and normal cases, yielding an effective result of 95% accuracy. Kumar et al. [21] introduced a model of a light-CNN with a modified-mel-frequency cepstral coefficient (M-MFCC) that was utilized to diagnose COVID-19 and other diseases like asthma, pertussis, and bronchitis. The crowdsourced data set of COVID-19 cough sound was gathered from Cambridge University. The highest accuracy of 92.32% was achieved by COVID-19.

Sabet et al. [22] introduced a model that was based on CNN. It was designed to produce an entirely efficient system. By doing the feature collection and using MFCC and data augmentation approaches, COVID-19 was identified. The CoughVid data set was applied for the training of the proposed CNN model and it achieved 94% accuracy. Lella et al. [23] proposed a deep CNN model to automatically identify the COVID-19 positive and negative cases of other respiratory diseases such as asthma, pertussis, bronchitis, and healthy cases by the utilization of voice, cough sound, and breathing. Crowdsourced COVID-19 was the sound data set that was collected from Cambridge University. This data set was applied to test the DCNN model; 95.45% accuracy was achieved by this model. Loey et al. [24] introduced a deep transfer learning (DTL) model that consisted of GoogleNet, ResNet18, ResNet50, ResNet101, MobileNetv2, and NasNetmobile. These were utilized to diagnose COVID-19. The binary classes were used in which COVID-19 and healthy patients were included. In this research, Coswara and Sarcos data sets were applied. By using audio sounds, the Coswara project was applied to detect COVID-19. There were 92 COVID-19-positive cases and 1,079 normal cases. In the Sarcos (SARS COVID-19 South Africa) data set, 8 cases of positive COVID-19 and 13 healthy patients were included. This model has achieved 94.9% accuracy.

Feng et al. [25] introduced machine learning (ML) approaches that were used to classify characteristics taken from the frequency and time domain. Support vector machine (SVM), K-nearest neighbor (K-NN), recurrent neural network (RNN), and random forest (RF) were applied in deep learning and the best outcomes were obtained using the RNN approach. When two separate data sets were utilized for the training and testing sets, the accuracy percentage of the system was determined to be 81.25% for RNN. Vijayakumar et al. [26] used SVM to distinguish between COVID-19, pertussis, typical cases, and pneumonia. There were 81 cough audio files in the databases, including 8 cases of COVID-19, 28 cases of pneumonia, 15 cases of pertusses, and 30 normal cough audio files. In this four-class analysis, the SVM classifier had an overall accuracy rate of 94%. Schuller et al. [27] employed a DL approach to identify COVID-19 patients by using CNN. They modified the CNN method, which assessed a patient's COVID-19 infection status using respiratory and coughing audio. The proposed method was roughly twice as effective as the standard method. A DL model was able to give the best results with the available data, while the CNN model had an accuracy of 80.7%.

Bansal et al. [28] introduced a CNN model based on mel-frequency cepstral coefficients (MFCC) for COVID-19 audio identification. Two methods for recognizing audio sounds were discussed in this study. In the initial stage, the MFCC technique took the spectrogram as an input. Second, a transfer learning-dependent technique and the VGG 16 architecture were used to evaluate the collection of image processing pipelines. The presented model's test accuracy and sensitivity were 70.58% and 81%, respectively, with an improved results strategy. Brown C et al. [29] created a model that employed respiratory sounds, and a support vector machine (SVM) was used by the researcher to classify COVID-19 and healthy cases using a binary classifier. The data set was gathered using a mobile application and a web browser. With the help of this model, a stunning accuracy of 82% was achieved. Imran et al. [30] first presented COVID-19, a smartphone app that records and analyzes three-second cough noises. Four different audio types were used in 328 cough recordings. Four categories were created from the 150 participants: COVID-19, pertussis, bronchitis, and healthy. Researchers employed the CNN approach to extract features, and this system achieved an accuracy of 92.64%. The categorization used the environmental sound categorization (ESC) data set. Dunne et al. [31] introduced a model based on CNN. It used cough sounds from smartphones and home sensors to examine the positive and negative COVID-19 cases. A good accuracy of 97.5% was attained.

The previously mentioned literature review identified multiple studies that employed chest X-rays and computed tomography (CT) to diagnose COVID-19 [32–36]. Few researchers studied the categorization of COVID-19 and normal individuals using cough, as shown in Table 12.1. Some of them researched the classification of COVID-19 and pneumonia using cough sounds. But some researchers believe that the symptoms of chest infections such as AST, COVID-19, NOR, PNEU, and TB are the same. That's why the gap in this chapter exists in which we need a DL-based classifier that can distinguish between AST, COVID-19, NOR, PNEU, and TB using the sound of coughing and breathing. The novel model presents several key points in which cough and breath sound files convert into spectrogram images. These spectrogram images employ the proposed model and six baselines DL models to predict the aforementioned chest infections. These baseline models are Vgg-16, Vgg-19, ResNet-152, MobileNet, Inception-ResNet-v2, and EfficientNet-B1.

12.3 MATERIALS AND METHODS

This section introduces the experimental procedure used to assess the proposed model. This section also presents six well-known deep CNN models: Vgg-16, Vgg-19, ResNet-152, MobileNet, Inception-ResNet-v2, and EfficientNet-B1.

12.3.1 Proposed Study Flow for the Diagnosis of Multiple Chest Infections

This chapter examines five types of chest infections, including AST, COVID-19, NOR, PNEU, and TB cases. The lungs, airways, alveoli, and trachea are all components of the respiratory system, which also includes the respiratory tract. Many chest infections that damage the lungs are characterized by coughing and breathing problems. Health professionals have a strong chance to stop the infection's growth and start treatment on time if chest infections are discovered in the initial stages. A huge change has come in the field of medical imaging through the use of artificial intelligence and image processing. Nowadays, in the field of medicine, image processing is utilized for analysis [24,38–40,49,50].

To categorize these infections, data sets obtained from a wide variety of sources were gathered. The duration of sounds was measured after the collection of data sets was completed. The sound duration was fixed at five seconds. These data sets employ sound files with the ".webm" file extension. This model required the ".wav" format to convert audio to images. To employ the online converter, the audio file was converted from "webm" to ".wav". After the sound file format was converted, a second online converter was used to turn the sound samples into spectrogram images. Graphical representations of an audio signal are known as spectrograms [41]. This converter is used to create spectrogram images from a huge quantity of audio recordings. The conversion from sound to spectrogram image is shown in Figure 12.1.

In this chapter, we propose an automated system for diagnosing multiple chest infections using deep learning called DMCIC_Net. Five classes of multiple chest infections were trained and tested by this system: AST, COVID-19, NOR, PNEU, and TB. The size of input images was determined at the resolution of 150 × 150 pixels. In addition, the data set was processed using the data normalization method to prevent the model from becoming overfitting. After that, to solve the issue of an unequal distribution data set, we applied the synthetic minority oversampling technique (SMOTE). This technique equalizes the number of samples contained within each class. After that, the data set was split into three parts: training, testing, and validation. Additionally, quick model training was accessible for the one shown in Figure 12.2 as well. In comparison to [28,31,42], the

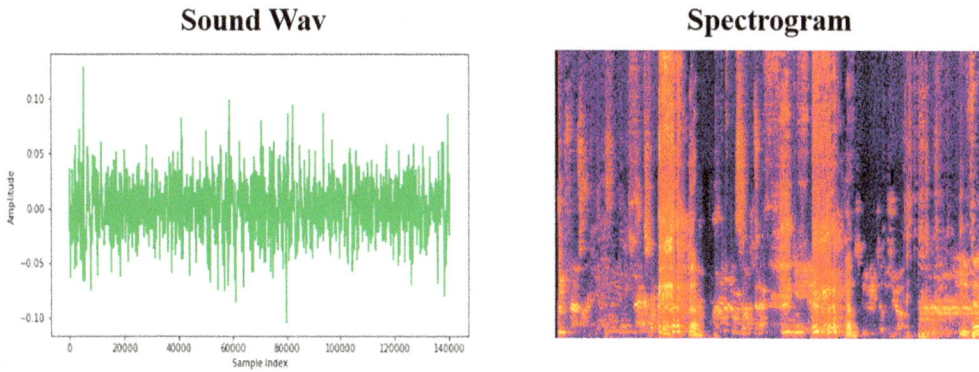

FIGURE 12.1   Convert sound into a spectrogram image.

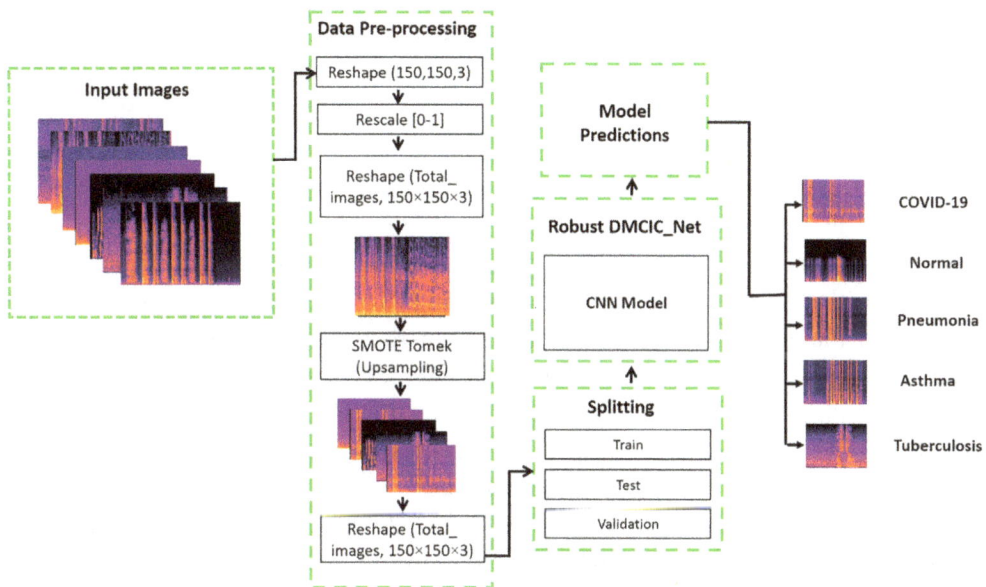

FIGURE 12.2   Work flow of the proposed DMCIC_Net.

size of input parameters was reduced. The experimental process had been fulfilled in the highest of 30 epochs. Following the completion of all epochs, the proposed DMCIC_Net achieved the desired maximum level of accuracy through validation and training. The performance of the proposed model DMCIC_Net was examined and distinguished from six pre-trained classifiers that were based on the confusion matrix: accuracy, recall, F1-score, AUC, precision, and loss.

12.3.2 Data Set Description

The COVID-19 cough audio data used in the Coswara Project originated from two other resources that were accessible on Kaggle [43] and this data was gathered by the Corswara web app (https://coswara.iisc.ac.in), which was responsible for both the gathering of information and the quality control of that data. The goal of the Coswara initiative was to

TABLE 12.2   A Detailed Description of the Cough and Breath Sound Data Set

Data set	COVID-19	Normal	Pneumonia	Asthma	Tuberculosis	Total
Coswara	120	48	-	-	-	168
Cough	19	151	-	-	-	170
Coughvid	2,073	1,664	-	-	-	3,737
CoronaHack	-	-	-	-	1,600	1,600
Respiratory	-	-	423	140	-	563
Total	2,212	1,863	423	140	1,600	6,238

create a database with open access that contained the respiratory sounds of COVID-19 patients as well as NOR case individuals. The project, which began on August 7, 2020, was an international effort to collect data on respiratory systems all over the globe. Over 1,600 people from all over the world took part in the survey, and their responses were compiled. The human ethics committee at the Indian Institute of Science (IISc) in Bangalore, India, gave their permission for the database. People who took part in the study had no personally identifying information collected from them, and the data they provided was thoroughly examined before being added to the collection. The database stored recordings of speech, cough, and respiration that were gathered through crowd-sourcing and the use of a dynamic website application designed specifically for mobile phones. Five to seven minutes was the average amount of time that a person spent interacting with the application. Before beginning the recording, all participants were instructed to maintain a 10 cm distance between their forehead and the device. Age, gender, geographic location, current health situation, and a history of any pre-existing medical conditions were also included in the documentation. There were 168 in the total, of which 120 COVID-19-positive and 48 NOR cases are included. Table 12.2 presents the complete statistics of the cough and breath sounds.

The Coughvid data set [44] was also used for COVID-19-positive and NOR cases, which included 3,737 named individual sound recordings in total and was accessed publicly. There were a total of 2,073 patients with COVID-19 positive and 1,664 patients with NOR. In the cough data set, [45] a total of 170 recordings of sound clips included 151 NOR patients and 19 COVID-19 patients. Corswara, after combining these three databases, had a total of 2,212 COVID-19-positive cases and 1,863 NOR cases. The respiratory sound data set [46] was a publicly accessible collection of AST and PNEU breathing sounds. There were 140 AST breath sounds and 423 PNEU breath sounds in the data set. The CoronaHack data set [47] was used for TB cough sounds in which 1,600 recordings of cough sounds were utilized. Figure 12.3 represents the cough and breath sound images of multiple chest infections.

12.3.3 Using SMOTE Tomek to Balance the Data Set

One of the most common approaches to resolving the problem of an unbalanced database is to increase the number of samples taken from the minority group. The objective is to replicate random samples from the minority group. This technique is commonly referred to as SMOTE, or synthetic minority oversampling technique. The up-sampling

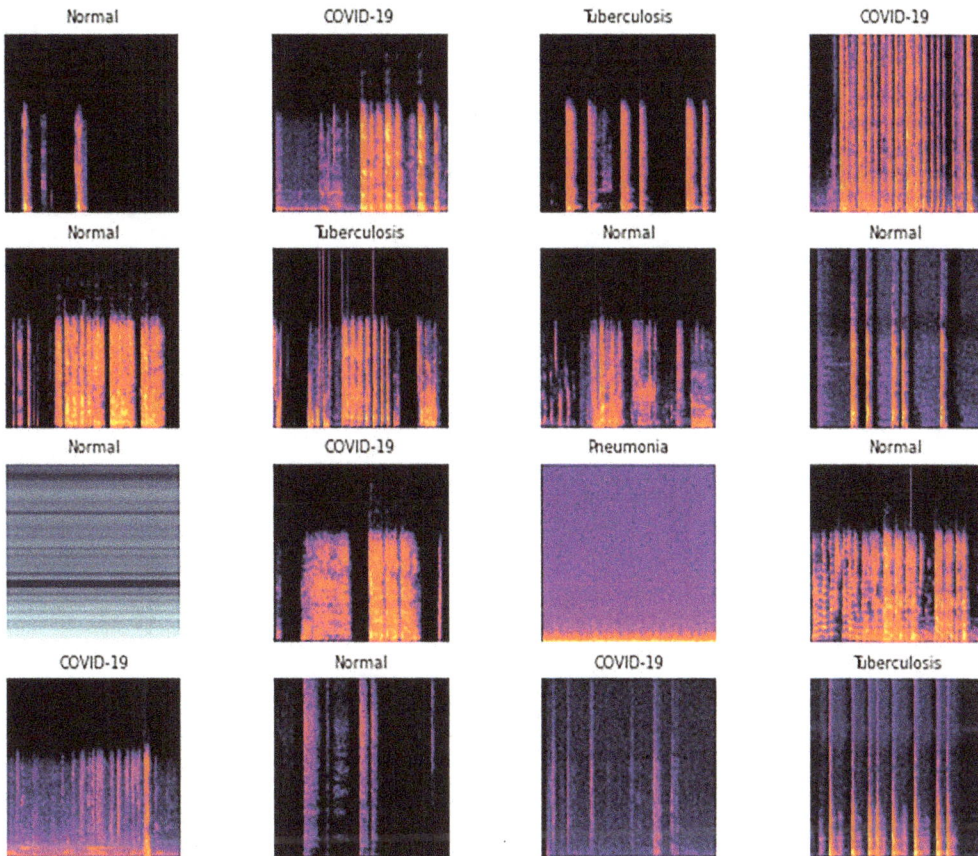

FIGURE 12.3 Original image samples of chest infections extracted from five data sets.

algorithm of SMOTE Tomek was used to get the samples for each class [19], as shown in Figure 12.4. The purpose was to integrate the SMOTE Tomek techniques to upgrade the effectiveness of dealing with unbalanced data sets. Synthetic points were created through the SMOTE to attain a balanced distribution. Before applying the method of up-sampling, the samples were expressed in Table 12.3, and after applying the technique of up-sampling, shown in Table 12.4.

12.3.4 Deep Learning Classifiers

The six DL classifiers were used for the detection of multiple chest infections. Vgg-16, Vgg-19, ResNet-152, MobileNet, Inception-ResNet-v2, and EfficientNet-B1 were included in these classifiers. By the utilization of ImageNet data, these classifiers were trained. This data set contained tens of thousands of distinct object categories, and classifiers were trained and assessed using it. VGG presented a model demonstrating that CNN's accuracy was greatly influenced by the depth of its network. In Vgg-16 [28], 18 architecture consisted of six distinct stages, with two convolutional layers and one max-pooling layer making up the first two phases. In the next three phases, three convolutional layers and one max-pooling layer were applied, and the final stage comprised three

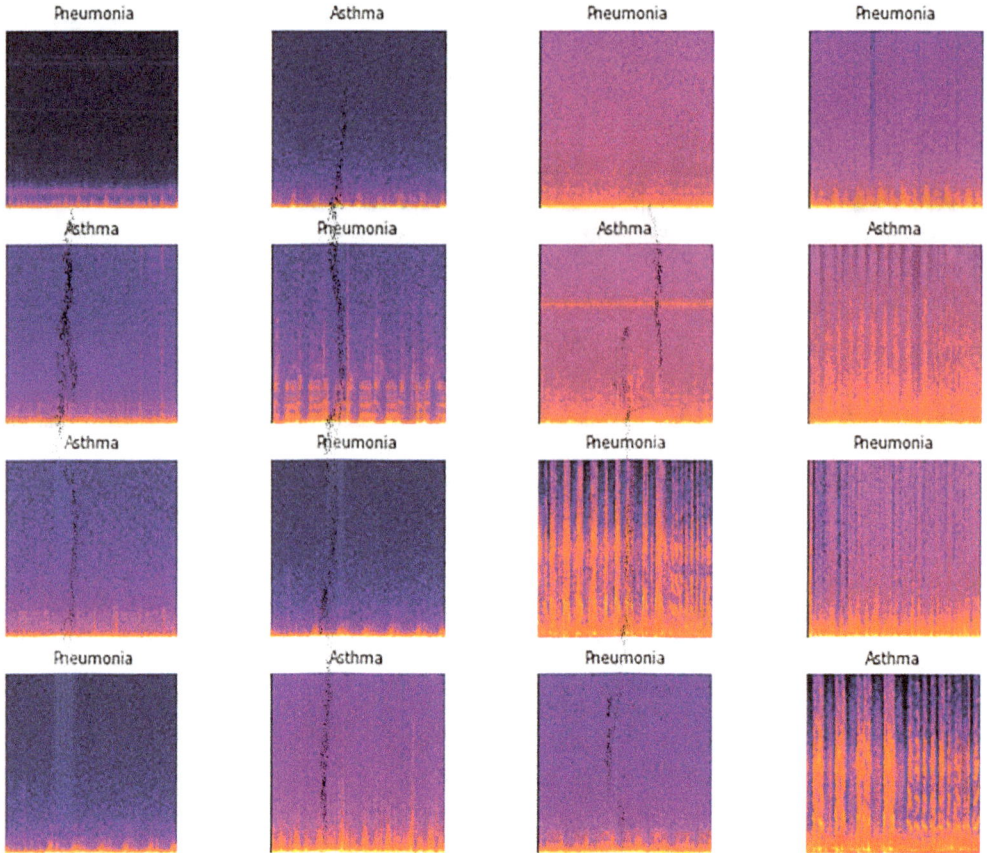

FIGURE 12.4 SMOTE Tomek creates samples of images to balance the classes.

TABLE 12.3 Before Up-sampling Distribution of Samples of Images

No. of classes	Class name	No. of images
0	AST	140
1	COVID-19	2,212
2	NOR	1,863
3	PNEU	423
4	TB	1,600

TABLE 12.4 After Up-sampling Distribution of Sample Images

No. of classes	Class name	No. of images
0	AST	2,186
1	COVID-19	2,212
2	NOR	1,863
3	PNEU	2,223
4	TB	1,600

fully connected layers. The VGG-19 was the enhanced version of the VGG-16. VGG-19 had 16 convolutional layers, 5 maximum pooling layers for feature extraction, and 3 fully connected layers for classification. ResNet-152 [24] was a residual system of 152 layers that consisted of 151 convolutional layers and 1 FC layer that was used to identify multiple chest infections. Inception-ResNet-v2 was a convolutional neural network trained on over 1 million photos from the ImageNet collection. The 164-layered network had the capability of assigning pictures to 1,000 different item categories. In Mobile-Net, [24] there were a total of 28: 27 of which were conversion layers, and 1 of which was the FC layer. EfficientNet-B1 was designed by utilizing a multi-objective neural network search that maximized both accuracy and floating-point operations.

12.3.5 Proposed Model

This section presents the proposed model and its architecture for multiple chest infections.

12.3.5.1 Structure of the Proposed DMCIC_Net

The CNN architecture is represented by the human brain's organic structure and is particularly beneficial for computer vision applications, such as object recognition, face detection, segmentation of images, and identification. In this study, we developed a robust DMCIC_Net that depended on the CNN model to correctly identify multiple chest infections. The DMCIC_Net model contained five convolutional blocks and a rectified linear unit (ReLU), one dropout layer, two dense layers, an activation function, and a softmax classification layer added in this model that is shown in Figure 12.5. Table 12.5 explains the proposed model for the identification of multiple chest infections with the succeeding layers.

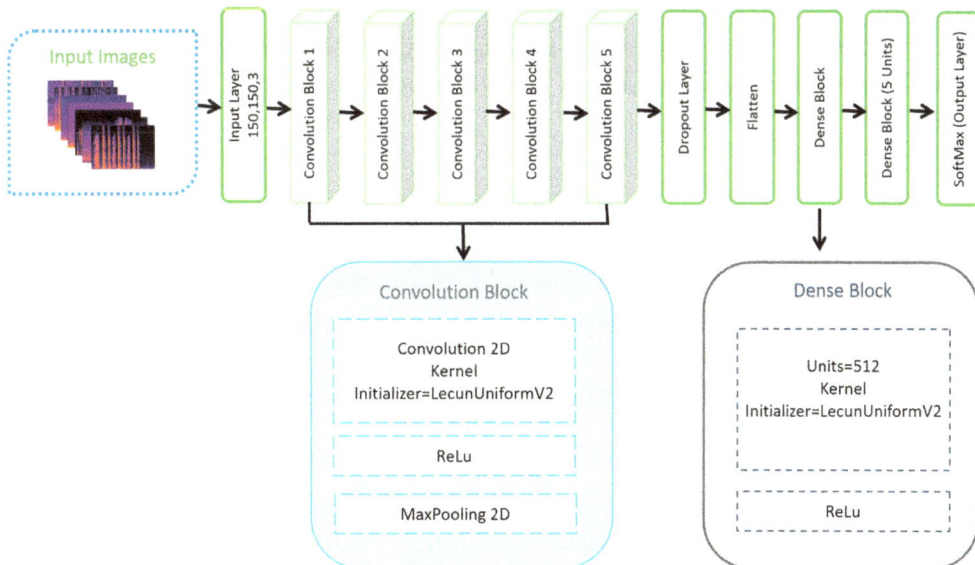

FIGURE 12.5   The architecture of the proposed DMCIC_Net to identify multiple chest infections.

TABLE 12.5 Total Number of Parameters Employed in the Proposed DMCIC_Net

Layer type	Outer shape	Parameters
Input Layer	(None, 150, 150, 3)	0
Block01	(None, 150, 150, 8)	224
Block02	(None, 75, 75, 16)	1,168
Block03	(None, 37, 37, 32)	4,640
Block04	(None, 18, 18, 64)	18,496
Block05	(None, 9, 9, 128)	73,856
Dropout_1	(None, 4, 4, 128)	0
Flatten	(None, 2048)	0
Dense_1	(None, 512)	1,049,088
ReLu	(None, 512)	0
Dense_2	(None, 5)	2,052
Output: SoftMax	(None, 5)	0
Total Parameters		1,150,037
Trainable Parameters		1,150,037
Non-Trainable Parameters		0

12.3.5.2 Convolutional Blocks of CNN Model

The fundamental building element of the proposed model is the convolutional block, but each convolutional block contains convolutional two dimensions, a ReLU, and a pooling two-dimensional with an average value. The initializer for the kernel layer to determine layer kernel weights, the LecunUniformV2 algorithm is first initialized. The gradient vanishing problem is resolved by the ReLU activation function, which enables the network to learn and operate faster.

The RGB colors are present in the image that is being input. The 371 convolutional layer is the name of the top layer in our design. This layer began with the procedure of adding filters, sometimes referred to as the 372 kernels. The kernel size is determined by two factors, as depicted by Eq (12.1).

$$Filter\ Size\,(FS) = f_w \times f_h \tag{12.1}$$

where f_w represents the width of the filter and f_h represents its height. In the course of our research, we decided to make the filter three times as large, so the first equation becomes FS = 3 × 3. These filters also referred to as feature identifiers help us to define low-level image features like sides and curves [5].

12.3.5.3 Flattened Layer

After the convolution layers and before the thick layers comes a layer that has been flattened out. Tensors are the sorts of data that can be used as input for convolution layers, whereas one-dimensional input is essential for dense layers [48]. As shown in Figure 12.6, a flattened layer turns the feature map into a vector so that it can be used by dense layers.

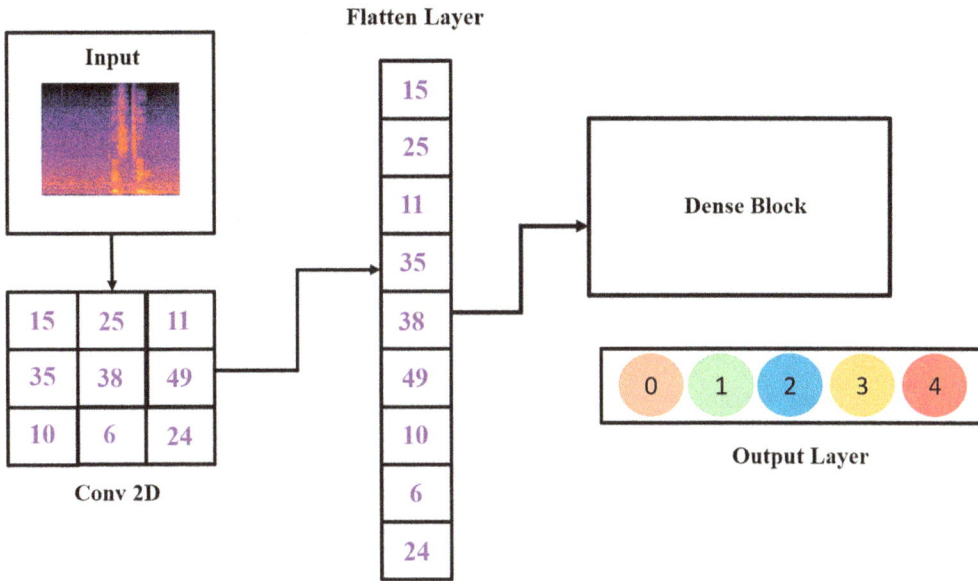

FIGURE 12.6   The fundamental structure of the flattened layer.

12.3.5.4 Dropout Layer

To shorten the amount of time required to train the model and to simplify the network, dropout layers switch the status of individual units. Dropout prevents models from becoming very accurate by turning off units in a random manner in which a probability distribution is used at the beginning of each epoch. As a consequence of this, the model picks up on all of the pertinent characteristics and fully avoids picking up any different characteristics during each iteration.

12.3.6 Dense Block of Proposed Model

In this chapter, we utilize two dense blocks with a lot of layers. The subsequent sections elaborate on the activation function of each block.

12.3.6.1 ReLU Activation

Numerical procedures known as activation functions determine whether or not the results from perception should be passed on to the upcoming layer. In essence, they enable and disable model nodes. Within the output layer, the activation function is called upon to activate the node, which gives back the label. It is subsequently allocated to the image once it has been analyzed by the model. Various functions activate it. We utilized ReLU in hidden layers due to its straightforward and time-efficient computation. Because our model is intended for multi-class classification, the output layer makes use of SoftMax, an activation function that is based on probability.

12.3.6.2 Dense Layer

The fully connected layer is another name for the dense layer. This layer takes only one vector as input and generates results according to its characteristics. Inside these layers, the images are diagnosed and given a label indicating their class. Backpropagation is the learning approach that is used for the model, and it is used in layers that are completely connected. The final output of the model identifies the image as having one of the five chest infections classifications. AST, COVID-19, NOR, PNEU, and TB are generated using a dense layer of five neurons and a softmax activation function. After a few layers, the softmax technique is used, with the number of neurons being equal to the number of classes. The total amount of parameters is 1,149,524, and they are divided into two categories: there are 1,149,524 trainable parameters and non-trainable parameters are zero.

12.3.7 Model Evaluations

In this chapter, we resolved the multi-classification problems where AST, COVID-19, NOR, PNEU, and TB were diagnosed accurately. Consequently, a confusion matrix was applied to examine the effectiveness of the model. Before training the model, the data set was divided into training and test sets. Using the test set, the model's performance was evaluated. We used a range of variables to assess the model's efficiency. The following evaluation metrics were used to evaluate the effectiveness of the proposed DMCIC_NET for multiple chest infection detection.

12.3.7.1 Accuracy

Accuracy was expressed as the proportion of correct identifies to all predictions using the following Eq (12.2):

$$Accuracy = \frac{TP + TF}{TP + FN + FP + TN} \tag{12.2}$$

where the values for TP, TN, FN, and FP stand for true positive, true negative, false negative, and accordingly false positive values.

12.3.7.2 Precision

The following Eq (12.3) was used to compute precision, which was defined as the ratio of correct positive predictions to all positive predictions:

$$Precision = \frac{TP}{TP + FP} \tag{12.3}$$

12.3.7.3 Recall

Recall is known as sensitivity or the proportion of actual positives. It is calculated by subtracting the total of false negatives from the sum of true positives. The recall was calculated using Eq (12.4).

$$Recall = \frac{TP}{TP + FN} \tag{12.4}$$

12.3.7.4 F1-Score

Generally, precision value of 1.0 and a recall value of 1.0 is considered a perfect case for a model of classification. The F1-score is defined as the harmonic mean of the recall and precision scores. It is unique in that it draws a distinct line for each class label on its graph, defined by the following Eq (12.5):

$$F1 - score = 2 \times \frac{Precision \times Recall}{Precision + Recall} \tag{12.5}$$

12.4 RESULTS AND DISCUSSION

We examined DMCIC-Net using the most recent deep network in the next section. Table 12.6 shows the comparisons between the proposed DMCIC_Net and six baseline neural networks. We implemented the same parameters to all of them to calculate the achievement of deep neural networks correctly.

12.4.1 Experimental Setup

A total of six models were applied by using keras in which six baseline models and one DMCIC_Net model was included. Furthermore, such approaches of programming that were not directly linked to convolutional networks were done in Python. A computer with Windows 10 and equipped with an 11 GB NVIDIA GPU and 32 GB of RAM was used for the experiment.

12.4.2 Accuracy Comparison of Proposed Model with Baseline Models

By applying the same data set, we compared our suggested and recently developed models, i.e., Vgg-16, Vgg-19, ResNet152, MobileNet, Inception-ResNet-v2, and EfficienNet-B1, with SMOTE Tomek. Before implementing SMOTE Tomek, we compared the suggested DMCIC_Net. The system gave the amazing result of the proposed model by using SMOTE Tomek, as shown in Figure 12.7. Table 12.6 shows the proposed DMCIC_Net with SMOTE Tomek, DMCIC_Net without SMOTE Tomek, Vgg-16, Vgg-19, ResNet152, MobileNet,

TABLE 12.6 Performance of DMCIC_Net Model Compared With Baseline Algorithms

Classifiers	Accuracy	Precision	Recall	F1-score	AUC
Vgg-16	93.02%	93.02%	93.02%	93.08%	99.33%
Vgg-19	88.15%	88.35%	87.39%	85.84%	98.73%
ResNet-152	92.85%	92.95%	92.74%	92.66%	99.28%
MobileNet	94.94%	94.93%	94.83%	94.97%	99.36%
Inception-ResNet-v2	90.62%	90.84%	90.36%	90.11%	99.32%
EfficientNet-B1	94.20%	94.48%	93.76%	93.99%	99.45%
Proposed Model **(With SMOTE Tomek)**	97.24%	97.24%	97.24%	97.27%	99.30%
Proposed Model **(Without SMOTE Tomek)**	64.94%	73.09%	54.62%	54.65%	91.77%

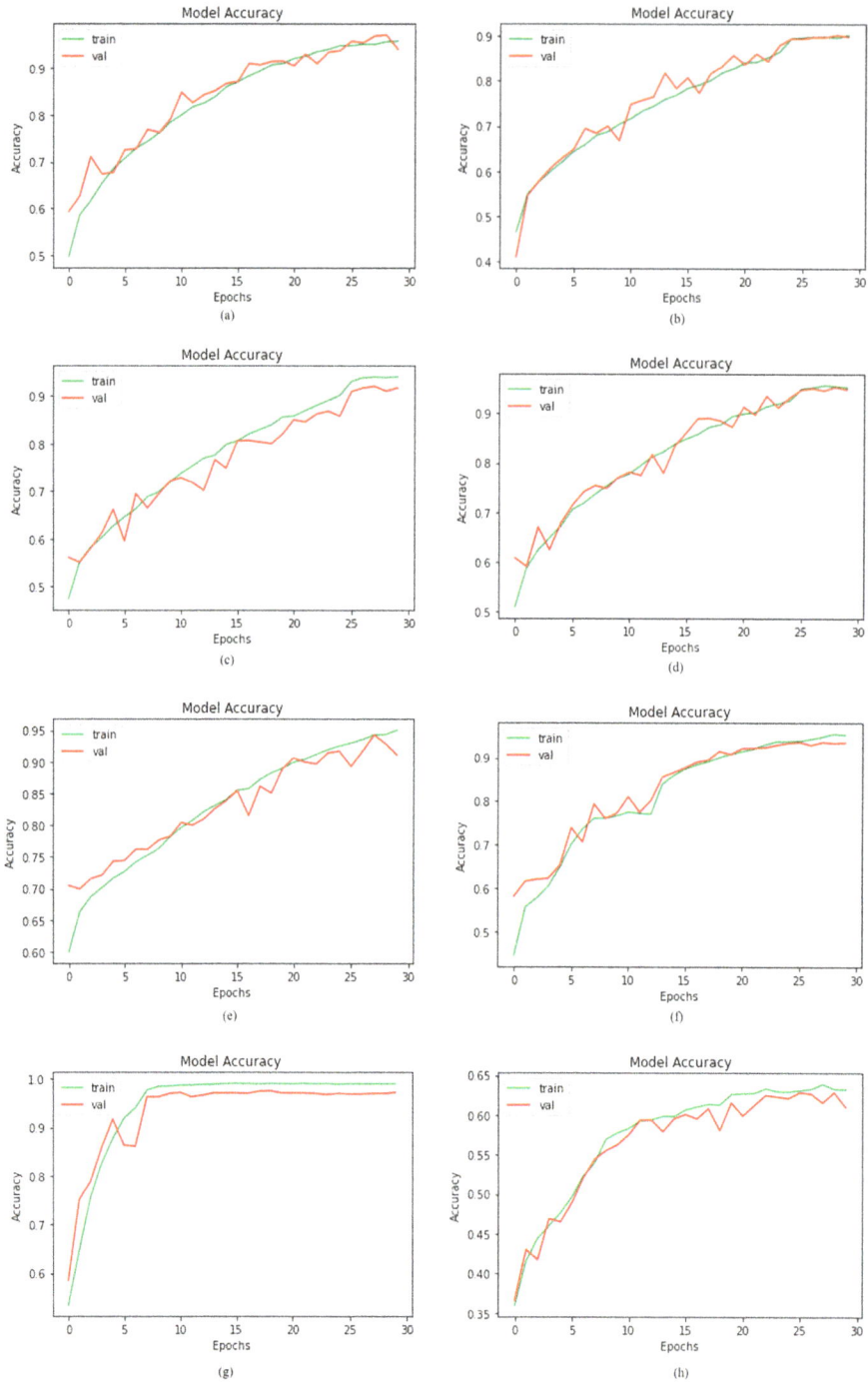

FIGURE 12.7   Stunning accuracy enhancement with or without SMOTE Tomek in the suggested model compared to other baseline deep neural networks: (a) Vgg-16, (b) Vgg-19, (c) ResNet152, (d) MobileNet, (e) Inception-ResNet-v2, (f) EfficientNet-B1, (g) proposed model with SMOTE Tomek, and (h) proposed model without SMOTE Tomek.

Inception-ResNet-v2, and EfficientNet-B1 achieved accuracies of 97.24%, 64.94%, 93.02%, 88.15%, 92.85%, 94.94%, 90.62%, and 94.20%, respectively.

12.4.3 AUC Comparison with Baseline Models

As previously described in this research, our suggested model was a deep CNN-based DMCIC-Net created of several blocks that were particularly successful in classifying the various classes of chest infections. We differentiated DMCIC_Net with six baseline deep neural networks to validate our suggested DMCIC_Net. These were six baseline neural networks, Vgg-16, Vgg-19, ResNet152, MobileNet, Inception-ResNet-v2 and EfficientNet-B1, that attained an AUC of 99.33%, 98.73%, 99.28%, 99.36%, 99.32%, and 99.45%, respectively. Figure 12.8 illustrates that the proposed DMCIC-Net with SMOTE Tomek and DMCIC-Net without SMOTE Tomek obtained AUC values of 99.30% and 91.77%, respectively. In the previous clarification, we observed that the performance of the proposed model, as measured by AUC, continued to be better and more reliable than that of other models.

12.4.4 Comparison with Baseline Models Using Precision

We estimated our proposed and recent models of Vgg-16, Vgg-19, ResNet152, MobileNet, InceptionResnet-V2, and EfficientNet-B1 using the SMOTE-Tomek to balance the data set. The system with SMOTE gave stunning outcomes for the proposed DMCIC_Net, along with various models. Precision values of 97.24%, 73.09%, 93.02%, 88.35%, 92.95%, 94.93%, 90.84%, and 94.48% were attained by the proposed DMCIC_Net with SMOTE Tomek, DMCIC_Net without SMOTE Tomek, Vgg-16, Vgg-19, ResNet152, MobileNet, Inception-ResNet-v2, and EfficientNet-B1, respectively. As the result of this study, we found that the performance of the proposed model DMCIC_Net was more precise and reliable than that of current deep models, as shown in Figure 12.9.

12.4.5 Comparison of DMCIC_Net with Baseline Models Using Recall

It is determined by splitting the amount of correct positive estimations through the actual amount of correct positives. Recall measures the model's capacity to identify positive samples. Higher recall values indicate that more positive samples were discovered. By the utilization of the recall curve, the proposed DMCIC_Net was differentiated between the baseline networks, as shown in Figure 12.10. The proposed DMCIC_Net with and without SMOTE Tomek and Vgg-16, Vgg-19, ResNet152, MobileNet Inception-ResNet-v2, and EfficientNet-B1 obtained the values of recall of 97.24%, 54.62%, 93.02%, 87.39%, 92.74%, 94.83%, 90.36%, and 93.76%, respectively. The results reveal that the proposed model depicts stunning performance.

12.4.6 F1-Score Comparison with Baseline Models

The suggested DMCIC_Net with and without SMOTE Tomek and Vgg-16, Vgg-19, ResNet152, MobileNet, Inception-ResNet-v2, and EfficientNet-B1 obtained the F1-score values of 97.27%, 54.65%, 93.08%, 85.84%, 92.66%, 94.97%, 90.11%, and 93.99%, respectively. The proposed DMCIC_Net with SMOTE Tomek achieved a higher value of the F1-score illustrated in Figure 12.11.

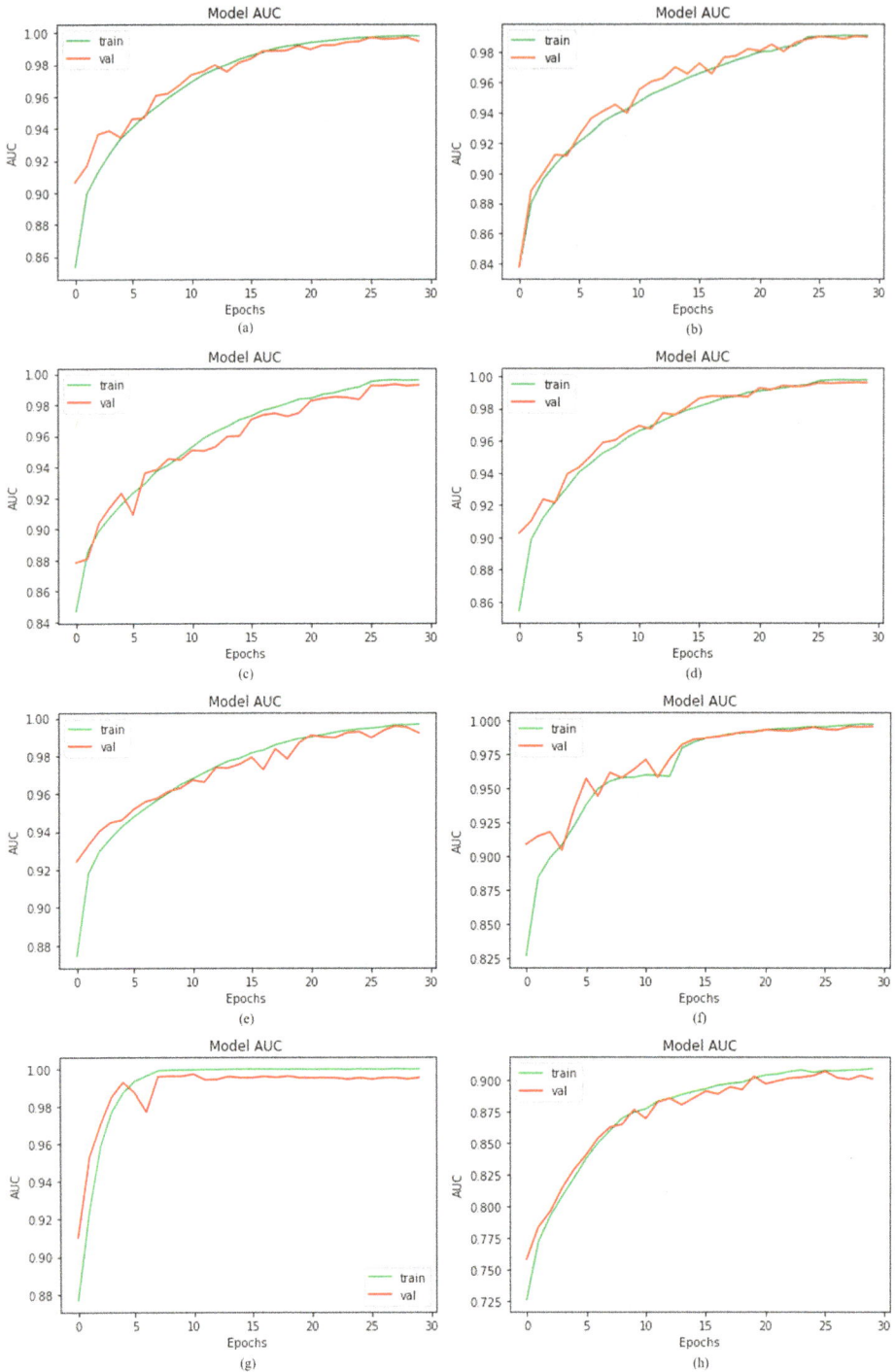

FIGURE 12.8 Classification result of multiple classes of proposed DMCIC_Net with or without SMOTE Tomek as presented by AUC: (a) Vgg-16, (b) Vgg-19, (c) ResNet152, (d) MobileNet, (e) Inception-ResNet-v2, (f) EfficientNet-B1, (g) proposed model with SMOTE Tomek, and (h) proposed model without SMOTE Tomek.

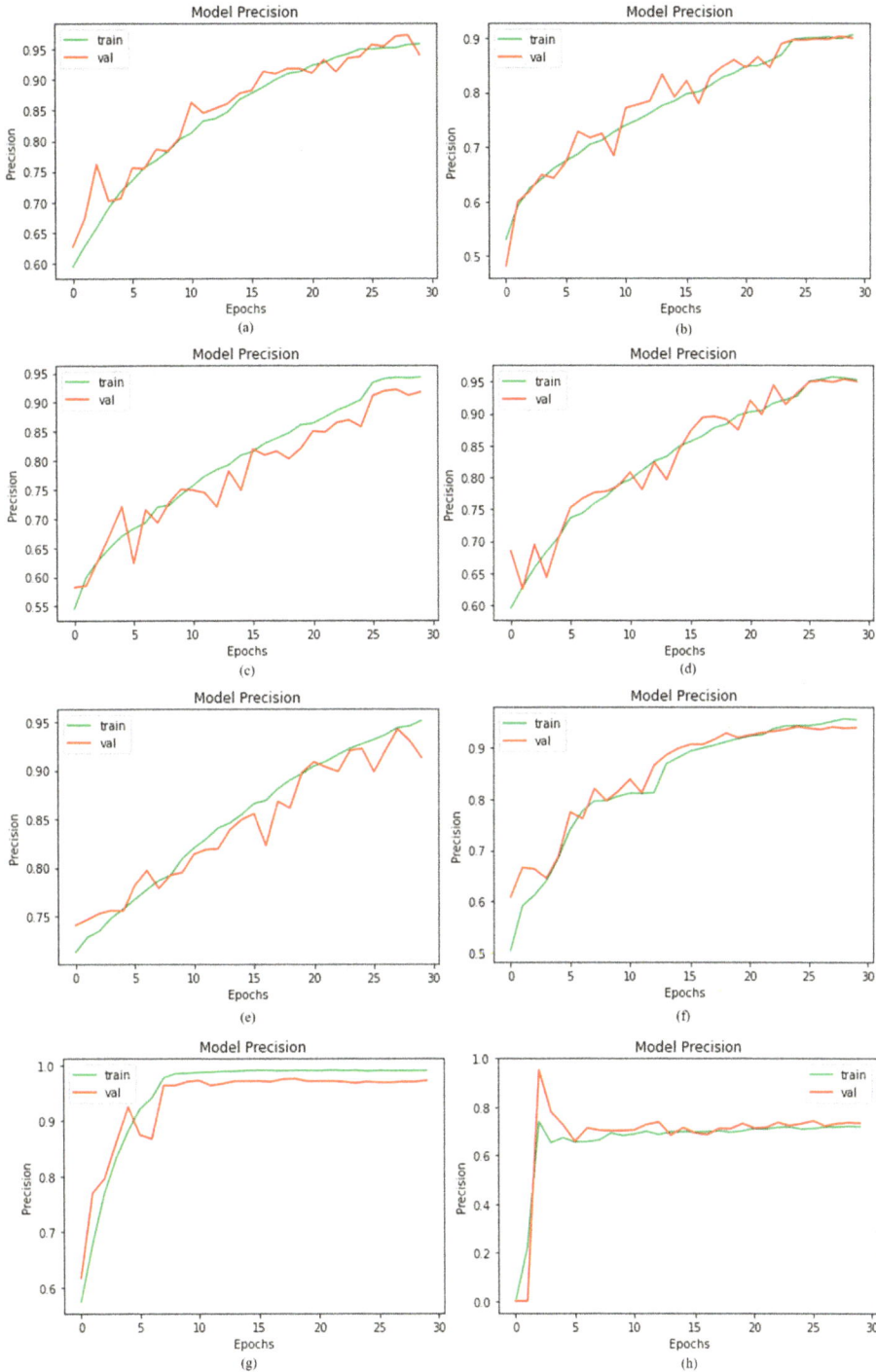

FIGURE 12.9   Precision results of proposed model DMCIC_Net and other baseline models: (a) Vgg-16, (b) Vgg-19, (c) ResNet152, (d) MobileNet, (e) Inception-ResNet-v2, (f) EfficientNet-B1, (g) proposed model with SMOTE Tomek, and (h) proposed model without SMOTE Tomek.

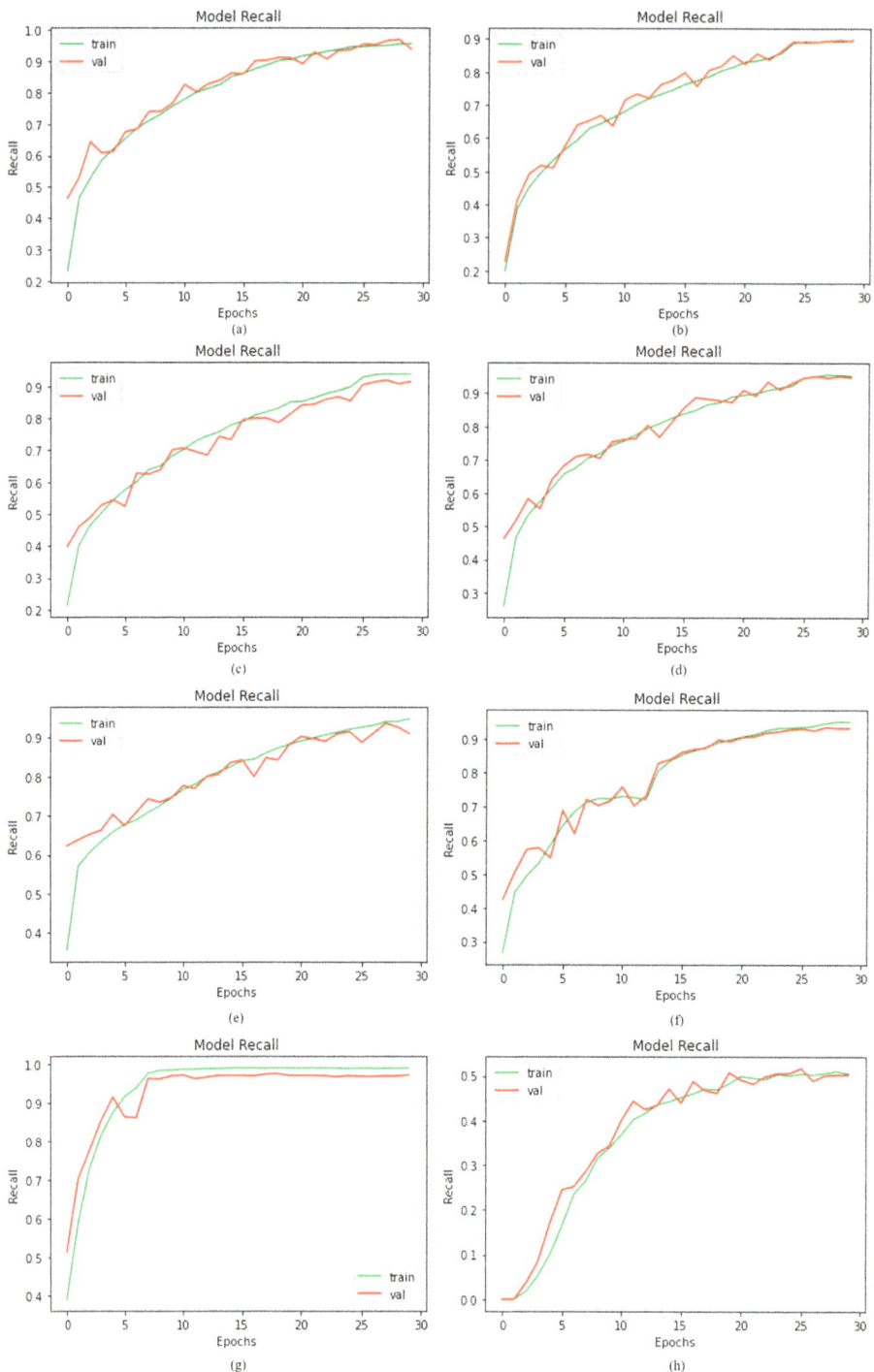

FIGURE 12.10 The recall analysis to give an overview of the recall between the correct positive and the positive estimate value for: (a) Vgg-16, (b) Vgg-19, (c) ResNet152, (d) MobileNet, (e) Inception-ResNet-v2, (f) EfficientNet-B1, (g) proposed model with SMOTE Tomek, and (h) proposed model without SMOTE Tomek.

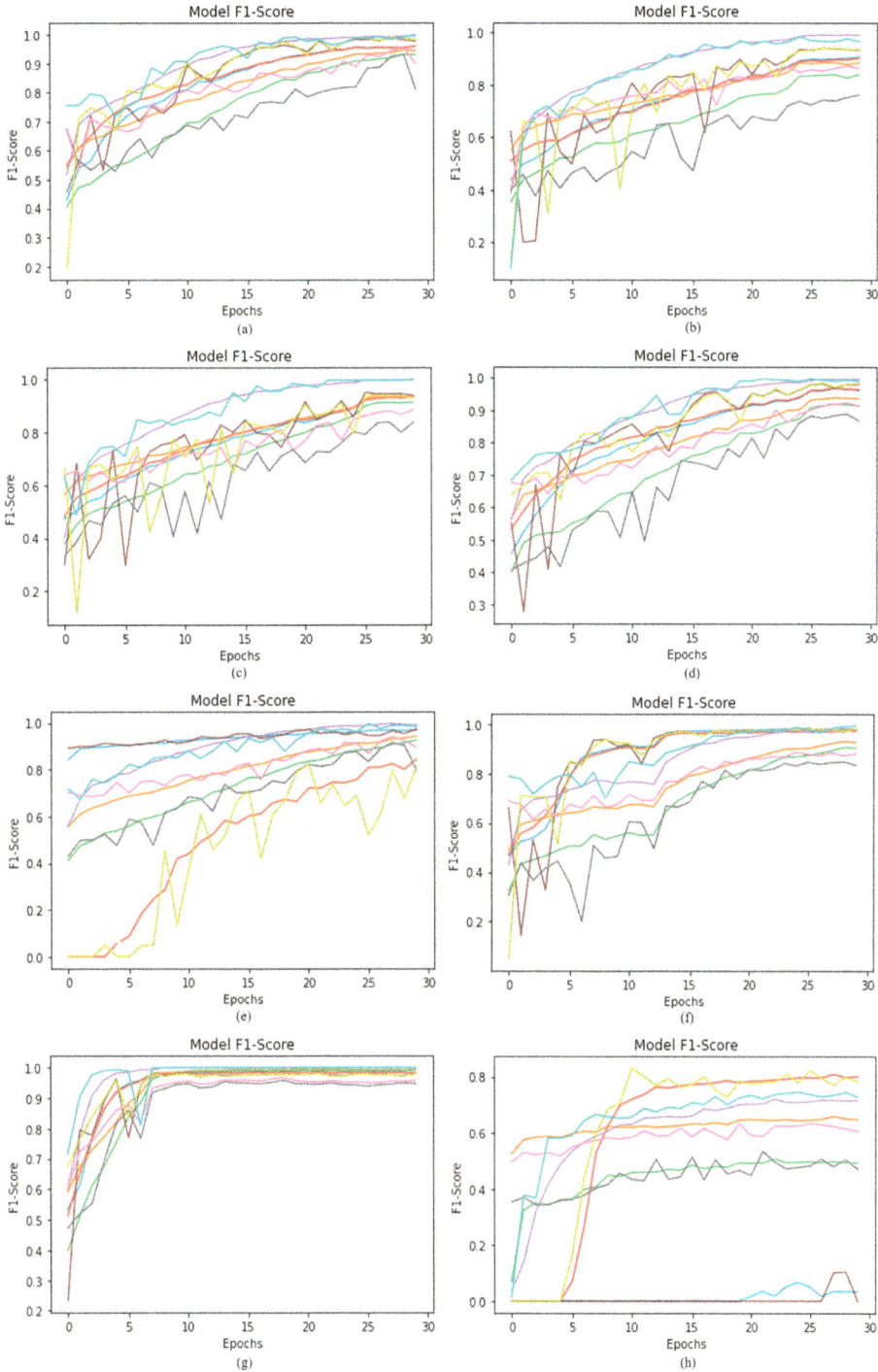

FIGURE 12.11 Compute the value of the F1-score among the suggested model and baseline models: (a) Vgg-16, (b) Vgg-19, (c) ResNet152, (d) MobileNet, (e) Inception-ResNet-v2, (f) EfficientNet-B1, (g) proposed model with SMOTE Tomek, and (h) proposed model without SMOTE Tomek.

12.4.7 Comparison of Proposed Model with Baseline Models Using Loss

The mathematical distinction between the expected amount and the actual amount was calculated by loss functions. For calculating loss, we employed a categorical cross-entropy approach. When the model was trained on up-sampled photos, the results were even more impressive. The suggested DMCIC_Net with and without SMOTE Tomek obtained the loss values of 0.15% and 0.77% while Vgg-16, Vgg-19, ResNet152, MobileNet, Inception-ResNet-v2, and EfficientNet-B1 obtained the loss values of 0.21%, 0.30%, 0.22%, 0.16%, 0.22%, and 0.17%, respectively. Figure 12.12 shows this remarkable improvement in the loss of the suggested DMCIC-Net with SMOTE Tomek.

12.4.8 Comparison of ROC with Current Models

The effectiveness of a classifier for binary or multi-classification was examined using a ROC curve when analyzing the results of clinical tests. A receiver operating characteristic (ROC) curve's area under the curve (AUC) was utilized to evaluate the performance of a classifier where a higher AUC often indicates a classifier's effectiveness. We examined the efficiency and accuracy of our suggested DMCIC-Net by using the ROC curve with and without SMOT Tomek. The suggested DMCIC_Net with SMOTE Tomek and without SMOTE Tomek was compared by the utilization of the ROC curve with baseline models. The proposed DMCIC_Net with and without SMOTE Tomek, Vgg-16, Vgg-19, ResNet152, MobileNet, Inception-ResNet-v2, and EfficientNet-B1 attained the ROC values of 0.95%, 0.65%, 0.86%, 0.84%, 0.89%, 0.90%, 0.83%, and 0.89%, respectively, as shown in Figure 12.13. The remarkable enhancement of the suggested DMCIC_Net with SMOTE Tomek and without SMOTE Tomek can be obvious in Figure 12.13.

12.4.9 AU(ROC) Extension for Multiclass Comparison Against Recent Models

The comparison of the suggested DMCIC-Net with baseline networks by utilizing the extension of the ROC curve is shown in Figure 12.14. After balancing the data set by using the SMOTE Tomek approach, the AUC greatly improved for the suggested approaches in contrast with baseline models. In the AUC, this important impact was also noted for proposed DMCIC_Net classes with and without SMOTE Tomek in which class 0 (AST), class 1 (COVID-19), class 2 (NOR), class 3 (PNEU), and class 4 (TB) are included. These enhancements in AUC illustrate the reliability of the SMOTE Tomek technique.

12.4.10 Comparison of DMCIC_Net with Six Models Using a Confusion Matrix

To verify our deep DMCIC-Net, we compared it to current deep networks Vgg-16, Vgg-19, ResNet152, MobileNet, Inception-ResNet-v2, and EfficientNet-B1. The system with SMOTE Tomek produced remarkable outcomes for the proposed model as well as other models, as shown in Figure 12.15.

The proposed model accurately classified 188 images out of 190 total images in AST cases, but incorrectly classified 2 images as PNEU. Out of 200 images, 197 images of COVID-19 were correctly identified, and 3 images were incorrectly classified as NOR. From the total of 163 images of NOR, 148 images were correctly recognized as such, while 15 images were misclassified as COVID-19. From 205 images of PNEU, 200 images

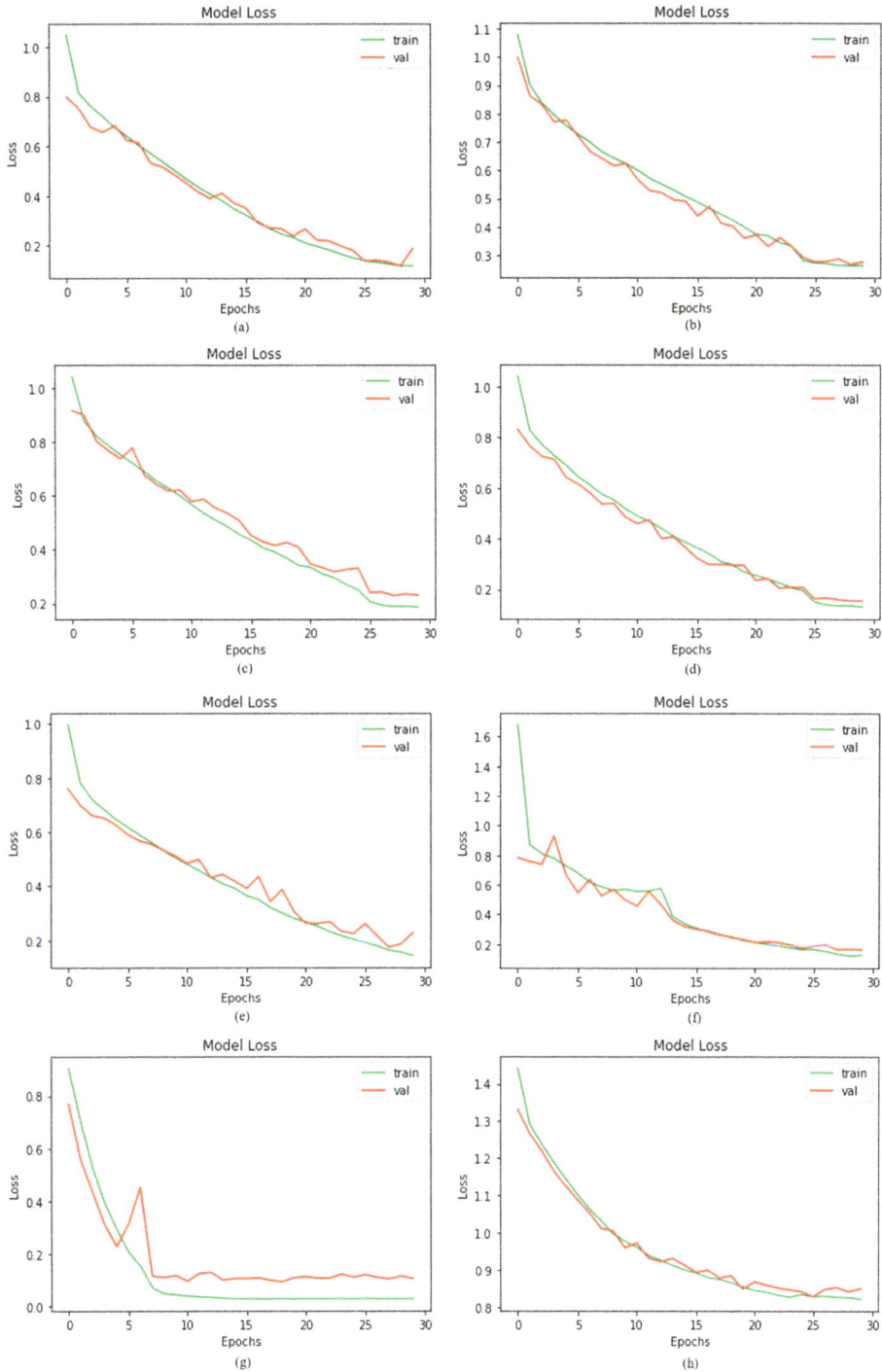

FIGURE 12.12   Loss value is calculated between proposed DMCIC_Net and baseline models: (a) Vgg-16, (b) Vgg-19, (c) ResNet152, (d) MobileNet, (e) Inception-ResNet-v2, (f) EfficientNet-B1, (g) proposed model with SMOTE Tomek, and (h) proposed model without SMOTE Tomek.

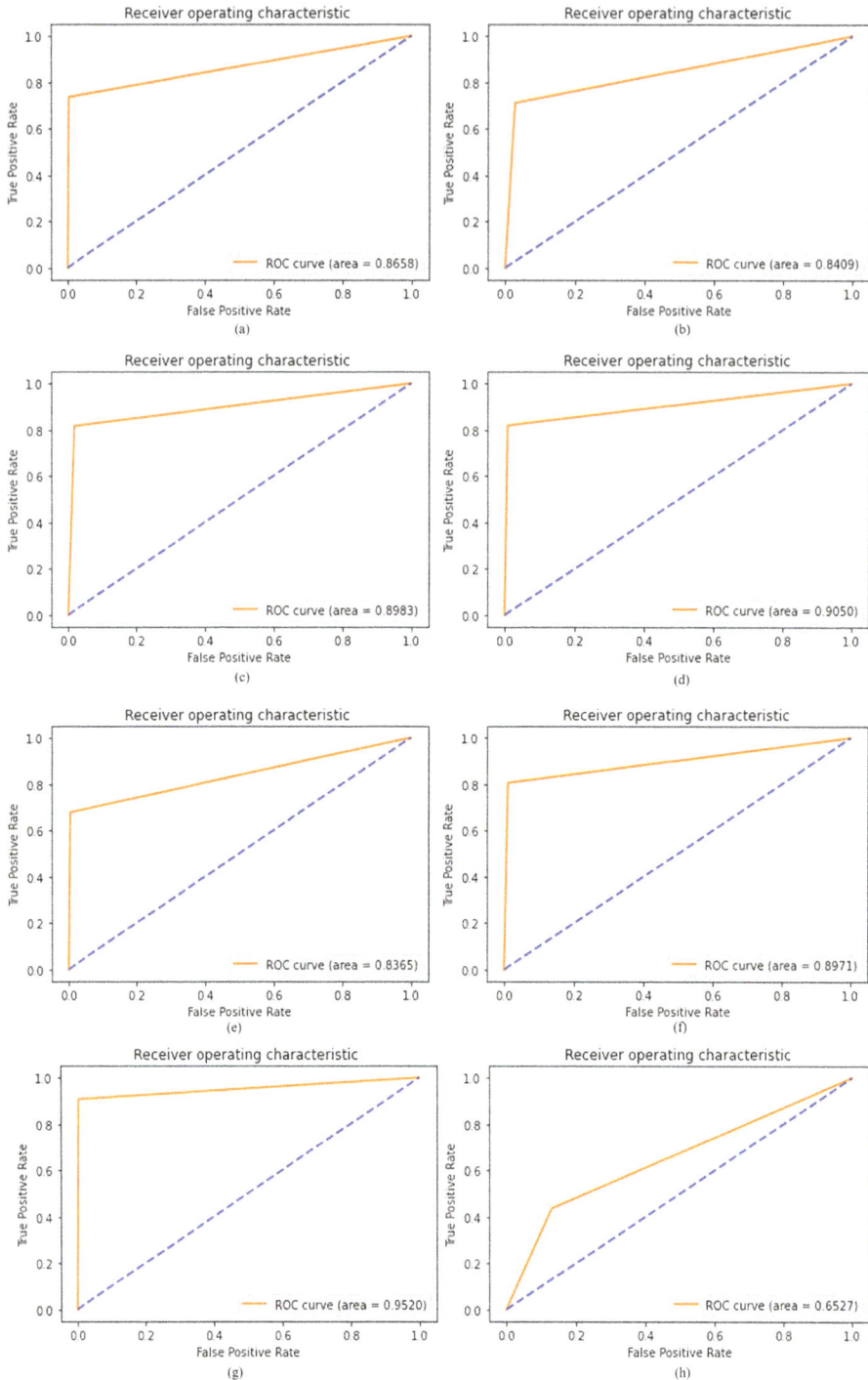

FIGURE 12.13 Comparison of the performance of ROC curve between proposed DMCIC and baseline models: (a) Vgg-16, (b) Vgg-19, (c) ResNet152, (d) MobileNet, (e) Inception-ResNet-v2, (f) EfficientNet-B1, (g) proposed model with SMOTE Tomek, and (h) proposed model without SMOTE Tomek.

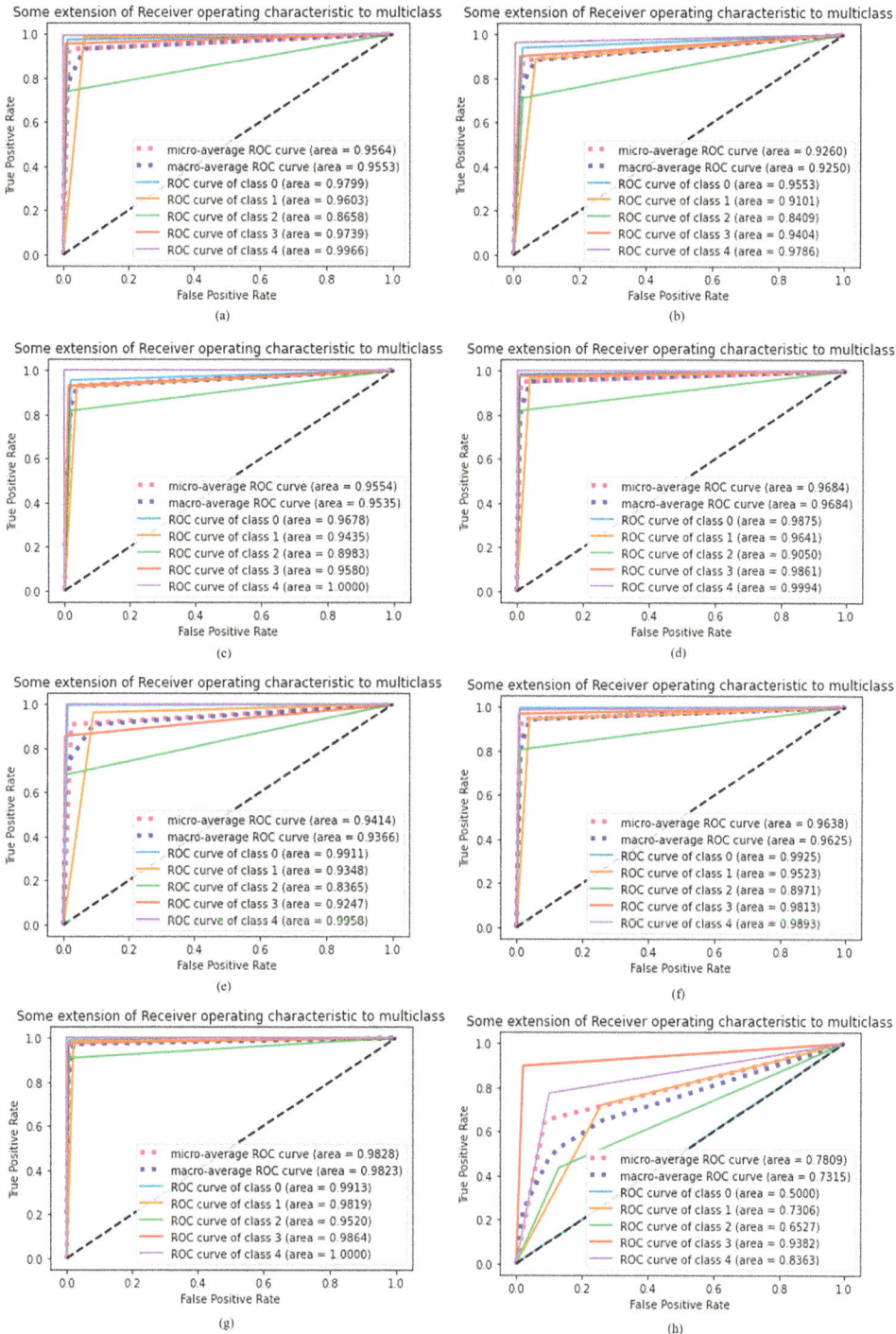

FIGURE 12.14 Analysis of AUC (ROC) curve with extension for suggested DMCIC_Net and baseline models: (a) Vgg-16, (b) Vgg-19, (c) ResNet152, (d) MobileNet, (e) Inception-ResNet-v2, (f) EfficientNet-B1, (g) proposed model with SMOTE Tomek, and (h) proposed model without SMOTE Tomek.

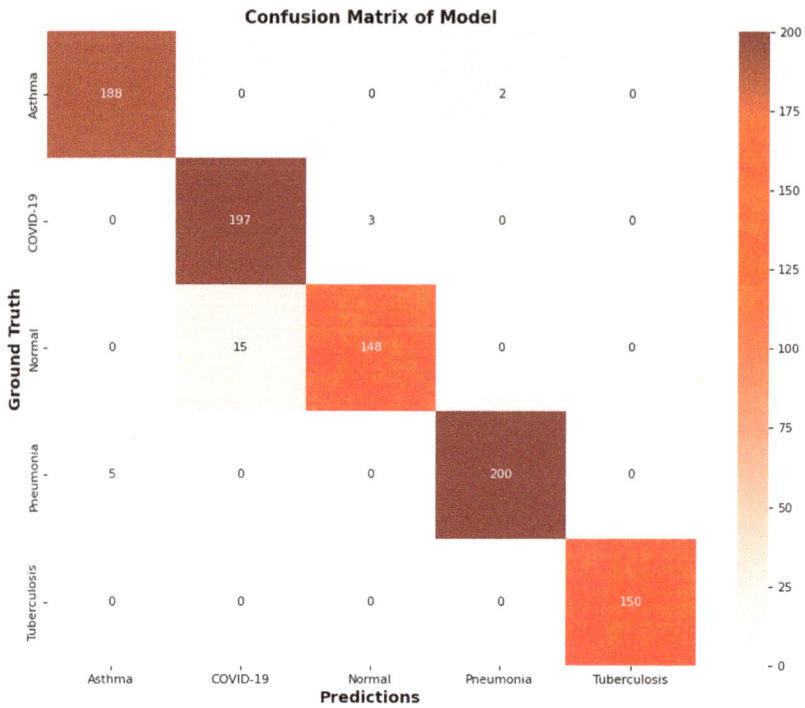

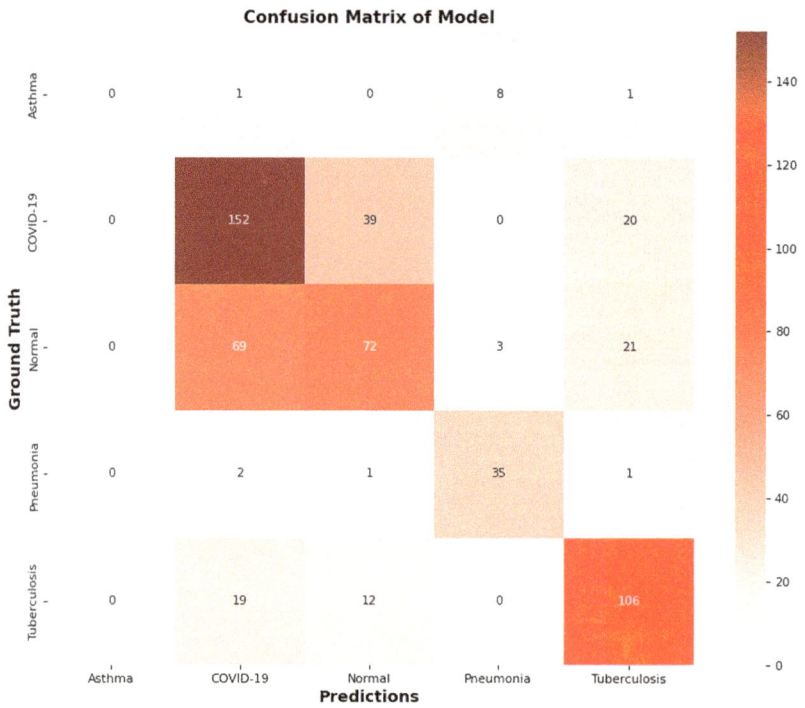

FIGURE 12.15 Confusion matrix: (a) proposed model with SMOTE Tomek and (b) proposed model without SMOTE Tomek.

TABLE 12.7   Comparison of the DMCIC_Net Model With SOTA

References	Models	Accuracy	Recall	Precision	F1-score
[17]	MSCCov19Net	74%	-	-	-
[18]	CNN + LSTM	95.01%	-	94.53%	95.01%
[19]	CNN	92.32%	-	-	93.48%
[20]	MLP + KNN	90%	0.74%	0.75%	-
[26]	SVM	94%	-	-	-
Proposed Model	**DMCIC_Net**	**97.24%**	**97.24%**	**97.24%**	**97.27%**

were correctly classified, whereas 5 images were misclassified as AST; 150 images as TB were classified out of 150 images and there was no misclassification.

12.4.11 Comparison of the Proposed Model with State of the Art

The proposed DMCIC_Net model was thoroughly evaluated in terms of numerous effectiveness metrics, including accuracy, recall, F1-score, and precision, using the most current state of the art (SOTA), as shown in Table 12.7.

12.4.12 Discussion

Spectrogram images can be employed to identify and categorize several different chest illnesses. The most accurate and efficient method for identifying whether the infection is an AST, COVID-19, NOR, PNEU, or TB is to use a spectrogram. An automated diagnostic method is needed to detect AST, COVID-19, NOR, PNEU, and TB due to a continuing rise in chest infection cases [23,38]. As a result, we developed a DMCIC_Net model that makes use of DL and is capable of accurately classifying a variety of chest conditions. The model allows medical professionals to start treating their patients early for infections like AST, COVID-19, NOR, PNEU, and TB. The five publicly available benchmark data sets were used to evaluate the performance of the proposed DMCIC_Net model. Six baseline models, including Vgg-16, Vgg-19, ResNet-152, MobileNet Inception-ResNet-v2, and EfficientNet-B1, have been used to compare the results of the proposed model [26]. Table 25.3 shows that the sounds obtained from data sets are unbalanced. Then, spectrogram visuals were created using the collected sounds. The uneven class of the images affected how well the model performed during training. We used the SMOTE Tomek approach to increase the minority class images in the data sets to tackle these issues. Figure 25.6 demonstrated that our proposed DMCIC_Net was capable of identifying instances of infection with the four categories of chest infections (AST, COVID-19, NOR, PNEU, and TB) and had the necessary training in these illnesses. According to Table 25.6, our DMCIC_Net classifier performs much better than the other six baseline classification techniques in classifying chest infections. The proposed DMCIC_Net model using the SMOTE Tomek method successfully classified AST, COVID-19, NOR, PNEU, and TB from spectrogram images with an accuracy of 97.24%. Furthermore, the accuracy of the DMCIC_Net model without the SMOTE Tomek method was 64.94%. The Vgg-16 model, on the other hand, achieved an accuracy of 93.02%. Similar to this, the accuracies of ResNet-152,

MobileNet, Inception-ResNet-v2, and EfficientNet-B1 were 92.85%, 94.94%, 90.62%, and 94.20%, respectively. Compared to other baseline models, the VGG19's performance in detecting chest infections was low.

12.5 CONCLUSION

Every year, cough and respiratory infections impose a significant strain on the personnel and resources of our healthcare system. Lung damage from these illnesses affects victims immediately. These infections cause various chest problems. These illnesses have an impact on populations worldwide. Numerous fatalities have been caused by poor and inefficient testing procedures, inadequate facilities, and a lack of earlier detection of chest infections. There are many possibilities, so a quick and efficient checking process is required. The proposed DMCIC_Net model for categorizing the five categories of multiple chest infections (AST, COVID-19, NOR, PNEU, and TB) was created and evaluated in this chapter. We proposed the DMCIC_Net model to distinguish between the five main types of chest infections. Each convolutional block in the modified structure is constructed using many layers and utilized to classify chest infections at their initial stages. The SMOTE Tomek algorithm is used to generate samples to address the data set imbalance issues and maintain a balanced amount of samples for each class. 97.24% accuracy, 97.24% recall, 97.27% F1-score, 97.24% precision, and 99.30% AUC was attained by our proposed DMCIC_Net. As a result, it can be concluded that DMCIC_Net is a valuable tool for medical professionals. In the future, we will incorporate more trained architectures with federated learning to ensure patient data privacy as well as the classification of chest diseases.

REFERENCES

1. Lu, H., Stratton, C. W., & Tang, Y. W. (2020). Outbreak of pneumonia of unknown etiology in Wuhan, China: The mystery and the miracle. *Journal of Medical Virology*, 92(4), 401.
2. Hui, D. S., Azhar, E. I., Madani, T. A., Ntoumi, F., Kock, R., Dar, O., ... & Petersen, E. (2020). The continuing 2019-nCoV epidemic threat of novel coronaviruses to global health—The latest 2019 novel coronavirus outbreak in Wuhan, China. *International Journal of Infectious Diseases*, 91, 264–266.
3. Eurosurveillance Editorial Team. (2020). Note from the editors: World Health Organization declares novel coronavirus (2019-nCoV) sixth public health emergency of international concern. *Eurosurveillance*, 25(5), 200131e.
4. Mahase, E. (2020). Coronavirus: Covid-19 has killed more people than SARS and MERS combined, despite lower case fatality rate.
5. Xu, X., Jiang, X., Ma, C., Du, P., Li, X., Lv, S., ... & Li, L. (2020). A deep learning system to screen novel coronavirus disease 2019 pneumonia. *Engineering*, 6(10), 1122–1129.
6. Update, C. (2020). Cases and deaths from COVID-19 virus pandemic-worldometer.
7. Saul, C. J., Urey, D. Y., & Taktakoglu, C. D. (2019). Early diagnosis of pneumonia with deep learning. *arXiv preprint arXiv:1904.00937*.
8. Kermany, D., Zhang, K., & Goldbaum, M. (2018). Labeled optical coherence tomography (oct) and chest X-ray images for classification. *Mendeley Data*, 2(2). *Konstantinos*.
9. Varshni, D., Thakral, K., Agarwal, L., Nijhawan, R., & Mittal, A. (2019, February). Pneumonia detection using CNN based feature extraction. In 2019 IEEE International Conference on Electrical, Computer and Communication Technologies (ICECCT) (pp. 1–7). IEEE.

10. Feng, S., Liu, Q., Patel, A., Bazai, S. U., Jin, C.-K., Kim, J. S., … others. (2022). Automated pneumothorax triaging in chest X-rays in the New Zealand population using deep-learning algorithms. *Journal of Medical Imaging and Radiation Oncology*, 66, 1035–1043.

11. World Health Organization. (2020). Clinical management of severe acute respiratory infection when novel coronavirus (nCoV) infection is suspected: Interim guidance, 25 January 2020 (No. WHO/nCoV/Clinical/2020.2). World Health Organization.

12. Floyd, K., Glaziou, P., Zumla, A., & Raviglione, M. (2018). The global tuberculosis epidemic and progress in care, prevention, and research: An overview in year 3 of the End TB era. *The Lancet Respiratory Medicine*, 6(4), 299–314.

13. Carfì, A., Bernabei, R., & Landi, F. (2020). Persistent symptoms in patients after acute COVID-19. *JAMA*, 324(6), 603–605.

14. Wang, D., Hu, B., Hu, C., Zhu, F., Liu, X., Zhang, J., … & Peng, Z. (2020). Clinical characteristics of 138 hospitalized patients with 2019 novel coronavirus–infected pneumonia in Wuhan, China. *JAMA*, 323(11), 1061–1069.

15. Pahar, M., Klopper, M., Reeve, B., Warren, R., Theron, G., & Niesler, T. (2021). Automatic cough classification for tuberculosis screening in a real-world environment. *Physiological Measurement*, 42(10), 105014.

16. Bateman, E. D., Hurd, S. S., Barnes, P. J., Bousquet, J., Drazen, J. M., FitzGerald, M., … & Zar, H. J. (2018). Global strategy for asthma management and prevention: GINA executive summary. *European Respiratory Journal*, 31(1), 143–178.

17. Ulukaya, S., Sarıca, A. A., Erdem, O., & Karaali, A. (2023). MSCCov19Net: multi-branch deep learning model for COVID-19 detection from cough sounds. *Medical & Biological Engineering & Computing*, 61, 1619–1629.

18. Nasab, K. A., Mirzaei, J., Zali, A., Gholizadeh, S., & Akhlaghdoust, M. (2023). Coronavirus diagnosis using cough sounds: Artificial intelligence approaches. *Frontiers in Artificial Intelligence*, 6, 1100112.

19. Sharma, G., Umapathy, K., & Krishnan, S. (2022). Audio texture analysis of COVID-19 cough, breath, and speech sounds. *Biomedical Signal Processing and Control*, 76, 103703.

20. Chowdhury, N. K., Kabir, M. A., Rahman, M. M., & Islam, S. M. S. (2022). Machine learning for detecting COVID-19 from cough sounds: An ensemble-based MCDM method. *Computers in Biology and Medicine*, 145, 105405.

21. Kranthi Kumar, L., & Alphonse, P. J. A. (2022). COVID-19 disease diagnosis with lightweight CNN using modified MFCC and enhanced GFCC from human respiratory sounds. *The European Physical Journal Special Topics*, 231(18), 3329–3346.

22. Sabet, M., Ramezani, A., & Ghasemi, S. M. (2022, March). COVID-19 detection in cough audio dataset using deep learning model. In 2022 8th International Conference on Control, Instrumentation and Automation (ICCIA) (pp. 1–5). IEEE.

23. Lella, K. K., & Pja, A. (2022). Automatic diagnosis of COVID-19 disease using deep convolutional neural network with multi-feature channel from respiratory sound data: Cough, voice, and breath. *Alexandria Engineering Journal*, 61(2), 1319–1334.

24. Loey, M., & Mirjalili, S. (2021). COVID-19 cough sound symptoms classification from scalogram image representation using deep learning models. *Computers in Biology and Medicine*, 139, 105020.

25. Feng, K., He, F., Steinmann, J., & Demirkiran, I. (2021, March). Deep-learning based approach to identify Covid-19. In SoutheastCon 2021 (pp. 1–4). IEEE.

26. Vijayakumar, D. S., & Sneha, M. (2021). Low cost Covid-19 preliminary diagnosis utilizing cough samples and keenly intellective deep learning approaches. *Alexandria Engineering Journal*, 60(1), 549–557.

27. Schuller, B. W., Coppock, H., & Gaskell, A. (2020). Detecting COVID-19 from breathing and coughing sounds using deep neural networks. *arXiv preprint arXiv:2012.14553*.

28. Bansal, V., Pahwa, G., & Kannan, N. (2020, October). Cough classification for COVID-19 based on audio MFCC features using convolutional neural networks. In 2020 IEEE International Conference on Computing, Power and Communication Technologies (GUCON) (pp. 604–608). IEEE.

29. Brown, C., Chauhan, J., Grammenos, A., Han, J., Hasthanasombat, A., Spathis, D., ... & Mascolo, C. (2020). *Exploring automatic diagnosis of COVID-19 from crowdsourced respiratory sound data.* In Proceedings of the 26th ACM SIGKDD International Conference on Knowledge Discovery & Data Mining (pp. 3474–3484).

30. Imran, A., Posokhova, I., Qureshi, H. N., Masood, U., Riaz, M. S., Ali, K., ... & Nabeel, M. (2020). AI4COVID-19: AI enabled preliminary diagnosis for COVID-19 from cough samples via an app. *Informatics in Medicine Unlocked*, 20, 100378.

31. Dunne, R., Morris, T., & Harper, S. (2020). High accuracy classification of COVID-19 coughs using Mel-frequency cepstral coefficients and a convolutional neural network with a use case for smart home devices.

32. Oh, Y., Park, S., & Ye, J. C. (2020). Deep learning COVID-19 features on CXR using limited training data sets. *IEEE Transactions on Medical Imaging*, 39(8), 2688–2700.

33. Hemdan, E. E. D., Shouman, M. A., & Karar, M. E. (2020). Covidx-net: A framework of deep learning classifiers to diagnose covid-19 in x-ray images. *arXiv preprint arXiv:2003.11055*.

34. Sethy, P. K., & Behera, S. K. (2020). Detection of coronavirus disease (covid-19) based on deep features.

35. Hemdan, E. E. D., Shouman, M. A., & Karar, M. E. (2020). Covidx-net: A framework of deep learning classifiers to diagnose Covid-19 in X-ray images.

36. Hilmizen, N., Bustamam, A., & Sarwinda, D. (2020, December). The multimodal deep learning for diagnosing COVID-19 pneumonia from chest CT-scan and X-ray images. In 2020 3rd International Seminar on Research of Information Technology and Intelligent Systems (ISRITI) (pp. 26–31). IEEE.

37. Bazai, S. U., Jang-Jaccard, J., & Wang, R. (2017). Anonymizing k-N N classification on MapReduce. International Conference on Mobile Networks and Management, (pp. 364–377).

38. Melek, M. (2021). Diagnosis of COVID-19 and non-COVID-19 patients by classifying only a single cough sound. *Neural Computing and Applications*, 33(24), 17621–17632.

39. Brown, C., Chauhan, J., Grammenos, A., Han, J., Hasthanasombat, A., Spathis, D., ... & Mascolo, C. (2020). Exploring automatic diagnosis of COVID-19 from crowdsourced respiratory sound data. In Proceedings of the 26th ACM SIGKDD International Conference on Knowledge Discovery & Data Mining, pp. 3474–3484.

40. Aamir, M., Bazai, S. U., Bhatti, U. A., Dayo, Z. A., Liu, J., & Zhang, K. (2023). Applications of Machine Learning in Medicine: Current Trends and Prospects. 2023 Global Conference on Wireless and Optical Technologies (GCWOT), (pp. 1–4).

41. Nessiem, M. A., Mohamed, M. M., Coppock, H., Gaskell, A., & Schuller, B. W. (2021, June). Detecting COVID-19 from breathing and coughing sounds using deep neural networks. In *2021 IEEE 34th International Symposium on Computer-Based Medical Systems (CBMS)* (pp. 183–188). IEEE.

42. Sattar, F. (2021). A fully-automated method to evaluate coronavirus disease progression with Covid-19 cough sounds using minimal phase information. *Annals of Biomedical Engineering*, 49(9), 2481–2490.

43. Sharma, N., Krishnan, P., Kumar, R., Ramoji, S., Chetupalli, S. R., Ghosh, P. K., & Ganapathy, S. (2020). Coswara--a database of breathing, cough, and voice sounds for COVID-19 diagnosis. *arXiv preprint arXiv:2005.10548*.

44. Orlandic, L., Teijeiro, T., & Atienza, D. (2021). The COUGHVID crowdsourcing data set, a corpus for the study of large-scale cough analysis algorithms. *Scientific Data*, 8(1), 156.

45. Morice, A. H., Millqvist, E., Bieksiene, K., Birring, S. S., Dicpinigaitis, P., Ribas, D. C., & Boon M. H. et. al. (2020). ERS guidelines on the diagnosis and treatment of chronic cough in adults and children. *European Respiratory Journal*, 55(1).

46. Sayers, E. W., Agarwala, R., Bolton, E. E., Rodney Brister, J., Canese, K., Clark, K., Connor, R. et al. (2019). Database resources of the National Center for Biotechnology Information. *Nucleic Acids Research* 47, no. Database issue, D23.

47. Albadr, M. A. A., Tiun, S., Ayob, M., and Al-Dhief, F. T. (2022). Particle swarm optimization-based extreme learning machine for COVID-19 detection. *Cognitive Computation*, 1–16.

48. Hameed, M., Yang, F., Bazai, S. U., Ghafoor, M. I., Alshehri, A., Khan, I., …Jaskani, F. H. (2022). Urbanization detection using LiDAR-based remote sensing images of Azad Kashmir using novel 3D CNNs. *Journal of Sensors*, 2022, 1–9.

49. Aamir, M., Rahman, Z., Ahmed Abro, W., Aslam Bhatti, U., Ahmed Dayo, Z., & Ishfaq, M. (2023). Brain tumor classification utilizing deep features derived from high-quality regions in MRI images. *Biomedical Signal Processing and Control*, 85, 104988. 10.1016/j.bspc.2023.104988.

50. Aamir, M., Rahman, Z., Dayo, Z. A., Abro, W. A., Uddin, M. I., Khan, I., Imran, A. S., Ali, Z., Ishfaq, M., Guan, Y., & Hu, Z. (2022). A deep learning approach for brain tumor classification using MRI images. *Computers and Electrical Engineering*, 101, 108105. 10.1016/j.compeleceng.2022.108105.

CHAPTER 13

Classifying Traffic Signs Using Convolutional Neural Networks Based on Deep Learning Models

Saira Akram, Sibghat Ullah Bazai, and Shah Marjan

Balochistan University of Information Technology, Engineering, and Management Sciences (BUITEMS), Quetta, Pakistan

13.1 INTRODUCTION

Traffic signs are considered one of the most significant entities within the network of roads as the basic objective of these signs is to be noticed, and warn and guide the pedestrians and drivers during the daytime as well as at night without any difficulty. Traffic signs are the combination of distinguishable features like text, multiple shapes, and sometimes vibrant color exhibiting meaningful information so it is very essential to interpret these signs correctly in order to avoid serious consequences. A survey conducted (Mishra & Goyal, 2022) cited that yearly more than 1.24 million deaths and life-threatening accidents are reported, which may occur due to many reasons including carelessness of drivers, fast driving, and the most hazardous is the misinterpretation of traffic signs. In view of these circumstances, the role of developing a robust and efficient system for the detection and classification of traffic signs has become one of the most considered areas of research in the last decade.

Machine learning, which is the emerging field of artificial intelligence, has been a great success story in the last few years (Bhatti et al., 2022). The machine learning aims to train the models in such a way that they would be able to anticipate the results based on their past experience and learning algorithms. The algorithms-based predictions are used to make decisions about security, investments, speech recognition, and many other critical applications. Not limited to this, machine learning contributed a major growth in the domain of computer vision (CV) for classifying images (Bhatti et al., 2023) object recognition (Guan et al., 2021), object detection (Aamir et al., 2023) and anomaly

DOI: 10.1201/9781003427674-13

detection (Noor et al., n.d.). These predictions are also used to strengthen the life-crucial applications like making predictions in healthcare (classifying the images as healthy or cancerous (Aamir et al., 2023)) and detecting and classifying the road signs.

Before diving into the depth of classifying traffic signs based on deep neural models, we first need to understand some basics of machine learning and how the machine learns.

13.2 HOW DOES A MODEL LEARN?

A simple machine learning model consists of two parts (shown in Figure 13.1). In the first stage, the model is trained on data, and during the training phase many examples of data (can be in any form text, image, voice, videos, and graphs) are given to the model. The model learns the pattern or extracts features from this data; once the model is trained on data, then in the second phase, unseen data is given to the model. On the basis of training, the model predicts an output for the new, unseen data.

Once the model predicted an output (Y') for the particular input (X), the loss is computed by using a loss function. To compute the loss, the loss function takes two values as input:

Y': Result predicted by the model for the input (X)
Y: Correct label corresponding to the input (X)

On the basis of these two values, the parameters of features are updated and sent back to the model for the new prediction. This learning is continuous until the model finds the feature values with minimum loss or the loss stops changing. Predictions done by any machine learning model are used in various applications, ranging from making predictions about investments to life-crucial applications like predicting traffic signs on roads. To make a prediction, these models are trained on an algorithm. These learning algorithms are grouped on the following properties:

1. Type of machine learning

2. Tasks performed by these models

3. Depth of machine learning

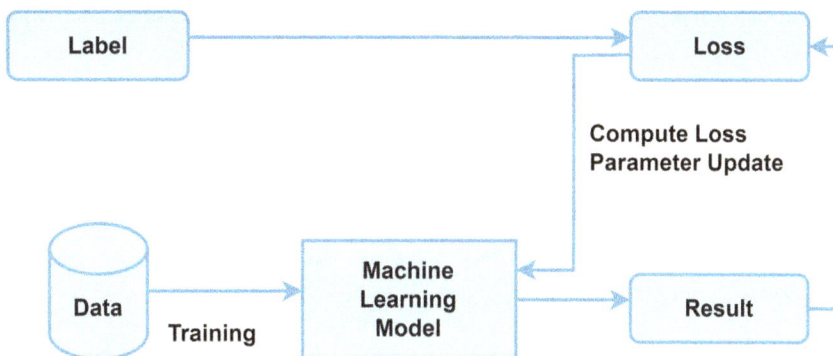

FIGURE 13.1 Trial and error process used by learning algorithms to train the model.

TABLE 13.1   Examples of Labeled Data

	Feature (X1)	Feature (X2)	Feature (X3)	Labels
Examples	7	1	1	1
	5	0	0	0
	8.5	1	1	1

TABLE 13.2   Examples of Unlabeled Data

	Feature (X1)	Feature (X2)	Feature (X3)	Labels
Examples	7	1	1	?
	5	0	0	?
	8.5	1	1	?

To have a clear understanding of these terms, it is necessary to understand the type of data on which these models get trained. In the framework of ML data or data set is the collection of multiple examples with distinguishable features that are used to train the model. There are two types of data (Wang et al., 2021):

1. *Labeled Data*: Labeled data includes both feature value(s) and a label for that particular feature value(s); for example, {Feature, Label}: {X, Y}. Examples of labeled data are shown in Table 13.1.

2. *Unlabeled Data:* Whereas unlabeled data contains only feature values, no label is given in unlabeled data. Examples of unlabeled data are shown in Table 13.2.

13.2.1 Types of Machine Learning

Types of machine learning or simply learning are dependent on the data set used in model training. This can be categorized in the following types:

1. *Supervised Learning*: In this learning, the model is trained on a data set having feature values and labels. The training data set consists of multiple examples; each example has a feature value and label. During the training process, the model learns and analyzes the data set and generates a map between the feature value (input: X) and the output (Y). Supervised learning is mostly used for solving the tasks of "classification" and "regression."

2. *Unsupervised Learning*: Here, the model is trained on the data with only feature values; no labeled input or output variable is provided to generate a relationship between input and output. Clustering is the most common application of unsupervised machine learning.

3. *Semi-Supervised Learning*: This learning comes between supervised and unsupervised learning. In this type of learning, the model is trained on labeled data, combined with

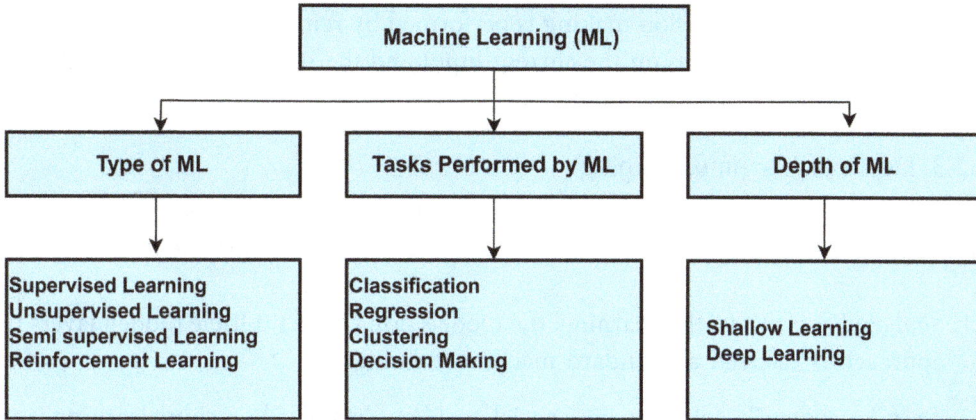

FIGURE 13.2 Dimensions of machine learning classification.

unlabeled data. The objective of combining labeled and unlabeled data set is to analyze the change in behavior and algorithm used in the training of model.

4. *Reinforcement Learning*: Reinforcement learning can be taken as a trial-and-error process as in this learning the model learns from the results of its actions. The model analyzes its past experience and tries to take the correct decision for the next action. In this learning, no labeled data is provided to the model, but actions are taken by the reinforcement agent (Figure 13.2).

13.2.2 Tasks Performed by Machine Learning

Tasks performed by models used in machine learning are categorized on the basis of the type of learning and the type of data used in the training process. These tasks are divided into the following types (Institute of Electrical and Electronics Engineers & Manav Rachna International Institute of Research and Studies, n.d.):

1. *Classification*: Classification can be done by using supervised learning. If the values predicted by model are discrete (categorial) in nature then this type of problem is called classification. For example:

   • Is the email spam or not spam?

   • Is this an image of a dog or cat?

2. *Regression*: If the values predicted by a supervised model are continuous in nature, then this task is classified as a regression problem. For example:

   • What is the value of house in Canada?

   • What is the probability of buying this ticket?

3. *Clustering*: Clustering is done by unsupervised learning. This is the process of arranging, organizing and labeling the data into one group with identical feature values; for example, grouping the leaves on the basis of their color (green, yellow, red).

4. *Decision Making*: Decision making is performed by reinforcement learning. In this, the output is dependent on the current input and the next provided input depends on the output of previous inputs.

13.2.3 Depth of Machine Learning

Models based on learning algorithms can also be classified on the basis of the number of hidden layers. As per this criterion, these models are categorized into the following types:

1. *Shallow Learning*: In this learning, the models don't have multiple hidden layers this approach is referred as standard machine learning.

2. *Deep Learning*: To solve the life-crucial problems, standard machine learning is not as suitable as deep learning can be because these models can extract feature with less intervention of human efforts. Deep models consist of a stack of multiple hidden layers that makes these models more suitable for making complicated decisions.

13.3 DEEP LEARNING

Machine learning is a vast and growing area of AI that focuses on training the machines in such a way that they would be able to make the decisions based on data and their past experience. Different types of algorithms are being used to train these models. The goal of these models is to replicate the human competencies and actions in making predictions on unseen scenarios. Machine learning models analyze the data on which these are trained and map a function or relation between the given input and labels and, on the basis of this training, the model makes a prediction for the unseen data.

Deep learning is asubset of machine learning that consists of NNs composed of more than one hidden layer between an input and output layer. This class of machine learning can perform and give satisfying results, even on unstructured data. The relationship between these terms is shown in Figure 13.3 (Aamir et al., 2019).

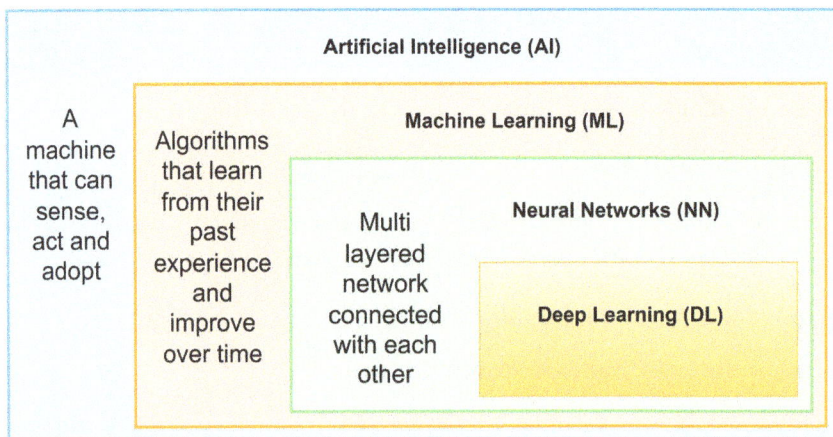

FIGURE 13.3   Where the deep learning fits in AI.

In standard machine learning, we have to train the model on structured data by tagging or giving labels to each different class; for example, to set the characteristics of a cat, dog, and cow, we have to assign tags or labels to each image to these three different classes of images, called feature extraction. Extracting features using human intervention can be very time consuming and also requires extra effort. To solve this issue, deep learning uses a diverse approach to make predictions or classifications. Deep models use a hierarchy of multiple layers connected with each other to process the image at each layer. The algorithms used in deep learning extract the useful information or features from the input data at each layer and then pass it to the next attached layer. In this way, these models are more efficient as the process of feature extraction even for unstructured data is done by a neural network itself (Ella Hassanien Roheet Bhatnagar Ashraf Darwish Editors, n.d.).

13.3.1 Training of Deep Learning Models

A deep model is composed of multi-layered neurons inspired by the human brain. In a brain, there are trillions of neurons interconnected with each other that pass information or electrical impulses from your brain to each part of your body so that any immediate action can happen. Similarly, a deep model consists of a neural network with many neurons in each layer. These neurons receive information from the first layer, "input layer", and pass to the next connected layer; this process is continued until the last layer, "output layer". The inbetween layer(s) are called hidden layer(s). To each neuron, a certain "weight" is assigned at each synapse that represents the strength of each connection.

The input value of each neuron is multiplied with the weight connected to it and summed with every product of neuron and weight; the result of the value of "bias" is added. After applying an "activation function" to this value, the output is passed to the next connected neuron. The output of the previous layer is the input for the next layer.

The algorithm that makes the deep model learns is the "perceptron" (Surya Engineering College & Institute of Electrical and Electronics Engineers, n.d.). A perceptron with no hidden layer is the single-layer perceptron shown in Figure 13.4, which is the easiest type of artificial neural network (ANN), whereas a perceptron with one or more hidden layers is called a multi-layered ANN used in deep learning, shown in Figure 13.5.

To train a deep learning model, the following two methods are used:

1. *Forward Propagation*: In this training, the input data travels from the input layer to the output layer. The input values are multiplied with the weight values and then summed up by the bias. After applying the activation function, the model gives the predicted value for the applied input. To find the difference between actual and predicted result, a "loss function" is used.

2. *Backward Propagation*: In this approach, if the predicted value is wrong, then to minimize this difference or error, the weight values are updated. In back propagation, to update the weights from the last to first layer, "optimizers" are used, which work on the principle of derivative. The weights are updated continuously until the desired accuracy is achieved.

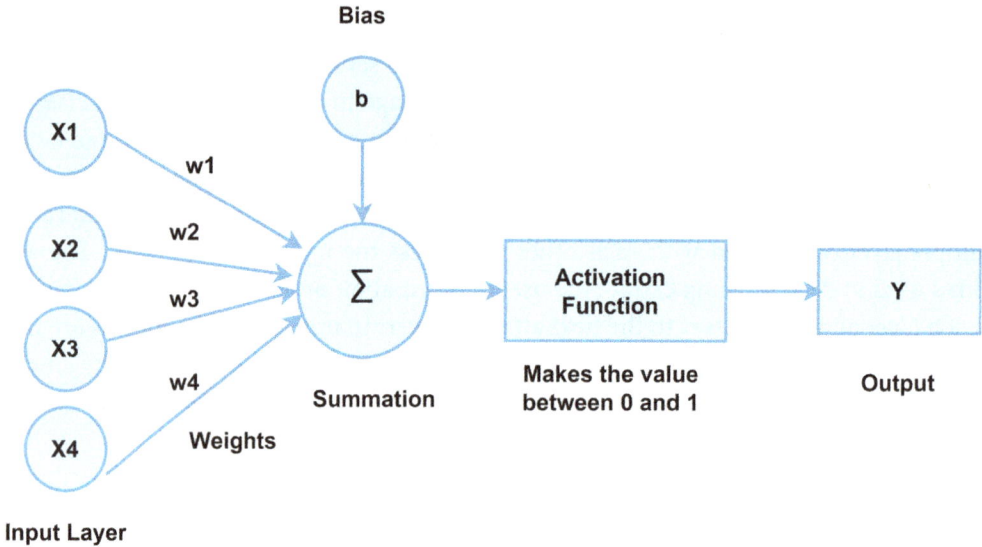

FIGURE 13.4   Single-layer neural network.

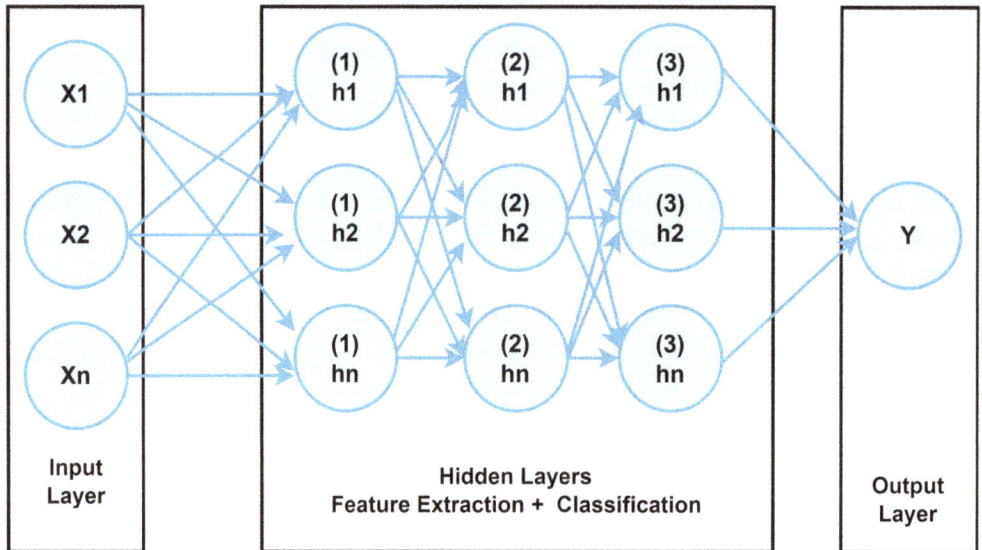

FIGURE 13.5   Multi-layered artificial neural network.

13.3.2 Algorithms Used to Train Deep Learning Models

Models based on deep learning use a variety of algorithms to classify the data. These algorithms are mainly divided into three classes based on the type of learning; supervised, unsupervised, and hybrid learning (Sarker, 2021), presented in Table 13.3.

TABLE 13.3   Algorithms Used to Train Deep Models on the Basis of Type of Learning

Algorithms Used in Deep Learning	
Type of learning	Algorithm
Supervised Learning	Multi-layered perceptron (MLP)
	Convolutional neural network (CNN)
	Recurrent neural network (RNN)
Unsupervised Learning	Deep belief network (DBN)
	Restricted Boltzmann machine (RBM)
	Self-organizing map (SOM)
	Auto encoders (AE)
	Generative adversarial networks (GAN)
Hybrid Learning	Deep transfer learning (DTL)
	Deep reinforcement learning (DRL)
	Auto encoders combined with support vector machine (SVM)
	Generative adversarial network combined with convolutional neural network

13.4 CLASSIFICATION OF IMAGES USING A CONVOLUTIONAL NEURAL NETWORK

Convolution-based networks are considered the most applied deep learning algorithm or method used in the area of CV for classifying images in supervised as well as in hybrid learning. The process of classifying the images can be divided into two types:

1. Classifying images using traditional methods

2. Image classification using CNN

13.4.1 Classifying Images Using Traditional Methods

Traditionally, classifying images is a two-step methodology. In the first step, the features of images like color and shape were selected. The process of feature selection was done by using hand-crafted methods like "histogram of oriented gradient (HOG)" and "localized binary model (LBM)". After feature extraction, in the second step, classification was done using simple ML algorithms like support vector machine (SVM), K-d tree, and random forest.

13.4.2 Image Classification Using CNN

Classifying images using a deep convolutional neural network provides more accuracy, as these models can extract meaningful features from an input image automatically without using any hand-crafted technique. In 1998, the first architecture of CNN was introduced to classify the handwritten digits named LeNet. The basic architecture of CNN is composed on a multi-layered neural network designed to achieve two main tasks; in the first step, features are extracted using convolution and pooling layers and then in the second phase, classification is done using fully connected and classification layers. Neurons with the specification of width, height, and depth and connected in each layer (Loukmane et al., 2020).

The convolution neural network, also called CovNet, processes the data in the form of a grid of numbers. To classify the images, it simply transforms the image into 2D grid or matrix of pixels. To differentiate the images of the model, it applies weights and biases to different objects in the image. The architecture of CNN involves multiple layers and each layer performs its own function. The input image is represented in a 2D matrix, so instead of applying matrix multiplication, it performs a "convolution operation". Mathematically, the convolution operation can be written as (Sai Kondamari Anudeep Itha, 2021):

$$y(t) = \int x(a)\, z(t - a)\, da$$

It can be simplified as:

$$y(t) = (x * z)(t)$$

In the above operation, x represents the input image applied to the model and any y(t) is the output class. The filter (also called as kernel) applied to the input pixels is represented by z. The kernel is a matrix of numbers representing different filters, which get convolved with the matrix of input data having stride various numbers. The size of a kernel is smaller than the size of the input matrix. After applying a convolution operation between the input and kernel, the obtained resultant matrix is called a "feature map". These feature maps are then used by the model to extract features that help in classification.

To classify the images using CNN, the following two types of approaches (Oztel et al., 2019) can be used to make the model learn or be trained on the data set (a data set is the collection of data of any type, including text, audio, images, videos, and graphs with multiple images of the same class or different classes); for example, a data set of class "cars" or a data set consists of multiple classes of "cars, trucks and cycles").

1. *Training from Scratch*: In this approach, we have to build the model from the first step or from the first layer. For example, if we want the model to classify the images as car, cycle, or truck, we have to collect approximately millions of images of these classes to train the model. During this process, the model learns to extract features using convolution and pooling layers and then classifies using the fully connected layers. This approach is time consuming and requires additional resources to build a model. In building the model from scratch, we can define the number of layers, number of filters, stride, type of activation function, and learning rate of our own choice.

2. *Transfer Learning*: In this learning, a pre-trained model (model trained on one data set) is used for another data set. By using this approach, the learning process of the model can be made easier by simply transferring the learned parameters (bias and weights) to a new problem. Transfer learning is used when the data set of our problem is too small; to achieve good results, the model must be trained on millions of images. For example, if a model is intended to classify the images between "car and truck", then we can use the model trained for the images of "car" and all we need

to do is to train this pre-trained model for the new small data set of "truck". One thing to note here is that using a model trained for recognizing text can give terrible results if the same model is used for classifying data in the form of images. To use a pre-trained model for new problem one must follow the given sequence of steps:

- Load the pre-trained model (trained for the same data type)

- Remove or reset the task-specific layers (fully connected or classifiers layers)

- Before starting training for new data, the feature extraction layers should freeze so the learned weights won't get updated

- Add the new classifier layers and train the model for a new data set

13.4.3 Overview of CNN Models Used for Image Classification

By using the basic architecture or layers of CNN, many architectures have developed a way to classify the images that can be used to solve the similar problem by using a transfer learning approach. The aim of discussing these methods is to provide a clear comparison among these methods so the readers or researchers can easily choose the model as per their requirements. Table 13.4 represents the summarized view of these models.

13.5 CLASSIFYING TRAFFIC SIGNS USING CONVOLUTIONAL NEURAL NETWORK

A traffic sign recognition system can be divided into classes: detecting the road sign and classify the traffic sign in to the correct class. This chapter focuses on the classification part using CNN, which is the class of deep learning models.

Before moving on to the classification part, the very first step is to train the model for the particular problem for which the model will be designed. In supervised learning, a properly labeled and organized data set containing millions of images will be given to the model so the model can extract useful information or features (like color, edges, shape, etc.) from the data that will be more helpful in classifying the image. The whole data set will be segmented in three parts: training set, validation set, and test data set. A training set utilizes model training when the model is fully trained and then it is evaluated by using a validation data set. During evaluation, the model's hyperparameters are adjusted to give more accurate predictions. Most commonly, hyperparameters are number of epochs and activation function. Once the model gets evaluated and parameters are adjusted, then finally the model will be tested on the test data set that makes the model ready to classify the unseen new images of the same classes.

13.5.1 Model Architecture

The whole architecture is divided into two parts. In the first part, there will be number of convolution layers, activation function, pooling layers, and dropout layer. In the second part of the model, there will be a dense or fully connected layer with a flatten layer. The CNN model architecture used to classify the traffic signs can have the following arrangement of layers, as shown in Figure 13.6 (Bailke, 2022).

TABLE 13.4 Summarized View of CNN-Based Architecture for Image Classification

Year	Model	Enhanced feature(s)	Depth	Data set	Input size	Error rate
2012	AlexNet	Use ReLU and activation function and Dropout	8	ImageNet	$227 \times 227 \times 3$	16.4
2014	VGG	Number of layers increased up to 16 and 19, having small kernel	16 19	ImageNet	$224 \times 224 \times 3$	7.3
2015	GoogLeNet	Enhanced number of layers, vary kernel size, concept of concatenation	22	ImageNet	$224 \times 224 \times 3$	6.7
2015	Inception-V3	Uses small kernel size	48	ImageNet	$229 \times 229 \times 3$	3.5
2016	Inception-V4	Concept of integration and transformation was divided	70	ImageNet	$229 \times 229 \times 3$	3.08
2016	ResNet	Introduced skip links based on symmetry mapping (reduced overfitting)	152	ImageNet	$224 \times 224 \times 3$	3.57
2016	WideResNet	Width was increased while decreasing the depth	28	CIFAR-10 CIFAR-100	$32 \times 32 \times 3$	3.89 18.85
2017	Xception	Convolution by depthwise afterwards pointwise convolution	71	ImageNet	$229 \times 229 \times 3$	0.055
2017	DenseNet	Block of layers	201	CIFAR-10 CIFAR-100 ImageNet	$224 \times 224 \times 3$	3.46 17.18 5.54
2018	CapsuleNet	Worked more on features relationship	3	MNIST	$28 \times 28 \times 1$	0.00855

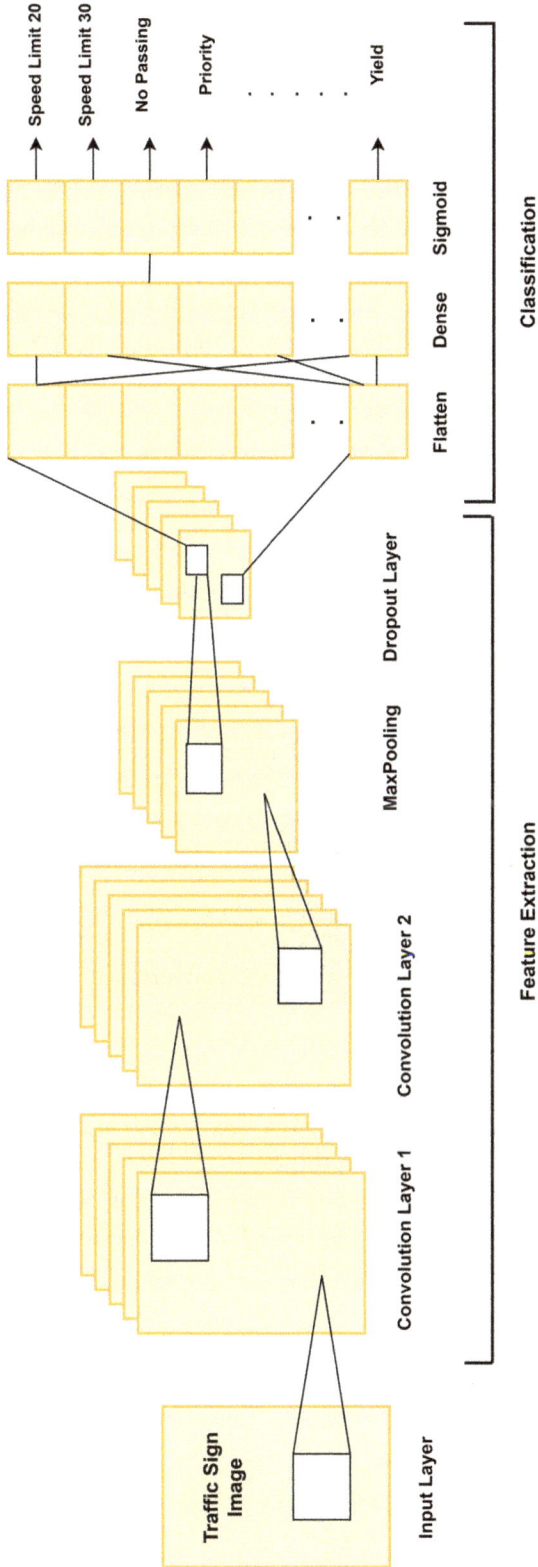

FIGURE 13.6 Layering sequence of deep CNN used to classify traffic signs.

13.5.1.1 Input Layer

Here, input data in the form of image is given to the model. Image data is represented in the 2D matrix of numbers that shows pixel values.

13.5.1.2 Convolutional Layer

At this layer, the convolution operation is performed between a 2D matrix of pixel values and a kernel value. Kernel is another name for filter (can also be understood as the feature detector). In this process, multiple filters are applied to the input image. Filters are a matrix of numbers of any size, but the size of filter must be smaller than the size of the input matrix. The filter matrix is convolved with the input matrix and the matrix produced as a result is called a feature map. These generated feature maps highlight the features of an input image.

13.5.1.3 Activation Function

To map the non-linearities between input and output class, an activation function is applied after each coevolution layer. Its function is to check the values of a feature map and convert all the negative values (smaller than 0) to 0 and if the values are greater than 0, they will be written as is in the new matrix. Some commonly used activation functions are the following:

- Sigmoid: This function receives a real number as input and the output value is constrained between 0 and 1.

- Tanh: It works like a sigmoid function as it takes input in real numbers, but the output is represented between −1 and 1.

- ReLU: To map the non-linearities in CNN, ReLU is the most generally used as an activation function. Whole input values are converted to positive values.

13.5.1.4 Pooling Layers

The feature maps generated by convolving the input and filters are of high dimensions. Pooling layer(s) are used in reducing the large-size feature maps into small-size feature maps. Some common techniques used at this layer are max pooling and average pooling. One main advantage of using this layer is it minimize the number of parameters and avoid overfitting.

13.5.1.5 Dropout Layer

During training of the model, the dropout layer sets the value of some neurons to 0 to avoid overfitting.

13.5.1.6 Flatten Layer

The 2D matrix obtained by the pooling layer will be converted into a single-column matrix or into a 1D matrix. After getting the flattened numbers, we can now make a fully connected dense neural network to classify the traffic sign.

13.5.1.7 Fully Connected Layer

This layer receives the vector (1D matrix) of numbers created by the flattened layer. The fully connected or dense layer of CNN works like the simple multi-layered neural network with neurons that are connected to each neuron present in the next layer; so-called a fully connected layer. It receives vectors of values from the previous layer and feeds these numbers into its neurons.

13.5.1.8 Output Layer

The last layer of the model is the output layer with neurons equal to the number of classes present in our data set. The model classifies the traffic sign and shows the assigned label of class as the output.

13.5.1.9 Loss Function

Finally, the loss will be calculated and apply the optimization technique. The loss function is used to calculate the error in the predicted result. This error is the difference between actual results and the predicted values. Many loss functions are used to find the loss values; the most commonly used are cross-entropy (softmax) loss function, Euclidean, and hinge loss function.

13.5.1.10 Training and Testing

During training of the model, optimizers like "stochastic gradient descent" are used to adjust the model's parameters (weight and bias). The values of these parameters are set to the point where the loss is minimized. Finally, the model is tested on the unseen data of the same classes to check the accuracy of the model.

The CNN model based on deep learning techniques automatically learns useful features from the input images of traffic sign and classify the class. There are a number of data sets available that can be used to train the model. The benchmark is the German Traffic Sign Recognition Benchmark because of its large number of images and all the images are organized properly. The sequence of steps illustrating the process of image classification is shown as the flowchart in Figure 13.7.

13.5.2 Benchmark on Traffic Sign Classification

A variety of models have been proposed to solve this particular problem of traffic sign classification based on deep learning CNN. To present the work done, a comparison is made on the basis of some parameters. This is summarized in Table 13.5.

13.6 DATA SET

The accuracy of predictions done by DL models is dependent on the data set. The data set is a collection of millions of examples having a great impact on the performance of any model. Data can be of any type; it may be in the form of text, images, audio, video, and graphs. A good data set must be relevant to the domain of problem, large enough, contain many examples of particular entity. In the background of ML, the data can be labeled or unlabeled (discussed in the Introduction).

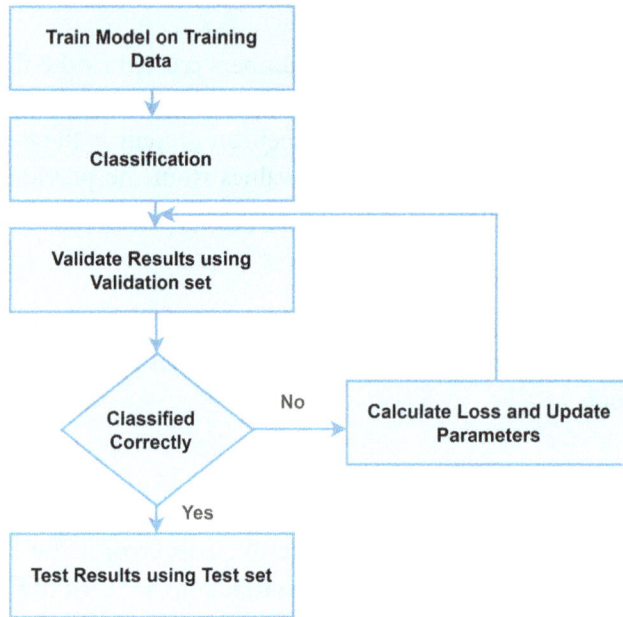

FIGURE 13.7 Sequence of steps from model training to model testing.

TABLE 13.5 Comparison of Multiple CNN-Based Methods Used to Classify Traffic Signs

References	Method	Data set	Accuracy
Saha et al., 2018	Residual CNN combined with hierarchical skip connections	GTSRB BTSD	99.33% 99.17%
Aslansefat et al., 2021	A model-agnostic technique based on deep CNN was presented, applied on any machine learning classifier, irrespective of its structure	GTSRB	0.9797%
Zhang et al., n.d.	Proposed a new method to train two CNN named Teacher and Student network; teacher network consisted of dense connectivity, whereas the student CNN composed of five convolved and one dense layer	CIFAR-10 (Teacher network) GTSRB (Student Network) BTSD (Student Network)	93.16% 99.61% 99.13%
Zaibi et al., 2021)	This study was done on enhanced variant of LeNet-5 using small number of parameters (0.38 million)	GTSRB BTSRD	99.84% 98.37%
Bouti et al., 2020	This research focused on two aspects: detecting signs and classification. For detection, SVM and HOG classification: LeNet based on CNN with modified features	GTSRB	96.85% (detecting) 96.23% (classifying)
Hechri & Mtibaa, 2020	Two stage methodology based on HOG and SVM for sign detection where as to classify the class CNN was used	GTSRB	99.37% (classification accuracy)

(Continued)

TABLE 13.5 (Continued)   Comparison of Multiple CNN-Based Methods Used to Classify Traffic Signs

References	Method	Data set	Accuracy
Hasegawa et al., 2020	To solve the issue of alteration, in contrast, during a different environment can greatly affect the classification. This study proposed a method based on CNN named Faster R-CNN and Single Shot multibox Detector (SSD)	Japanese Road Sign	Results are mentioned for each class achieved accuracy up to 91%
Bangquan & Xiong, 2019	Proposed two methods, ENet and EmdNet, for classification and detection, respectively	GTSRB	98.6%
Chinese Association of Automation et al., n.d.	The research proposed both detection and classification of "circular shaped signs". Hough transform method is used to detect the sign and for classification CNN is applied	GTSRB	98.2%
(Université Yahia Fares de Médéa & Institute of Electrical and Electronics Engineers, n.d.)	In this study, enhanced LeNet-5 based on CNN is used to classify the traffic sign after feature detection	GTSRB	97.1% validation accuracy

Creating a data set is a difficult and time-consuming task. In the case of supervised learning, along with large amount of data, the data must be labeled correctly on the basis of these labels associated with feature values. The models create a relationship and draw a pattern. To learn a relationship between input and output, the data on which the model is supposed to train must be of high quality. In the case of image data, all the images must be captured clearly and edges should not be cropped because low-quality and cropped images can affect the predictions in an unanticipated way.

Despite using a standard ML approach in the process of detecting and classifying traffic signs, the use of DL-based CNN models proved their remarkable accuracy (also shown in Table 13.5). And this can only be achieved by using large amounts of data to train the model and learn the pattern of relationship between inputs and labels associated with input features. So, we can say that the accuracy of any DL model is dependent on availability of good and large amounts of data. In order to achieve maximum accuracy, the data set must be trustworthy, well-organized, and labeled properly (in case of supervised learning); images are of high dimensions and must be captured from every angle. To detect or classify traffic signs, the data set can be created manually or can be downloaded and used that are publicly available for the researchers.

There are many data sets available for detecting and classifying traffic signs, but the most commonly used is the GTSRB, which is also considered the benchmark among all available data sets.

TABLE 13.6   Specifications of GTSRB

Data set	GTSRB
Origin	Germany
Total Number of Groups	4
Prohibitory Signs	12
Danger Signs	15
Mandatory Signs	8
Other Signs	8
Total Number of Signs (Classes)	43
Images in Training Set	39,209
Images in Test Set	12,630
Image Size (Pixels)	$15 \times 15 \times 15$ to $250 \times 250 \times 250$
Highest Accuracy	99.3%

13.6.1 German Traffic Sign Recognition Benchmark

This data set can be used to detect as well as for classification purposes. The data set is considered the benchmark among all publicly available data sets because of the following reasons (Bailke, 2022):

- It contains a large number of images.

- The images are captured in different environments like background and different colors.

The whole data set is divided into four groups and each group has different numbers of signs and each sign has multiple images. The total number of classes in this data set is 43 (Wen & Jo, 2018). The complete specification of GTSRB is summarized in Table 13.6 (Nadeem et al., 2022).

This data set can be easily downloaded from authenticated Internet sources. The classes in the data set are labeled from 0 to 42.

13.6.2 Comparison With Other Data Sets

Other than GTSRB, some data sets are also developed to train the model for detection and classification of traffic signs. A clear comparison among these developed data sets is presented in Table 13.7 (Nadeem et al., 2022).

TABLE 13.7   Specifications of Other Data Sets Used to Detect and Classify Traffic Signs

Data set	Origin	Classes	Number of images	Image size (pixels)	Accuracy
BTSD	Belgium	100+	13,444	$1628 \times 1236 \times 1236$	99.72%
STSD	Sweden	7	3,488	$1280 \times 960 \times 960$	82%
RSTD	Russia	140	80,000 +	$1280 \times 720 \times 720$	92.90%
PTSD	Pakistan	35	359	$64 \times 64 \times 64$	54.8%
JRSD (Hasegawa et al., 2020)	Japan	7,160	4,477	$1093 \times 615 \times 615$	94.64%

13.7 EXPERIMENTAL SETUP

The following steps will be taken to start building the model for classifying traffic signs:

1. **Impot Libraries:** Libraries used to create ML-based model can be imported by using the "import" command in Python.

2. **Download Data Set:** You can download the benchmark data set GTSRB from GitHub, Kaggle, and any other Internet source.

3. **Data Pre-processing:** Apply techniques for pre-processing the data like one hot coding and normalize the data set.

4. **Model Building:** Here, you create your model by defining layers. With each layer, you have to give parameter values.

5. **Training and Evaluation:** Now you can train your model by defining the epoch and batch size of your own choice.

13.8 CONCLUSION

The primary function of traffic signs is to warn, guide, and be noticed easily during daytime, nighttime, as well as under any weather. These signs are the combination of distinguishable features like text, multiple shapes, and vibrant color exhibiting meaningful information so it is very essential to interpret these signs correctly in order to avoid serious consequences. In view of these circumstances, the need of developing a robust system to detect and classify these signs prior to any terrible event is essential. Before DL techniques, standard ML algorithms were used to detect and classify these signs, but the process was time consuming and accuracy was not as satisfied as they had to be. By applying DL models, we can achieve promising predictions in any field, even for the detection and classification of traffic signs. CNN is one of the most commonly used algorithms based on DL, used for the applications of computer vision, including detection and classification of images. CNNs are based on a multi-layered deep neural network consisting of many hidden layers that can extract useful information from the input image so the results predicted by these models are more accurate than a traditional ML. This chapter efficiently describes all the concepts from a simple ML to experimental setup that will assist the researchers to understand every important terminology need to build an efficient model to classify traffic signs. Models based on CNN are compared to provide guidelines to researchers in selecting the model as per the need of their problem. Additionally, the importance of data sets and types of data like labeled and unlabeled data are discussed thoroughly along with the benchmark data set used in detection and classification of traffic signs. A complete model architecture, including functions of each layer of CNN, is presented. Finally, experimental setup is given as a basic guidance to assist the researchers in creating their own problem-solving models and get promising predictions by applying the concepts discussed in this chapter.

REFERENCES

Aslansefat, K., Kabir, S., Abdullatif, A., Vasudevan, V., & Papadopoulos, Y. (2021). Toward improving confidence in autonomous vehicle software: A study on traffic sign recognition systems. *Computer, 54*(8), 66–76.

Aamir, M. (2023). A progressive approach to generic object detection: A two-stage framework for image recognition. *Computers, Materials & Continua, 75*, 6351–6373. 10.32604/cmc.2023. 038173.

Aamir, M., Rahman, Z., Ahmed Abro, W., Aslam Bhatti, U., Ahmed Dayo, Z., & Ishfaq, M. (2023). Brain tumor classification utilizing deep features derived from high-quality regions in MRI images. *Biomedical Signal Processing and Control, 85*, 104988. 10.1016/j.bspc.2023.104988.

Aamir, M., Rahman, Z., Ahmed Abro, W., Tahir, M., & Mustajar Ahmed, S. (2019). An optimized architecture of image classification using convolutional neural network. *International Journal of Image, Graphics and Signal Processing, 10*(10), 30.

Bhatti, U. A., Huang, M., Neira-Molina, H., Marjan, S., Baryalai, M., Tang, H., … & Bazai, S. U. (2023). MFFCG–Multi feature fusion for hyperspectral image classification using graph attention network. *Expert Systems with Applications, 229*, 120496.

Bhatti, U. A., Zeeshan, Z., Nizamani, M. M., Bazai, S., Yu, Z., & Yuan, L. (2022). Assessing the change of ambient air quality patterns in Jiangsu Province of China pre-to post-COVID-19. *Chemosphere, 288*, 132569.

Bailke, P. (2022). Traffic sign classification using CNN. *International Journal for Research in Applied Science and Engineering Technology, 10*(2), 198–206. 10.22214/ijraset.2022.40224

Bangquan, X., & Xiong, W. X. (2019). Real-time embedded traffic sign recognition using efficient convolutional neural network. *IEEE Access, 7*, 53330–53346. 10.1109/ACCESS.2019.2912311

Bouti, A., Mahraz, M. A., Riffi, J., & Tairi, H. (2020). A robust system for road sign detection and classification using LeNet architecture based on convolutional neural network. *Soft Computing, 24*(9), 6721–6733. 10.1007/s00500-019-04307-6

Chinese Association of Automation, IEEE Systems, M., & Institute of Electrical and Electronics Engineers. (n.d.). *Proceedings, 2019 Chinese Automation Congress (CAC2019): Nov. 22–24, 2019, Hangzhou, China.*

Guan, Y., Aamir, M., Hu, Z., Dayo, Z. A., Rahman, Z., Abro, W. A., & Soothar, P. (2021). An object detection framework based on deep features and high-quality object locations. *Traitement du Signal, 38*, 719–730. 10.18280/ts.380319.

Hassanien, A. E., Bhatnagar, R., & Darwish, A. Editors (n.d.). *Advances in Intelligent Systems and Computing 1141 Advanced Machine Learning Technologies and Applications Proceedings of AMLTA 2020.* http://www.springer.com/series/11156

Hasegawa, R., Iwamoto, Y., & Chen, Y.-W. (2020). Robust Japanese road sign detection and recognition in complex scenes using convolutional neural networks. *Journal of Image and Graphics, 8*(3), 59–66. 10.18178/joig.8.3.59-66

Hechri, A., & Mtibaa, A. (2020). Two-stage traffic sign detection and recognition based on SVM and convolutional neural networks. *IET Image Processing, 14*(5), 939–946. 10.1049/iet-ipr. 2019.0634

Institute of Electrical and Electronics Engineers, & Manav Rachna International Institute of Research and Studies. (n.d.). *Proceedings of the International Conference on Machine Learning, Big Data, Cloud and Parallel Computing: Trends, Prespectives and Prospects: COMITCON-2019: 14th–16th February, 2019.*

Loukmane, A., Grana, M., & Mestari, M. (2020, October 21). A model for classification of traffic signs using improved convolutional neural network and image enhancement. *4th International Conference on Intelligent Computing in Data Sciences, ICDS 2020.* 10.1109/ ICDS50568.2020.9268761

Mishra, J., & Goyal, S. (2022). An effective automatic traffic sign classification and recognition deep convolutional networks. *Multimedia Tools and Applications*, *81*(13), 18915–18934. 10.1007/s11042-022-12531-w

Nadeem, Z., Khan, Z., Mir, U., Mir, U. I., Khan, S., Nadeem, H., & Sultan, J. (2022). Pakistani traffic-sign recognition using transfer learning. *Multimedia Tools and Applications*, *81*(6), 8429–8449. 10.1007/s11042-022-12177-8

Noor, S., Marjan, S., Bazai, S. U., Akram, S., Ghafoor, M. I., & Ali, F. (n.d.). *Generative Adversarial Networks for Anomaly Detection: A Systematic Literature Review.*

Oztel, I., Yolcu, G., & Oz, C. (2019). Performance comparison of transfer learning and training from scratch approaches for deep facial expression recognition. *UBMK 2019 - Proceedings, 4th International Conference on Computer Science and Engineering*, 290–295. 10.1109/UBMK.2019.8907203

Saha, S., Kamran, S. A., & Sabbir, A. S. (2018). *Total Recall: Understanding Traffic Signs using Deep Hierarchical Convolutional Neural Networks.* 10.1109/ICCITECHN.2018.8631925

Sai Kondamari Anudeep Itha, P. (2021). *A Deep Learning Application for Traffic Sign Classification.* www.bth.se

Sarker, I. H. (2021). Deep learning: A comprehensive overview on techniques, taxonomy, applications and research directions. In *SN Computer Science* (Vol. 2, Issue 6). Springer. 10.1007/s42979-021-00815-1

Surya Engineering College, & Institute of Electrical and Electronics Engineers. (n.d.). *Proceedings of the 3rd International Conference on Computing Methodologies and Communication (ICCMC 2019)*: 27–29, March 2019.

Université Yahia Fares de Médéa, & Institute of Electrical and Electronics Engineers. (n.d.). *Proceedings of the 2018 International Conference on Applied Smart Systems (ICASS): ICASS-2018: Médéa*, Algeria, 24–25 November 2018.

Wang, C., Wu, Y., Qian, Y., Kumatani, K., Liu, S., Wei, F., Zeng, M., & Huang, X. (2021). *UniSpeech: Unified Speech Representation Learning with Labeled and Unlabeled Data.* https://github.com/

Wen, L., & Jo, K. H. (2018). Traffic sign recognition and classification with modified residual networks. *SII 2017-2017 IEEE/SICE International Symposium on System Integration*, 2018-January, 835–840. 10.1109/SII.2017.8279326

Zaibi, A., Ladgham, A., & Sakly, A. (2021). A lightweight model for traffic sign classification based on enhanced LeNet-5 network. *Journal of Sensors*, *2021*. 10.1155/2021/8870529

Zhang, J., Wang, W., Lu, C., Wang, J., & Sangaiah, A. K. (n.d.). *Lightweight deep network for traffic sign classification.* 10.1007/s12243-019-00731-9/Published

Cloud-Based Intrusion Detection System Using a Deep Neural Network and Human-in-the-Loop Decision Making

Hootan Alavizadeh[1] and Hooman Alavizadeh[2]

[1]*Department of Computer Science and Engineering, Wright State University, Dayton, OH, USA*

[2]*Department of Computer Science and Information Technology, School of Computing, Engineering and Mathematical Sciences, La Trobe University, Melbourne, Australia*

14.1 INTRODUCTION

Cyber-attack identification is one of the most significant challenges of many network contexts, such as cloud [1]. Various types of cyber-attacks can be launched against the clouds' resources and infrastructures from outside the cloud, such as compromised Internet of Thing (IoT) devices, sensors, or online social networks (OSNs) [2]. The attackers can launch multiple types of cyber-attacks, especially distributed denial of service (DDoS) attacks toward the crucial cloud components. They would be able to have normal behaviors to either deceive the security mechanisms (i.e., IDS) used in the cloud or gain required information from the targeted such as vulnerabilities, list of critical VMs and servers, and IP address of critical targets and so forth through launching probing attacks.

Many security mechanisms have so far been proposed to secure the cloud environment. However, most of the current methods are designed by automated approaches such as machine learning-based IDS and anomaly detection methods that may not be able to defend against some subtle attacks. Even a capable IDS using strong attack classification methods have limitations that are typical of any machine learning system, such as inability to detect some threats, enhance the learning abilities by getting feedback about its mistakes,

DOI: 10.1201/9781003427674-14

and lack of evaluation using experts in the loop. To address these shortcomings, cloud environments need to be secured with a comprehensive IDS that utilizes strong attack classification techniques alongside the knowledge of experts to be able to monitor the traffic, detect various attack types, and evaluate the capabilities of both attackers and defenders. Thus, human-in-the-loop systems can be well mingled with IDS systems to both improve the learning capabilities of the IDS and increase the attack classification accuracy, and also make decisions about the undetected attacks and deal with uncertainties.

In this chapter, we proposed a cloud-based IDS equipped with a human-in-the-loop module leveraging a game theory model for decision making. The IDS module is able to detect various types of attacks and decision making is able to evaluate the system for undetected attacks (or those classified wrongly by IDS) through formulating a non-zero-sum game model. The main contributions of this chapter, which to the best of our knowledge have not already been proposed by other works, are listed as follows:

- We propose a cloud-based IDS for using a convolutional neural network (CNN) to classify four attack classes, such as DoS, probe, privilege, and access, using a NSL-KDD data set. We further evaluate the performance of our proposed CNN model using performance metrics.

- We propose a HITL module to detect attacks that are either not detected by our IDS module or classified wrongly. Our designed HILT module is able to perform decision making to accept the traffic or send it to honeypot for deceiving the attackers.

- We incorporate a game theory model into our HITL module to evaluate various attack scenarios as a game between the defender (IDS) and the attacker.

- We provide the formal mathematical definitions for game model and payoff functions, and also provide the formulated extensive form of game based on two types of games.

The rest of the chapter is organized as follows. Section 14.2 presents the related works. Section 14.3 introduces some background and definitions together with cloud and attack model definitions. Section 14.4 describes our proposed IDS and related evaluations such as CNN model and game model. Section 14.5 provides the discussion and limitation of this study. We conclude our chapter in Section 14.6.

14.2 RELATED WORK

A number of studies have been proposed to develop and evaluate the IDS systems for various contexts, such as cloud computing [1]. Machine learning techniques have been widely studied in IDS and anomaly detection in recent years [3]. Many new cybersecurity IDS techniques have been proposed by leveraging either traditional machine learning approaches [4] or deep learning techniques [3].

In [5,6], the authors proposed a VM-based intrusion detection system (IDS) for cloud computing. They proposed a solution using the Dempster Shafer theory (DST) [5] operations using fault-tree analysis (FTA) to detect and analyze DDoS attacks in cloud computing environments. However, their proposed method is basically a host-based approach and is unable to handle various attacks launched to cloud computing and the network infrastructure. A deep learning approach for anomaly detection and classification was proposed in [7]. The authors proposed a deep learning approach to adapt self-learning for selecting required features for network anomaly detection. Their classification algorithm was able to separate legitimate traffic from other malicious ones such as flooding, injection type, and impersonation attacks. However, their proposed method was conducted basted on a small data set and was not able to detect crucial attacks such as distributed of service (DoS) or probing attacks. However, convolutional neural networks (CNN) have been utilized in many studies to perform malware classification in IDS. Based on various studies, CNN approaches have a high potential for IDS and can be considered as one of the most efficient techniques with high accuracy and results [8]. Compared with other deep neural network techniques, CNN has fewer parameters with the same number of hidden units that enhance the learning process [9,10]. In [8], the authors proposed a novel IoT malware traffic analysis method using CNN. Their method was able to detect five attack categories and visualize both malicious and normal network traffics.

However, various relevant research introduced the game theory and studied the different strategies of network attack and defense. To perform the security assessment and active protection of the network, in [11], the authors proposed a network optimal active defense method based on the offensive and defensive game models. Similarly, in [12], the authors introduced a network security defense decision-making method based on the offensive and defensive differential game to analyze the change process of the security state of the network system and construct the differential game model of attack and defense. In [13], the authors presented a network security optimal attack and defense decision-making method based on the non-cooperative non-zero-sum game model. It generates an optimal attack and defense strategy by analyzing the attack and defense interactions of attackers and defenders. In [14], the authors suggested an algorithm for selecting the optimal defense strategy based on the static Bayesian game. The proposed algorithm calculates the effectiveness of the defense strategy based on probability and gives the optimal active defense strategy selection algorithm for network intrusion detection. In [15], the authors recommended a wireless sensor network intrusion detection approach. In [16], a game theory method is used to analyze the security combination model of a firewall, intrusion detection system (IDS), and vulnerability scanning technology and gave the optimal intrusion detection calculation method.

In this chapter, we leverage a non-zero-sum game model in our proposed human-in-the-loop (HITL) module to evaluate the possible strategies for the defender (IDS) based on the attacker strategies that may deceive the defender.

14.3 PRELIMINARIES

In this section, we define some related notations and mathematical formalisms for our cloud and threat model, intrusion detection system, and game model. The defined notations will further be used throughout this chapter.

14.3.1 Cloud and Attack Model

We modeled a cloud system consisting of four virtual machines (VMs) hosting a web server (WS), two databases (noted as DB-S1 and DB-S2), and a honeypot (HP). The cloud model is demonstrated in Figure 27.1. The cloud system is equipped with a central cloud-based intrusion detection system (IDS), which is able to monitor and inspect the incoming traffic toward the cloud. Moreover, the IDS system utilizes a deep learning module to train the system to capture different threats such as DoS, Probe, Privilege, Access, and other attacks. The cloud-based IDS includes a decision-making module equipped with a HITL module for evaluating the correctness of a trained neural network classifier based on real network traffic and also forwarding normal traffic to the destination and malicious ones to the honeypot. The role of the HITL module is crucial as the feedback from the HITL module will update the training module and will be sent to the game-theory module for evaluation of IDS. We assume that the attacker is able to use various reconnaissance tools and techniques to gain enough information about the targets in the cloud system. The attacker has some information regarding the cloud hosts, VMs, and network gained through the various network and vulnerability scanning tools such as Nessus, open vulnerability assessment scanner (OpenVAS), and so forth. The attacker can also launch passive attacks (such as probe attacks) to gain information about the targets and later launch active attacks such as DoS toward the targets (i.e., DBs). Figure 14.1

14.3.2 Deep Neural Network

In this chapter, we utilize a deep neural network in our proposed IDS module which is able train the model based on CNN to detect and classify different attack categories. We use NSL-KDD data set to train our CNN model. A CNN contains different layers that are basically input, convolution layer, batch normalization, activation, pooling, flatten, and

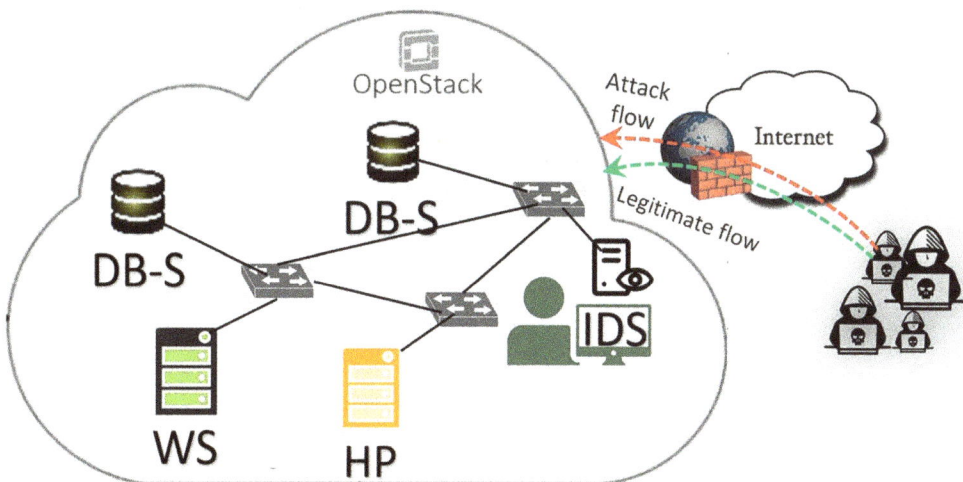

FIGURE 14.1 Cloud system model equipped with cloud-based IDS utilizing HITL and GT modules.

TABLE 14.1   The Specific Implementation Operations of Data Enhancement

Categories	Notation	Definitions
Normal	N	Normal traffics and activities by attacker
DoS	D	Denial of service attack, users' services can be interrupted by attackers
Probe	Pb	Attacker plans to collect information by scanning the cloud and find vulnerabilities
Privilege	Pr	Attacker tries to get user privilege to the victim's machine using local access
Access	A	Attacker sends packets to the target host to get access without a local account

other layers [17]. Convolutional layers are the seminal component of a CNN architecture that are based on a hierarchical structure. They can help the model to learn complex features by generating a number of feature maps. A CNN also utilizes the activation function that is a non-linear and element-wise function, such as sigmoid or rectified linear unit (ReLU) [3]. ReLU is responsible for activating all units of the feature maps with a non-linear rectification function as MAX (X, 0) so that all the negative values in the matrix X will be set as zeros and make all other values a constant number.

One of the most commonly used data sets used for the intrusion detection techniques for training and testing the models is a NSL-KDD data set. It has been utilized in various studies to evaluate different machine learning and deep learning–based algorithms for IDS [18,19]. The NSL-KDD data set contains 41 features where each is labeled as a normal traffic or specific attack type such as denial-of-service (DoS), probe, root to local (R2L), and so forth. Table 14.1 defines the attack classes for NSL-KDD that we consider in our study.

14.3.3 Performance Metrics

We use the following measurements to evaluate the performance of our proposed DNN models used in the IDS module. We utilized accuracy, precision, recall, and F1-score as follows.

14.3.1.1 Accuracy

Accuracy is generally used to measure a model's accuracy using true positive (TP), true negative (TN), false positive (FP), and false negative (FN). Accuracy measures the total correct predictions numbers against all the predictions, as in Equation (14.1).

$$\text{Accuracy} = \frac{TP + TN}{TP + FP + TN + FN} \tag{14.1}$$

14.3.1.2 Precision

A precision indicator measures the rate of positive instances against the total predicted positive instances. It finds out how much the model is right when it says it is right; see Equation (14.2).

$$\text{Precision} = \frac{TP}{TP + FP} \tag{14.2}$$

14.3.1.3 Recall

Recall or sensitivity indicates the rate of percentage of positive instances against the total actual positive instances. It shows how much extra right ones the model missed when it showed the right ones. See Equation (14.3).

$$\text{Recall} = \frac{TP}{TP + FN} \tag{14.3}$$

14.3.1.4 F1-Score

The F1-score is considered the harmonic mean of both precision and recall values. A model can be considered a good one if it has a higher F1-score. Equation (14.4) shows the F1-score formula.

$$\text{F1 score} = \frac{2 * \text{Precision} * \text{Recall}}{\text{Precision} + \text{Recall}} \tag{14.4}$$

14.4 INTRODUCTION CLOUD-BASED IDS MODULE

In this section, we explain the proposed cloud-based IDS module, which is designed based on different seminal components for intrusion detection and evaluation of the proposed model, including a model training module using CNN, decision-making module using HITL phase, and a game theory module for evaluation of the IDS capabilities on detection of attacks. Figure 14.2 shows our proposed IDS model and interactions between the main modules of the system.

14.4.1 CNN Model Training

We leverage a deep learning model based on a convolutional neural network (CNN) to detect the attacks. To train our CNN model, we use the NSL-KDD data set, including 41 features. A NSL- KDD data set will be fed into a pre-processing module in our proposed IDS model (as in Figure 14.2) which extracts the related features from the data set and trains the model based on the related features [20]. We also used a database (DB) to keep and update the NSL-KDD records based on the output from the decision-making module and the further evaluation will be performed by the HITL module. This can enhance the accuracy of our CNN model for further classification.

We implemented the attack classifier using the CNN architecture that can be described by six layers, such as (i) the input layer that holds the NSL-KDD features; (ii) convolution layers that include a set of learnable kernels for producing feature maps; (iii) dropout layer that is used to prevent overfitting and provide generalization [21]; (iv) max-pooling that selects the feature values and reduce the feature maps; (v)

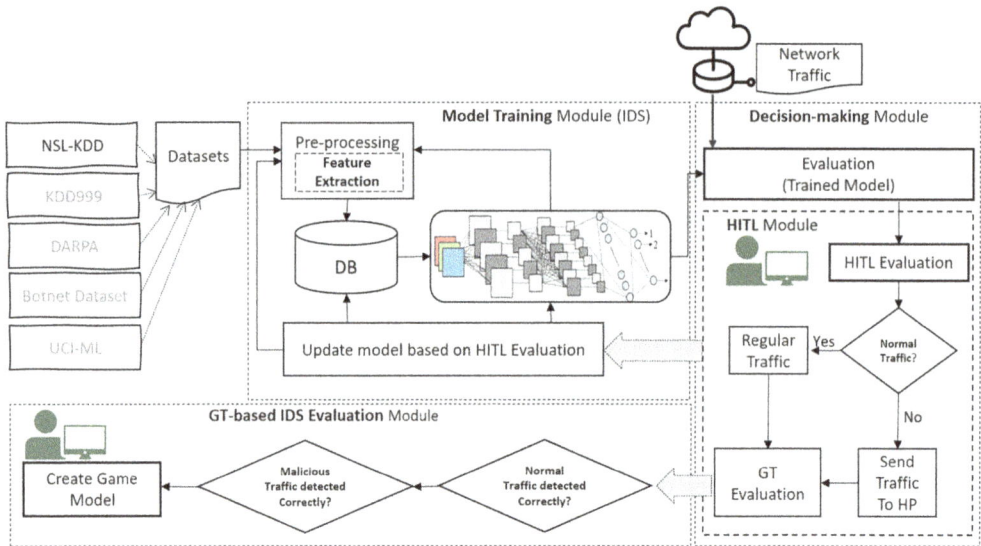

FIGURE 14.2 Cloud-based IDS model designed for training, classifying, and evaluation of different attacks for the cloud model.

flatten layer that creates a vector from the max-polling results as the final feature vector. The output of flatten layer is passed to a fully connected sigmoid layer for final classification; and (vi) dense layers providing the output based on six different classes. The CNN model can be trained to classify five attack types, as demonstrated in Table 14.1, alongside normal activities. The trained CNN model based on the NSL-KDD data set will further be used for evaluation in the decision-making module.

14.4.2 Decision Making and Evaluation

In this section, we describe the decision-making module of our proposed method that can evaluate the trained CNN model based on either testing the data set or real incoming network traffic. The incoming network traffic will be sent to the CNN evaluation module equipped with a trained model to measure the performance of the trained CNN model based on various metrics explained in Section 14.3.3. The main part of the evaluation is performed using a HITL module in which the experts evaluate the correctness of CNN classification based on the incoming traffic, either from trained data set or real traffic. HITL experts evaluate the input data based on the results of CNN evaluation and accept the normal traffic to enter the cloud and avoid the malicious traffic and send them to the honeypot for deception. Moreover, the HITL module needs to categorize the incoming traffic based on the CNN evaluation outcome to decide about the nature of the attacks based on two types: active and passive. The HITL module further generates a game model based on the correctness of evaluation for both active and passive attacks. The game model is described in Section 14.4.4.1.

14.4.3 Performance Evaluation

The proposed CNN models were implemented using an open-source machine learning library (tensor-flow) utilizing Keras. Experiments were performed in a Google Colab environment using GPUs running. We evaluate the classification accuracy of our proposed CNN model described in Figure 14.2 using the performance metrics. We run 20 epochs to train the CNN model using the features obtained from the NSL-KDD data set. Our experiments are designed to evaluate the ability of our proposed CNN model to detect the network intrusions based on the attacks described in Table 14.1. We divided the NSL-KDD to two training and testing data sets. We used 20% of the NSL-KDD data set for testing and evaluation. Figure 14.3 demonstrates the distributions of attack classes and normal activities for both training and testing data sets. However, there is a lower rate for both probing and access attacks in the data set. Moreover, the privilege attack class has the lowest value for both training and testing data sets. This may impact the accuracy of our classification model for those attack types with a lower number of samples in the data set. However, by utilizing the HITL module, the new identified attacks will be added to the training database for the next rounds of training to improve the accuracy of the CNN model gradually. In this study, we only run our evaluation based on one run of experiments due to a lack of real attack samples and network traffic. However, we plan to leverage more data sets and real traffic collected from various sources such as honeypots to improve our CNN model and evaluate our model based on a real testbed in our future work.

Figure 14.4 shows the accuracy and loss of the purposed CNN model. The model was trained based on 20 epochs taken around 29 seconds per epoch/step on the GPU in Google Colab. We observed a high performance for our proposed CNN model. At the end of the 20th epoch, the model reached around 99% accuracy for training and around 97% for testing. Moreover, the loss functions for both training and testing values were decreased gradually and reached the minimum value. However, we set the training epochs at a maximum of 20 steps to avoid overfitting. The loss values for both training and testing reached around 0.4% on epoch 20. We further evaluated the performance of classification for each individual defined class based on accuracy, precision, F1-score, and recall on the testing data set.

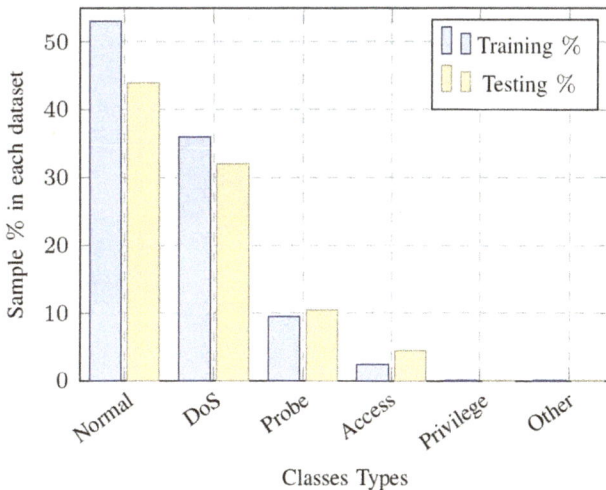

FIGURE 14.3 Attack class distribution for training and testing.

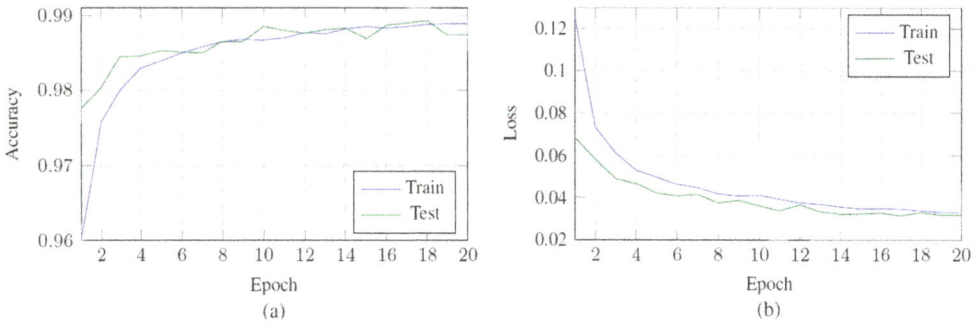

FIGURE 14.4 Comparing the (a) accuracy and (b) loss values of CNN for both training and testing phases based on 20 epochs.

Figure 14.5 demonstrates the confusion matrix for our proposed CNN model based on six different categories, including both normal and attack traffic based on our testing data set. The main diagonal values show the correctly classified TP or TN based on both values and percentages. However, the off-diagonal cells demonstrate the percentages and total numbers for incorrectly classified values of FN or FP. Considering the confusion matrix, we can observe that the true-positive rate for normal, DoS, and probe classes have the values of 99%, 100%, and 99%, respectively. However, true-positive values for privilege and access attack categories are 64% and 70%. Any other attack types are identified by the true-positive rate of 70%. However, the results for the minority class of access and privilege are lower because of unbalanced distributions of the number of samples for each class.

FIGURE 14.5 Confusion matrix based on the attack categories and CNN model parameters.

14.4.4 GT-Based Evaluation

In this section, we propose our game theory model in our GT-based evaluation module equipped with HITL interaction for evaluation of the possible strategies for the defender (IDS) based on the attacker strategy. in some cases, the attacker may use various attack strategies and technique to deceive the IDS. Even capable IDS systems may be deceived by different attack strategies, such as having less suspected activities by attackers. To this end, we leverage a GT model to evaluate the effectiveness of out IDS module to detect the attacks by defining a game model to evaluate different attack scenarios.

14.4.4.1 Game Model Definition

Table 14.1 shows the notation and definitions used to design the game theory model for capturing and evaluating our proposed IDS model based on two types of attacks: active and passive attacks. We define the non-zero-sum game model as G =< (P), (S), (U) >, where P = {A, D} denotes two players as attacker and defender, represented by A and D. Each player ($p \cdot \in P$) can choose a (pure) strategy, s_p, from the strategy space, $S_p = \{s_1, \ldots ., s_j\}$. The strategy for the defender is to either accept the traffic as regular (R) or avoid the traffic and send it to honeypot (H) to deceive the attacker. Likewise, the attacker can either launch active or passive attacks. Please note that the attacker can have normal activities as well at anytime to deceive the defender. However, the HITL module needs to make an evaluation based on the CNN evaluation module and categorize the attack types based on active and passive attacks, as follows.

- Type I: Passive attack (avoid detection), which includes normal behaviors by attackers to deceive the IDS and suspicious, such as probing attack.

- Type II: Active attack, which consists of normal activities for deception and malicious attacks, such as DoS, access, and privilege.

We assume that the defender does not know the attack's type (whether if it is type I or type II) and also, we assume that the attacker has no information about the honeypot used by the defender. Thus, the attacker may launch attacks, even if the attacks are detected by the defender and sent to the honeypot.

14.4.4.2 Payoff Function

The payoff function (u) for a player for a given strategy can be defined as $u_p(s_{PA}, s_{PD})$. Computing the quantitative values for payoff function and computing the game outcome directly depends on the strategy selection. Therefore, it is necessary to quantify the payoffs of the strategies of both sides accurately. In this chapter, we used different metrics to quantify payoff, rewards, costs, and loss. Table 14.2 demonstrates the used notations for metrics to calculate payoff matrix. Both attacker and defender reward functions are defined based on the equations shown in Figure 14.6.

As we have attack sets for each attacker type, the general reward functions for the proposed game for different scenarios are defined based on Figure 14.6. $R_{\tau,\alpha}^A$ and $R_{\tau,\alpha}^D$

TABLE 14.2   The Notations and Definitions Used for the
GT Model

A/D	Suspicious (M)
B_s	Successful detection gain
I_d	Damage on system attack (impact)
G_a	Gain for successful attack (reward)
G_{pa}	Gain for successful probing attack (reward)
C_a	Attacker cost for unsuccessful attack
C_{pa}	Attacker cost of unsuccessful probing attack
C_{ids}	Defender's cost for successful attack
C_{ud}	Defender's cost for unnecessary defense L_{ids}
L_{ids}	Defender's loss for probed target
e	Extra gain/costs
G_d	Gain for successful defend (reward)
G_{da}	Defender's award (normal behaviors)

are the rewards for the attacker type and the defender, respectively, when the attacker type τ = I uses an attack action against the system; the attack can either be detected and sent to the honeypot or not detected or behave as normal activity. This is similar for the attacker type τ = II. Tables 14.3 and 14.4 demonstrate the payoff matrix or both attack types based on the variables Id, Ga, Gda, Gpa, Cud, Cpa, and Cids, defined in Table 14.1. Note that these metrics can be obtained through various resources such as the national vulnerability database (NVD) for impact, exploitability score, and attack costs to compute the payoff matrix.

$$R^{\mathcal{A}}_{\tau_I,\alpha} = \begin{cases} -G_a & \text{if attacker's } \textit{normal} \text{ activity is not detected} \\ G_{pa} & \text{if attacker's } \textit{malicious} \text{ activity is not detected} \\ 0 & \text{if attacker's } \textit{normal} \text{ activity is detected by } IDS \\ C_{pa} & \text{if attacker's } \textit{malicious} \text{ activity is detected by } IDS \end{cases}$$

$$R^{\mathcal{D}}_{\tau_I,\alpha} = \begin{cases} G_{da} & \text{if attacker's } \textit{normal} \text{ activity is not detected} \\ -L_{ids} & \text{if attacker's } \textit{malicious} \text{ activity is not detected} \\ -C_{ud} & \text{if attacker's } \textit{normal} \text{ activity is detected by } IDS \\ 0 & \text{if attacker's } \textit{malicious} \text{ activity is detected by } IDS \end{cases}$$

$$R^{\mathcal{A}}_{\tau_{II},\alpha} = \begin{cases} G_a + e & \text{if attacker's } \textit{normal} \text{ activity is not detected} \\ G_a & \text{if attacker's } \textit{malicious} \text{ activity is not detected} \\ -C_a & \text{if attacker's } \textit{normal} \text{ activity is detected by } IDS \\ -C_a & \text{if attacker's } \textit{malicious} \text{ activity is detected by } IDS \end{cases}$$

$$R^{\mathcal{D}}_{\tau_{II},\alpha} = \begin{cases} -I_d - C_{ids} - e & \text{if attacker's } \textit{normal} \text{ activity is not detected} \\ -I_d - C_{ids} & \text{if attacker's } \textit{malicious} \text{ activity is not detected} \\ G_d + e & \text{if attacker's } \textit{normal} \text{ activity is detected by } H \\ G_d & \text{if attacker's } \textit{malicious} \text{ activity is detected by } H \end{cases}$$

FIGURE 14.6   Game reward functions equations.

TABLE 14.3   Type I Payoff Matrix

A/D	Normal (N)	Suspicious (M)
Regular (R)	$-G_a$, G_{da}	G_{pa}, $-L_{ids}$
Honeypot (H)	0, $-C_{ud}$	C_{pa}, 0

TABLE 14.4   Type II Payoff Matrix

A/D	Normal (N)	Suspicious (M)
Regular (R)	$G_a + e$, $-I_d - C_{ids} - e$	G_a, $-I_d - C_{ids}$
Honeypot (H)	$-C_a$, $G_d + e$	C_a, G_a

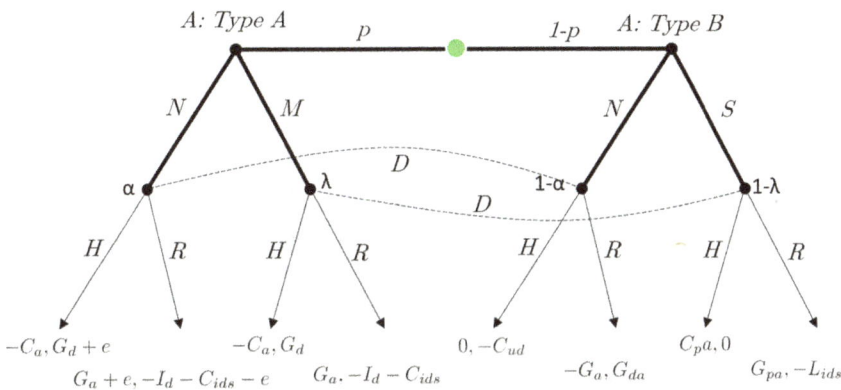

FIGURE 14.7   Extensive form of game based on two types of games.

We formulate the extensive form of the game to show the explicit representation of an attacker's and defender's possible moves, their choices at every decision point, the (possibly imperfect and incomplete) information each player, and the chance of events based on the types and beliefs, and their payoffs for all possible game outcomes. The extensive form of the game is represented in Figure 14.7. However, in this chapter, we only formulate the game model for our purposed IDS system based on both attacker and defender to obtain the extensive form of the game and did not include numerical values for the game theory parameters and metrics to solve the game based on the realistic values. However, we plan to evaluate the GT-based module based on realistic values that can be obtained from various resources, such as NVD and CVSS scores such as attack impact, cost, and exploitability values based on the vulnerability of our cloud components in our future work.

14.5 DISCUSSION AND LIMITATIONS

We proposed a CNN model as the main part of our IDS that is able to classify the different types of attacks, such as DoS, probe, privilege, and access together with other attacks based on the NSL-KDD data set. Table 14.5 shows the performance metrics of a CNN model. Based on the observed results, our trained CNN model has a very high

TABLE 14.5   Evaluation Metrics for the CNN Model

Metric	Attack Categories					
	Normal	DoS	Probe	Privilege	Access	Other
Precision	0.98	1	0.98	0.64	0.96	0.78
Recall	0.99	1	0.99	0.64	0.70	0.70
F1-Score	0.99	1	0.99	0.64	0.81	0.74
Accuracy	0.99	0.99	0.99	0.99	0.99	0.99

TABLE 14.6   Evaluation Metrics for the CNN Model

Approach	Reference	Dataset	Accuracy
SOM	Ibrahim et al. [23]	NSL-KDD	75%
RF		NSL-KDD	74%
BiLSTM	Jiang et al. [24]	NSL-KDD	79%
CNN-BiLSTM		NSL-KDD	83%
CNN	Our model	NSL-KDD	99%

performance for classifying DoS and probe attacks [22] for all computer performance metrics. Moreover, we compared our CNN model performance with other machine learning–based IDS proposed in other studies in Table 14.6. As it can be seen, our model has a higher accuracy compared with the other approaches with a rate of 99%. However, the worst accuracy obtained by the SVM approach is about 68%. Both BiLSTM and CNN- BiLSTM hybrid approaches have high accuracies of 79% and 83%, respectively. While it is very difficult for the IDS system to detect and classify most of the attack classes, one can still benefit from using human experts to judge the performance of the trained model and make decisions based on different attack and defense scenarios using formulating and solving the game theory models. However, the main limitations of our proposed cloud-based IDS system are as follows: (i) Lack of implementation of our proposed method in a realistic cloud environment. We plan to implement our proposed method in this chapter to a realistic private cloud and test our trained CNN model using real attack traffics. (ii) Training the CNN model using more data sets and updating the training database and our neural network learning parameters using the feedback obtained from the human experts in the HITL module. (iii) We only proposed a game theory model to formalize the attacker and defender strategies based on various metrics such as attack impact and cost. However, we did not solve the formulated game using numerical values to obtain the optimal defender's strategies by using different methods such as the Nash equilibrium. We plan to solve our game model based on numerical values that can be obtained using NVD and CVSS scores in our future work.

14.6 CONCLUSION

In this chapter, we presented a cloud-based IDS module that was capable of detecting different attacks toward the cloud from outside, such as DoS, probe, privilege, access, and some other attack categories. We observed that our proposed CNN model can have higher performance compared with other machine learning algorithms. Although our proposed CNN model can detect those attack categories with a high accuracy, there is still a chance for the attackers to leverage various deception techniques to avoid the detection. To address this problem, we equipped our IDS framework with a decision-making module that leveraged the human-in-the-loop system to evaluate the undetected or wrongly classified attacks using the human expert's judgment. Our HITL module evaluated various attack and defense strategies by formulating a non-zero-sum game model.

REFERENCES

1. Lonea, A. M., Popescu, D. E., and Tianfield, H. Detecting DDOS attacks in cloud computing environment. *International Journal of Computers Communications & Control* 8, 1 (2012), 70–78.
2. Faghani, M. R., and Nguyen, U. T. Mobile botnets meet social networks: design and analysis of a new type of botnet. *International Journal of Information Security* 18, 4 (2019), 423–449.
3. Abu Al-Haija, Q., and Zein-Sabatto, S. An efficient deep-learning-based detection and classification system for cyber-attacks in IoT communication networks. *Electronics* 9, 12 (2020), 2152.
4. Bhattacharyya, D. K., and Kalita, J. K. *Network anomaly detection: A machine learning perspective*. CRC Press, 2013.
5. Wang, Z., Wang, R., Gao, J., Gao, Z., and And Liang, Y. Fault recognition using an ensemble classifier based on Dempster Shafer theory. *Pattern Recognition* 99 (2020), 107079.
6. Yang, K., Liu, J., Zhang, C., and And Fang, Y. Adversarial examples against the deep learning-based network intrusion detection systems. In *MILCOM 2018-2018 IEEE Military Communications Conference (MILCOM)* (2018), IEEE, pp. 559–564.
7. Thing, V. L. IEEE 802.11 network anomaly detection and attack classification: A deep learning approach. In *2017 IEEE Wireless Communications and Networking Conference (WCNC)* (2017), IEEE, pp. 1–6.
8. Shire, R., Shiaeles, S., Bendiab, K., Ghita, B., and And Kolokotronis, N. Malware squid: A novel IoT malware traffic analysis framework using convolutional neural network and binary visualisation. In *Internet of Things, Smart Spaces, and Next Generation Networks and Systems*. Springer, 2019, pp. 65–76.
9. Aamir, M., Rahman, Z., Ahmed Abro, W., Tahir, M., and Mustajar Ahmed, S. An optimized architecture of image classification using convolutional neural network. *International Journal of Image, Graphics and Signal Processing*, 11, 10 (2019), 30–39.
10. Ullah, R., Bazai, S. U., Aslam, U., and Shah, S. A. A. Utilizing blockchain technology to enhance smart home security and privacy. In Proceedings of International Conference on Information Technology and Applications: ICITA 2022, 2023.
11. Liang, X., and Xiao, Y. Game theory for network security. *IEEE Communications Surveys & Tutorials*, 15, 1 (2013), 472–486.
12. Zhang, H., Li, T., and Huang, S. Network defense decision-making method based on attack-defense differential game. *Acta Electronica Sinica* 46, 6 (2018), 1428–1435.
13. Liu, G., Zhang, H., and Li, Q. M. Network security optimal attack and defense decision-making method based on game model. *Journal of Nanjing University of Science and Technology*, 38, 1 (2014), 12–21.
14. Zhang, H., Wang, J., Yu, D., Han, J., and Li, T. Active defense strategy selection based on static Bayesian game. In Third International Conference on Cyberspace Technology (CCT 2015), pp. 1–7. IET, 2015.

15. Shen, S., Li, Y., Xu, H., and And Cao, Q. Signaling game-based strategy of intrusion detection in wireless sensor networks. *Computers & Mathematics with Applications* 62, 6 (2011), 2404–2416.

16. Li, Q., Hou, J., Meng, S., and Long, H. GLIDE: A game theory and data-driven mimicking linkage intrusion detection for edge computing networks. *Complexity*, 2020 (2020), 1–18.

17. Lin, X., Zhao, C., and Pan, W. Towards accurate binary convolutional neural network. *Advances in Neural Information Processing Systems* 30 (2017).

18. Da Costa, K. A., Papa, J. P., Lisboa, C. O., Munoz, R., and De Albuquerque, V. H. C. Internet of Things: A survey on machine learning-based intrusion detection approaches. *Computer Networks* 151 (2019), 147–157.

19. Liu, H., and Lang, B. Machine learning and deep learning methods for intrusion detection systems: A survey. *Applied Sciences*, 9, 20 (2019), 4396.

20. Shi, W.-C., and And Sun, H.-M. Deepbot: a time-based botnet detection with deep learning. *Soft Computing* 24 (2020), 16605–16616.

21. Srivastava, N., Hinton, G., Krizhevsky, A., Sutskever, I., and Salakhutdinov, R. Dropout: a simple way to prevent neural networks from overfitting. *The Journal of Machine Learning Research* 15, 1 (2014), 1929–1958.

22. Tareen, S., Bazai, S. U., Ullah, S., Ullah, R., Marjan, S., and Ghafoor, M. I. Phishing and intrusion attacks: An overview of classification mechanisms. In 2022 3rd International Informatics and Software Engineering Conference (IISEC), 2022.

23. Ibrahim, L. M., Basheer, D. T., and Mahmod, M. S. A comparison study for intrusion database (Kdd99, Nsl-Kdd) based on self organization map (SOM) artificial neural network. *Journal of Engineering Science and Technology*, 8, 1 (2013), 107–119.

24. Jiang, K., Wang, W., Wang, A., and Wu, H. Network intrusion detection combined hybrid sampling with deep hierarchical network. *IEEE Access*, 8 (2020), 32464–32476.

Index

Pages in *italics* refer to figures and pages in **bold** refer to tables.

For Product Safety Concerns and Information please contact our EU
representative GPSR@taylorandfrancis.com
Taylor & Francis Verlag GmbH, Kaufingerstraße 24, 80331 München, Germany

www.ingramcontent.com/pod-product-compliance
Lightning Source LLC
Chambersburg PA
CBHW080935220326
41598CB00034B/5789